Advanced Principles of Virology

Advanced Principles of Virology

Editor: Harvey O'Brien

R CALLISTO REFERENCE

www.callistoreference.com

Callisto Reference,
118-35 Queens Blvd., Suite 400,
Forest Hills, NY 11375, USA

Visit us on the World Wide Web at:
www.callistoreference.com

ISBN: 978-1-64116-127-5 (Hardback)

Cataloging-in-Publication Data

Advanced principles of virology / edited by Harvey O'Brien.
    p. cm.
Includes bibliographical references and index.
ISBN 978-1-64116-127-5
1. Virology. 2. Microbiology. I. O'Brien, Harvey.
QR360 .A38 2019
579.2--dc23

Table of Contents

Preface..VII

Chapter 1   **Regulation of Viral Replication, Apoptosis and Pro-Inflammatory Responses by 17-AAG during Chikungunya Virus Infection in Macrophages**..........................1
Tapas K. Nayak, Prabhudutta Mamidi, Abhishek Kumar,
Laishram Pradeep K. Singh, Subhransu S. Sahoo, Soma Chattopadhyay and
Subhasis Chattopadhyay

Chapter 2   **Feline Panleucopenia Virus NS2 Suppresses the Host IFN-β Induction by Disrupting the Interaction between TBK1 and STING**..........................22
Hongtao Kang, Dafei Liu, Jin Tian, Xiaoliang Hu, Xiaozhan Zhang, Hang Yin,
Hongxia Wu, Chunguo Liu, Dongchun Guo, Zhijie Li, Qian Jiang, Jiasen Liu and
Liandong Qu

Chapter 3   **PreC and C Regions of Woodchuck Hepatitis Virus Facilitate Persistent Expression of Surface Antigen of Chimeric WHV-HBV Virus in the Hydrodynamic Injection BALB/c Mouse Model**..........................34
Weimin Wu, Yan Liu, Yong Lin, Danzhen Pan, Dongliang Yang, Mengji Lu
and Yang Xu

Chapter 4   **Virus Resistance is not Costly in a Marine Alga Evolving under Multiple Environmental Stressors**..........................46
Sarah E. Heath, Kirsten Knox, Pedro F. Vale and Sinead Collins

Chapter 5   **Isolation and Characterization of a Double Stranded DNA Megavirus Infecting the Toxin-Producing Haptophyte *Prymnesium parvum***..........................64
Ben A. Wagstaff, Iulia C. Vladu, J. Elaine Barclay, Declan C. Schroeder,
Gill Malin and Robert A. Field

Chapter 6   **Change in *Emiliania huxleyi* Virus Assemblage Diversity but not in Host Genetic Composition during an Ocean Acidification Mesocosm Experiment**..........................75
Andrea Highfield, Ian Joint, Jack A. Gilbert, Katharine J. Crawfurd and
Declan C. Schroeder

Chapter 7   **A Student's Guide to Giant Viruses Infecting Small Eukaryotes: From *Acanthamoeba* to *Zooxanthellae***..........................90
Steven W. Wilhelm, Jordan T. Bird, Kyle S. Bonifer, Benjamin C. Calfee,
Tian Chen, Samantha R. Coy, P. Jackson Gainer, Eric R. Gann,
Huston T. Heatherly, Jasper Lee, Xiaolong Liang, Jiang Liu, April C. Armes,
Mohammad Moniruzzaman, J. Hunter Rice, Joshua M. A. Stough,
Robert N. Tams, Evan P. Williams and Gary R. LeCleir

Chapter 8   **Porcine Rotaviruses: Epidemiology, Immune Responses and Control Strategies**..........................108
Anastasia N. Vlasova, Joshua O. Amimo and Linda J. Saif

Chapter 9   **Structure of Ty1 Internally Initiated RNA Influences Restriction Factor Expression** 135
Leszek Błaszczyk, Marcin Biesiada, Agniva Saha, David J. Garfinkel and
Katarzyna J. Purzycka

Chapter 10   **Envelope Protein Mutations L107F and E138K are important for Neurovirulence Attenuation for Japanese Encephalitis Virus SA14-14-2 Strain** 155
Jian Yang, Huiqiang Yang, Zhushi Li, Wei Wang, Hua Lin, Lina Liu,
Qianzhi Ni, Xinyu Liu, Xianwu Zeng, Yonglin Wu and Yuhua Li

Chapter 11   **Novel Approach for Isolation and Identification of Porcine Epidemic Diarrhea Virus (PEDV) Strain NJ using Porcine Intestinal Epithelial Cells** 167
Wen Shi, Shuo Jia, Haiyuan Zhao, Jiyuan Yin, Xiaona Wang, Meiling Yu,
Sunting Ma, Yang Wu, Ying Chen, Wenlu Fan, Yigang Xu and Yijing Li

Chapter 12   **A Glimpse of Nucleo-Cytoplasmic Large DNA Virus Biodiversity through the Eukaryotic Genomics Window** 180
Lucie Gallot-Lavallée and Guillaume Blanc

Chapter 13   **A Point Mutation in a Herpesvirus Co-Determines Neuropathogenicity and Viral Shedding** 194
Mathias Franz, Laura B. Goodman, Gerlinde R. Van de Walle,
Nikolaus Osterrieder and Alex D. Greenwood

Permissions

List of Contributors

Index

Preface

Viruses are submicroscopic parasitic particles. They can infect all organisms, from bacteria and archaea to plants and animals, and cause a variety of diseases. The scientific study of viruses, their structure, classification and evolution is known as virology. The studies of the processes of infection and the resultant viral diseases are also a focus of this science. Research in virology has significant applications in medicine, neuroscience, materials science and nanotechnology. This book strives to provide detailed information about the theories and concepts of virology to help develop a better understanding of the latest advances within this field. From theories to research to practical applications, case studies related to all contemporary topics of relevance to this field have been included in this book.

This book is a result of research of several months to collate the most relevant data in the field.

When I was approached with the idea of this book and the proposal to edit it, I was overwhelmed. It gave me an opportunity to reach out to all those who share a common interest with me in this field. I had 3 main parameters for editing this text:

1. Accuracy – The data and information provided in this book should be up-to-date and valuable to the readers.

2. Structure – The data must be presented in a structured format for easy understanding and better grasping of the readers.

3. Universal Approach – This book not only targets students but also experts and innovators in the field, thus my aim was to present topics which are of use to all.

Thus, it took me a couple of months to finish the editing of this book.

I would like to make a special mention of my publisher who considered me worthy of this opportunity and also supported me throughout the editing process. I would also like to thank the editing team at the back-end who extended their help whenever required.

Editor

Regulation of Viral Replication, Apoptosis and Pro-Inflammatory Responses by 17-AAG during Chikungunya Virus Infection in Macrophages

Tapas K. Nayak [1,†], Prabhudutta Mamidi [2,†], Abhishek Kumar [2], Laishram Pradeep K. Singh [1,‡], Subhransu S. Sahoo [1], Soma Chattopadhyay [2,*] and Subhasis Chattopadhyay [1,*]

[1] School of Biological Sciences, National Institute of Science Education & Research, Bhubaneswar, HBNI, Jatni, Khurda, Odisha 752050, India; tapas.sbsniser@gmail.com (T.K.N.); laishrampk@gmail.com (L.P.K.S.); subhransusahoo87@gmail.com (S.S.S.)

[2] Infectious Disease Biology, Institute of Life Sciences, (Autonomous Institute of Department of Biotechnology, Government of India), Nalco Square, Bhubaneswar, Odisha 751023, India; rinku.prabhu@gmail.com (P.M.); abhishekbt13@gmail.com (A.K.)

*   Correspondence: sochat.ils@gmail.com (So.C.); subho@niser.ac.in (Su.C.);

†   These authors contributed equally to this paper.

‡   Current address: Department of Zoology, University of Kalyani, Kalyani, Nadia District, West Bengal 741235, India.

Academic Editor: Andrew Mehle

Abstract: Chikungunya virus (CHIKV) infection has re-emerged as a major public health concern due to its recent worldwide epidemics and lack of control measures. Although CHIKV is known to infect macrophages, regulation of CHIKV replication, apoptosis and immune responses towards macrophages are not well understood. Accordingly, the Raw264.7 cells, a mouse macrophage cell line, were infected with CHIKV and viral replication as well as new viral progeny release was assessed by flow cytometry and plaque assay, respectively. Moreover, host immune modulation and apoptosis were studied through flow cytometry, Western blot and ELISA. Our current findings suggest that expression of CHIKV proteins were maximum at 8 hpi and the release of new viral progenies were remarkably increased around 12 hpi. The induction of Annexin V binding, cleaved caspase-3, cleaved caspase-9 and cleaved caspase-8 in CHIKV infected macrophages suggests activation of apoptosis through both intrinsic and extrinsic pathways. The pro-inflammatory mediators (TNF and IL-6) MHC-I/II and B7.2 (CD86) were also up-regulated during infection over time. Further, 17-AAG, a potential HSP90 inhibitor, was found to regulate CHIKV infection, apoptosis and pro-inflammatory cytokine/chemokine productions of host macrophages significantly. Hence, the present findings might bring new insight into the therapeutic implication in CHIKV disease biology.

Keywords: Chikungunya virus; Alphavirus; macrophage; MHC; TNF; HSP90; apoptosis; 17-AAG

1. Introduction

Chikungunya virus (CHIKV), a mosquito borne re-emerging Alphavirus, belonging to the Togaviridae family, is endemic mainly in Africa, India, China and many other parts of Asia [1–4]. It was first isolated from Tanzania (formerly Tanganyika), Africa in 1952 during an epidemic of dengue-like illness [5]. Chikungunya fever (CHIKF) is often characterized by sudden appearance of high fever, headache, nausea, vomiting, rashes over skin [6] and followed by polyarthralgia, myalgia [7,8] and gastrointestinal complaints [9]. In some cases, complications such as myopericarditis [10], retrobulbar neuritis [11,12], nephritis [13], myocarditis, pericarditis [14], meningoencephalitis and death, have also been reported in CHIKV infection [9,15,16]. CHIKV is mainly transmitted to the vertebrate host by *Aedes* mosquitoes and maintained in two distinct transmission cycles: urban cycle between human and mosquitoes and sylvatic cycle within forest dwelling mosquitoes and non-human primates [17]. It is an enveloped virus, containing 11.8 kb long single stranded positive sense RNA genome with two open reading frames (ORF). The 5′ ORF codes for non-structural proteins, nsP1-4, mainly involved in viral replication and 3′ ORF codes for three major structural proteins, capsid, E1 and E2 [18].

Like other viral infections, following inoculation, CHIKV induces strong and quick type I interferon (IFN) response. Initially, it has been suggested that wild type adult mice but not the neonates are resistant to CHIKV infection and in neonates' disease severity is age dependent. Adult mice with IFN-α/βR$^{+/-}$ or IFN-α/βR$^{-/-}$ develop a mild or severe CHIKV infection respectively [19]. Moreover, it has been reported that CHIKV nsP2 can suppress anti-viral pathway by inhibiting IFN-α/β receptor signaling [20]. CHIKF is also associated with increased level of pro-inflammatory cytokines IL-1β, IL-6, IL-12, TNF, IFN-γ [21–24], GM-CSF [25,26], chemokines IL-8, and MCP-1 [27–29], as well as decreased level of pro-inflammatory chemokine RANTES [30].

CHIKV targets a wide range of immune and non-immune cells for its replication, propagation and dissemination. Among them, antigen presenting cells (APCs) such as macrophages/monocytes are known to play important roles towards modulating adaptive immune response against pathogens [31–35]. Among blood leukocytes, monocytes are known to be the major host cells for acute CHIKV infection in humans [36]. Previous studies have shown that macrophages could also be infected with CHIKV, both in vivo [37] and in vitro [22]. In both mouse and macaque models, it has been found that CHIKV induces predominant infiltration of monocytes, macrophages and NK cells with the production of MCP-1, TNF and IFN-γ at the site of inoculation, suggesting a strong immune activation [38]. This productive infection of CHIKV in macrophages could be associated with arthritis, tenosynovitis and myositis [38] despite a robust immune activation [39].

Recent reports have shown that viral infection often induces expression of various intracellular stress related proteins. The heat shock proteins (HSP) are molecular chaperones which bind and stabilize misfolded or unfolded polypeptides to ensure their proper folding and assembly with other polypeptides to decipher the normal protein function [40–43]. Induction of HSPs has been reported in both RNA and DNA viral infections, however, the type of HSP involved in a viral infection depends on the kind of virus and the type of host cells associated to the infection [44,45]. Recent studies have shown the functional requirement of HSP90 for Human cytomegalo virus (HCMV), Hepatitis C virus (HCV), Herpes Simplex virus-1 (HSV-1), Human Immunodeficiency virus-1 (HIV-1), Hepatitis E virus (HEV), Epstein Barr virus (EBV), Vaccinia virus and rotavirus infections [46–53]. The protein expression of HSP90 usually does not change during viral infection in macrophages [54], however macrophage immune responses for antigen presentation, phagocytosis and inflammatory responses are affected by functional HSP90.

Recently, it has been reported that Geldanamycin (GA) [55] and two other newly synthesized HSP90 inhibitory drugs (HS-10 and SNX-2112) [56] can regulate CHIKV infection both in vitro and in vivo. However, a functional role of HSP90 in CHIKV replication and associated immune modulation in macrophages during infection remains obscured. Accordingly, we have tested whether 17-AAG, a HSP90 functional inhibitor and a proposed therapeutic drug [57], has any regulatory role in CHIKV infection, apoptosis and the altered host immune response in macrophages.

2. Materials and Methods

2.1. Cells and Viruses

The Indian outbreak strain of CHIKV, DRDE-06 (accession no. EF210157.2), CHIKV prototype strain S 27 (accession no. AF369024.2) and Vero cells (African green monkey kidney epithelial cell line) were kind gifts from Dr. Manmohan Parida, Defence Research & Development Establishment (DRDE), Gwalior, India. The mouse monocyte/macrophage cell line, Raw264.7 (ATCC® TIB-71™)

was maintained in RPMI-1640 (HiGlutaXL™ RPMI-1640) supplemented with 2.0 mM L-glutamine, Penicillin 100 U/mL, Streptomycin 0.1 mg/mL (Himedia Laboratories Pvt. Ltd., Mumbai, India), 10% Fetal bovine serum (FBS; PAN Biotech, Aidenbach Germany) at 37 °C under a humidified incubator with 5% CO_2. Vero cells were maintained in Dulbecco's modified Eagle's medium (DMEM; PAN Biotech) supplemented with 5% FBS, Gentamycin (Sigma-Aldrich, St. Louis, MO, USA). The enzyme free cell dissociation reagent (ZymeFree™; Himedia Laboratories, Pvt. Ltd., Mumbai, India) was used for subculturing the cells.

2.2. Antibodies and Reagents

The mouse anti-CHIKV-nsP2 monoclonal antibody, used in this study was developed by us [58]. The anti-CHIKV-E2 monoclonal antibody was a kind gift from Dr. Manmohan Parida, DRDE, Gwalior, India. HRP linked secondary antibodies, H-2k^d PE, I-A/I-E PE, isotype PE, isotype APC and HSP90 antibodies were purchased from BD Biosciences (San Jose, CA, USA). CD86 APC and CD80 APC were purchased from eBiosciences (San Diego, CA, USA). The monoclocal antibodies for cleaved caspase-3 (Asp175), cleaved caspase-8 (Asp387) and caspase-9 (C9) were purchased from cell signaling technology (Danvers, MA, USA). The anti-mouse Alexa Fluor 488 was purchased from Invitrogen (Carlsbad, CA, USA). Mouse IgG1 isotype control, GAPDH and β-actin were purchased from Abgenex India Pvt. Ltd. (Bhubaneswar, India). Saponin, Anisomycin and Bovine serum albumin fraction V were purchased from Sigma-Aldrich. 17-Allylaminogeldanamycin (17-AAG) and Z-VAD-FMK were purchased from Merck Millipore (Billerica, MA, USA).

2.3. MTT Assay

MTT assay was performed to assess cytotoxicity of 17-AAG and Z-VAD-FMK using EZcount™ MTT cell assay kit (Himedia Laboratories Pvt. Ltd., Mumbai, India) according to the manufacturer's instructions. Briefly, the Raw264.7 cells were seeded in 96 well plate at a density of 5×10^3 cells per well before drug treatment. The cells were then washed in $1\times$ PBS and incubated with different concentrations of drugs in triplicate. The DMSO was taken as solvent control. After 24 h of treatment, the cells were incubated with the MTT reagent to a final concentration of 10% of the total volume. Then, the cells were incubated for 2 h in the incubator for the formation of visible crystals. Later, the media was removed carefully and 100 μL of solubilization solution was added per well followed by incubation for 15 min at room temperature (RT). The absorbance of the solution was taken at 550 nm by Microplate Reader (Bio-Rad, Hercules, CA, USA). Next, the percent viable cells were calculated in comparison to the control cells of the same plate in triplicate.

2.4. CHIKV Infection in Macrophage

The Raw264.7 cells of low passage number were seeded in six well plate before 18–20 h of infection with 60%–70% confluency. The cells were infected with DRDE-06 strain of CHIKV with different multiplicity of infection (MOI) as described earlier [59]. Briefly, the cells were washed in $1\times$ PBS and the virus was added over confluent monolayer for 2 h in the incubator with manual shaking at an interval of 15 min. Then, the virus inoculum was washed in $1\times$ PBS and the cells were maintained at 37 °C in complete RPMI-1640 media. The infected cells and the supernatants were collected at different time points and subjected to further processing according to the assay. Further, the CHIKV infected and the mock cells were examined under microscope ($20\times$ magnification) and pictures were taken at different hours post infection (hpi) to observe the cytopathic effect (CPE).

Vero cells with at least 90% confluency were seeded in 35 mm^2 cell culture dishes in complete media. Next day, the cells were infected with different strains of CHIKV as described previously [59]. The cells and supernatants were collected at different hpi and processed according to the assay.

For 17-AAG or Z-VAD-FMK treatment, the infection was carried out for 2 h in the presence of solvent control, DMSO or the drug. The cells were washed thoroughly with $1\times$ PBS after 2 h and cultured in serum free media containing the drug for 3 h. Then, serum was added to the cells and maintained in the incubator until harvesting [55].

2.5. Plaque Assay

For the determination of the virus titer, plaque assay was performed on the Vero cells as described previously [59]. Briefly, after infecting the Vero cells using cell culture supernatants collected from CHIKV infected Raw cells, the cells were overlaid with complete DMEM containing methyl cellulose and maintained in 37 °C incubator. After development of the visible plaques (4–5 days), the cells were fixed in formaldehyde at room temperature, washed gently in distilled water and stained with crystal violet. Then, the numbers of plaques were counted manually under white light.

2.6. Flow Cytometry (FC)

Flow cytometric assay was performed as mentioned elsewhere [60]. In brief, mock and CHIKV infected Raw264.7 cells were harvested by scraping, fixed in 4% paraformaldehyde for 10 min at RT and were washed twice in ice cold $1\times$ PBS to remove excess paraformaldehyde. Then, the cells were resuspended in FACS buffer ($1\times$ PBS, 1% BSA, 0.01% NaN_3) and stored at 4 °C until staining. For intra cellular staining (ICS) of CHIKV antigens, the cells were permeabilized in permeabilization buffer ($1\times$ PBS + 0.5% BSA + 0.1% Saponin + 0.01% NaN_3) followed by blocking in 1% BSA (in permeabilization buffer) for 30 min at RT. Then, the cells were washed with permeabilization buffer, incubated with anti-CHIKV-nsP2 or E2 antibodies [58], washed two times in permeabilization buffer followed by incubation in Alexa Fluor® 488 conjugated chicken anti-mouse IgG (H + L) secondary antibody. Both the primary and secondary antibodies were diluted in permeabilization buffer. The mouse IgG was taken as an isotype control during ICS. For the surface staining (CS), the fixed cells were washed with ice cold FACS buffer and then, fluorophore conjugated antibodies against different immune markers were added (diluted in FACS buffer). The mock and CHIKV infected cells were incubated with antibodies for 30 min on ice and washed twice with ice cold FACS buffer to remove un-bound antibodies. The fluorophore compatible isotype control antibodies were used as an isotype control for CS. The FcR blocking reagent (Miltenyi biotec, Bergisch Gladbach, Germany) was used at a dilution of 1:20 prior to the primary antibody incubation to prevent non-specific binding of antibodies to the Fc receptors on macrophages. Then, the cells were acquired by the BD FACS Calibur™ flow cytometer (BD Biosciences) and analyzed by the CellQuest Pro software (BD Biosciences). A total of approximately ten thousand cells were acquired per sample.

2.7. Sandwich ELISA for Cytokine Analysis

Cytokine production by the macrophages was analyzed from the cell culture supernatants as mentioned in methods by the BD OptEIA™ sandwich ELISA kit (BD Biosciences) according to the manufacturer's instructions. The cytokine concentrations in the test samples were calculated in comparison with the corresponding standard curve that was constructed using different concentrations of the recombinant cytokines in pg/mL.

2.8. Annexin V Staining

To detect the apoptotic cells, FC was carried out by using BD Annexin V Detection Kit I (BD Biosciences). Briefly, both mock and CHIKV infected cells were detached from the cell culture dishes by trypsin-EDTA treatment. The cells were washed twice in ice cold $1\times$ PBS and then resuspended in 100 μL of $1\times$ Annexin V binding buffer at a density of 1×10^6 cells/mL. Then, 2.5 μL each of APC conjugated Annexin V and 7-AAD cocktail was added per sample, gently mixed by vortexing and incubated at RT for 15 min in the dark. After that, 400 μL of $1\times$ Annexin V binding buffer was added per tube, samples were acquired immediately by the BD FACS Calibur™ and analyzed by the CellQuest Pro software. A total of approximately five thousand cells were acquired per sample.

2.9. Western Blot Analysis

Protein expression was assessed by Western blot analysis according to the protocol described earlier [55]. In brief, both the mock and CHIKV infected cells were washed once with ice cold $1\times$ PBS and the whole cell lysate (WCL) was prepared by Radio Immuno Precipitation Assay (RIPA) lysis buffer (150 mM NaCl, 1% NP-40, 0.5% sodium deoxycholate, 0.1% SDS, 50 mM Tris, pH 8.0, supplemented with protease inhibitor cocktail) and centrifuged at 13,000 rpm for 15 min at 4 °C. The protein concentration was quantified by the Bradford reagent (Sigma-Aldrich). Equal amount of protein was loaded in the 10%–12% SDS-PAGE after mixing with $2\times$ sample buffer (130 mM Tris-Cl, pH 8.0, 20% (v/v) Glycerol, 4.6% (w/v) SDS, 0.02% Bromophenol blue, 2% DTT) at a ratio of 1:1 and blotted on to a PVDF membrane (Millipore, MA, USA). Then the transferred membranes were blocked with 3% BSA followed by overnight incubation with the different primary antibodies, HSP90 (1:1000), cleaved caspase-3 (1:1000), cleaved caspase-8 (1:1000), caspase-9 (C9) (1:1000), β-actin (1:1000) and GAPDH (1: 1000). Then, the membranes were thoroughly washed with TBST for five times and incubated with the HRP conjugated secondary antibodies for 2 h at RT. After washing with TBST for three times, the membranes were subjected to chemiluminescence detection (Immobilon Western Chemiluminescent HRP substrate, Millipore or SuperSignal West Femto reagent, Thermo Scientific, Waltham, MA, USA) by the Bio-Rad gel doc with the Quantity One software (Bio Rad). For band intensity quantification, Western blot images were subjected to further analysis by the Quantity One 1-D analysis software while normalizing to the corresponding loading control.

2.10. RT-PCR

For quantitation of CHIKV RNA inside the host cells, Raw264.7 cells were infected with DRDE-06 strain of CHIKV with and without 17-AAG as mentioned previously. Cells were harvested at 8 hpi and the total RNA was isolated using Trizol reagent (Invitrogen). The RT was performed with equal amount of RNA (1 µg) by using First Strand cDNA synthesis kit (Thermo Scientific, Waltham, MA, USA) as per the manufacturer's instructions. This cDNA was used to amplify viral non-structural gene (NSP2) using primers (F)-5′CGAGGATCCACTGAATGAAATATGC-3′ and (R)-5′CGACTCGAGTTAACATCCTGCTCGGGTGG-3′; structural gene (E1) using primers (F)-5′TGCCGTCACAGTTAAGGACG3′ and (R)-5′CCTCGCATGACATGTCCG3′ through RT-PCR, along with GAPDH as housekeeping gene in Eppendorf master cycler pro S. The RT-PCR products were subjected to 1.5% agarose gel electrophoresis and GAPDH served as internal amplification control. Relative band intensities were calculated with the help of the Image J software (NIH, Bethesda, MD, USA) for the respective amplification.

2.11. Statistical Analysis

Statistical Analysis was performed using the GraphPad Prism 5.0 software (GraphPad Software Inc., San Diego, CA, USA). Data are represented as Mean $\pm$ SEM. The comparison between the groups was performed by the 2 way ANOVA with Bonferroni post-hoc test unless otherwise mentioned. Data presented here are representative of at least three independent experiments. $p < 0.05$ is considered as statistically significant difference between the groups.

3. Results

3.1. Determination of CHIKV Infection in Macrophages

In order to establish CHIKV infection, Raw264.7 and CHIKV strain, DRDE-06 were used at MOI 5. The infected cells along with the supernatants were harvested at an interval of 4 h from 0 to 24 hpi. The cells were processed for ICS to detect CHIKV specific antigens, nsP2 and E2 at different time points. Next, the expression pattern of these two CHIKV proteins was assessed by FC and it was observed that, both nsP2 and E2 were detected as early as 4 hpi (nsP2: 2.0 ± 0.24 and E2: 2.05 ± 0.22), while the highest level of proteins were noticed at 8 hpi (nsP2: 10.19 ± 2.04 and E2: 14.47 ± 0.17) followed by the gradual decrease at later time points (Figure 1A,B). Accordingly, Raw cells were harvested only at 8 hpi to assess CHIKV infection for subsequent experiments and representative FC data have been given towards the CHIKV replication and infection at 8 hpi (Figure 1C,D). Our current observation indicates that the virus could infect and replicate actively in Raw264.7 cells, with a peak of nsP2 and E2 levels around 8 hpi.

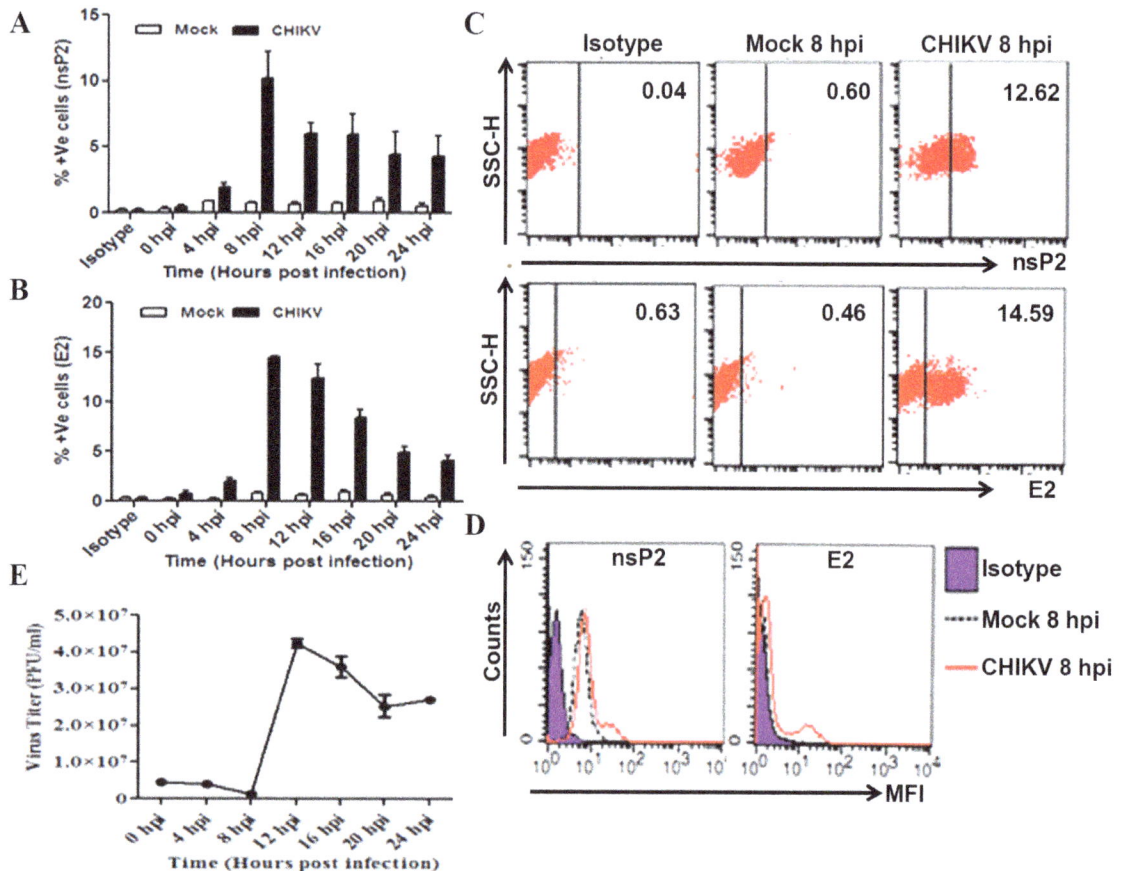

Figure 1. Infection of Chikungunya virus in macrophages. The Raw264.7 cells in 6 well plate were in vitro infected with CHIKV at MOI 5 for different time points or as described in the materials and methods section. The infected cells were harvested and stained with respective antibodies against the viral proteins followed by FC analysis. Representative bar diagram showing the time kinetics of percent positive cells for nsP2 (**A**) and E2 (**B**) in mock and CHIKV infected Raw cells at an interval of 4 hpi. (**C**) Representative dot plot analysis showing the expression of nsP2 (upper) and E2 (lower), along with isotype control (left), mock (middle) and CHIKV infected (right) macrophages at 8 hpi. (**D**) Mean fluorescence intensity (MFI) of nsP2 and E2 at 8 hpi representing the isotype (purple filled), mock (black dashed line) and CHIKV infected (red solid line) macrophage cells. (**E**) Line diagram of viral plaque forming units (PFU/mL), determined by plaque assay from CHIKV infected macrophage cell culture supernatants collected at different time points. Data represent mean $\pm$ SEM of three independent experiments. $p < 0.05$ was considered as statistically significant difference between the groups.

Next, to determine the release of the new infectious virus particles, plaque assay was performed using the supernatants collected at different hpi from the above experiments. Almost no virus particle was found to be released at 8 hpi, while the viral titer increased significantly to $4.25 \times 10^7 \pm 0.12 \times 10^7$ PFU/mL at 12 hpi indicating effective replication of CHIKV and release of the newly synthesized virus particles (Figure 1E). Subsequently, the viral count went down slowly at 16, 20 and 24 hpi. Since the supernatant was collected at every 4-h interval, the viral titer reflects the newly generated virus particles that were released in that particular period. Taken together, our data show that CHIKV could successfully infect and actively replicate in mouse macrophage cells in vitro.

Since susceptibility of CHIKV infection varies among cell lines with different MOIs in vitro [61], the efficacy of the DRDE-06 strain of CHIKV to infect Raw264.7 cell was also tested at different MOIs (0.1, 1, 5 and 10) at 8 hpi. It was observed that Raw264.7 cell was not susceptible to CHIKV infection at the MOI 0.1 and 1 in vitro as no significant E2 percent positive cells were detected with respect to the mock. However, at the MOI 5 and 10, E2 positive cells were found to be around 15% and 30% respectively (Figure S1A–C). Thus, MOI 5 was used for the rest of our experiments as an optimal virus load to the cells.

Earlier, it was reported that DRDE-06 strain of CHIKV replicates faster than S 27 strain in the Vero cell line [59]. Thus, here the infectivity of the S 27 strain was tested in the Raw264.7 cell line, to demonstrate that CHIKV infection in Raw cells is not strain specific. It was observed that S 27 also infects and replicates in Raw264.7 cells with less infectivity as compared to DRDE-06 strain (Figure S2C). Immune cells are generally found to be resistant to viral infections as compared to other somatic cells [61]. Thus, here the permissiveness of DRDE-06 was compared in Vero cell line. The expression pattern of CHIKV proteins was assessed by FC analysis. It was observed that both nsP2 and E2 were detected as early as 4 hpi (DRDE-06 nsP2: 31.79 ± 2.30 and E2: 85.62 ± 4.67), while the highest level of proteins were noticed at 8 hpi (DRDE-06 nsP2: 74.83 ± 0.52 and E2: 83.84 ± 5.39) followed by the gradual decrease at the later time points (Figure S2A,B). Taken together, our data show that different strains of CHIKV could successfully infect and actively replicate in mouse macrophage cells in vitro. Moreover, the macrophages (Raw264.7 cells) are found to be less permissible to CHIKV than Vero cells.

3.2. CHIKV Infection Induces Apoptosis in Macrophages

Apoptosis of virus infected cell is known to bear one of the important consequences towards viral replication, dissemination of the virus particle to the neighboring host cells as well as antigen presentation [62–64]. CHIKV infection has been recently reported to induce apoptosis in host epithelial cells [55,61,64]. However, apoptosis in host macrophages during CHIKV infection has not been reported yet. To assess whether mouse macrophages undergo apoptosis post CHIKV infection, the Raw cells were inoculated with CHIKV and processed at different time intervals for subsequent analysis. The bright field microscopic images showed the development of CPE at 8 hpi in CHIKV infected Raw cells (Figure 2A). Furthermore, visible tiny cell blebbings were observed at 12 hpi followed by few rounding and detachment of cells at 24 hpi. However, no such morphological changes were observed in the mock. The observed characteristic features of the cells indicate that the cells might have undergone apoptotic process after CHIKV infection. In order to confirm apoptosis in macrophages during CHIKV infection, the mock and the infected cells were stained with Annexin V and 7-AAD at different hpi. The FC analysis depicted a very small fraction of cells to be positive for Annexin V in both the mock (6.2 ± 0.5) and the CHIKV (7.55 ± 0.36, $p > 0.05$) infected cells at 8 hpi (Figure 2B). However, a significant increase in the population of Annexin V positive cells was observed at 12 hpi (Mock; 5.6 ± 0.9 vs. CHIKV; 8.67 ± 0.49, $p < 0.01$) and at 24 hpi (Mock; 5.3 ± 0.06 vs. CHIKV; 17.7 ± 0.05, $p < 0.001$) (Figure 2B). Interestingly, both the Annexin V and the 7-AAD dual positive cells were not increased significantly with time in the CHIKV infected macrophages as compared to the mock (Figure 2C), which confirms that CHIKV induces apoptotic marker, Annexin V without inducing necrosis. To further confirm the induction of apoptosis, the infected cells were harvested at 0, 4, 8, 12 and 24 hpi. Western blot analysis was performed to detect apoptosis through cleaved caspase-3. Surprisingly, the induction of cleaved caspase-3 was observed as early as 4 hpi in the case of CHIKV infected samples. Moreover, the expression of cleaved caspase-3 at 8 and 12 hpi in the CHIKV infected cells was significantly higher than the corresponding mock (Figure 2D). Hence, our data suggest that CHIKV infection induces apoptosis in macrophages in a time dependent manner.

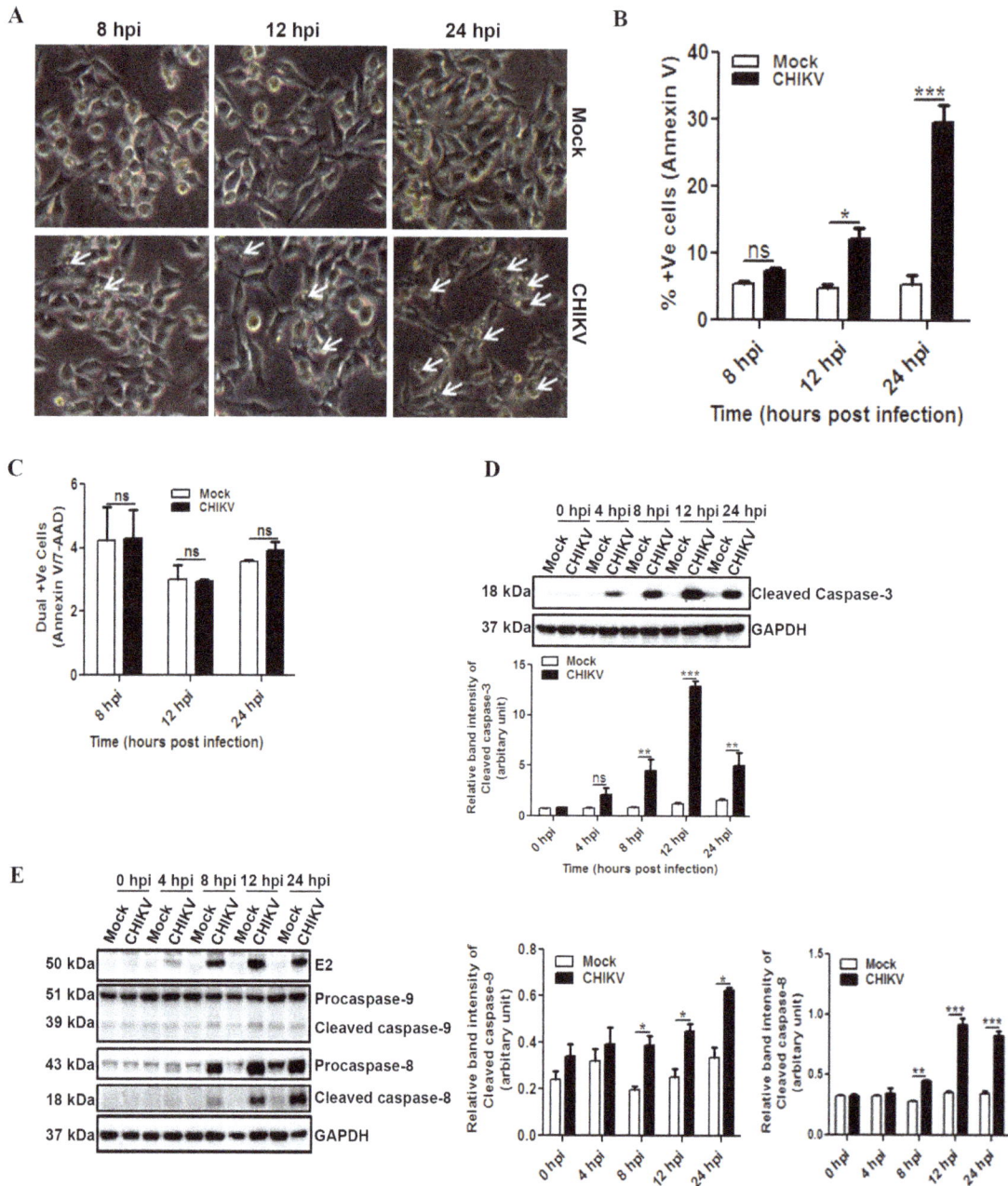

Figure 2. Induction of apoptosis following CHIKV infection in Macrophages. Raw264.7 cells were infected with CHIKV at MOI 5 for different time points as described earlier. Infected cells were viewed under bright field microscope or processed for FC analysis. (**A**) Bright field microscopic images taken at 8, 12 and 24 hpi with $20\times$ magnification. White arrows indicate observed morphological changes in the infected cells. (**B**) Graphical representation showing percent positive cells for Annexin V in mock (white bar) and CHIKV infected samples (dark bar) at 8, 12 and 24 hpi. (**C**) Bar diagram showing percent dual positive cells for Annexin V and 7-AAD in mock (white bar) and CHIKV infected samples (dark bar) at 8, 12 and 24 hpi. (**D**) CHIKV infected Raw cells were harvested, lysed and separated by SDS-PAGE. Western blot was performed and expression pattern of cleaved caspase-3 was assessed at different hpi as shown (upper), bar diagram showing relative band intensity of cleaved caspase-3 (lower). GAPDH was used as loading control. (**E**) Western blot analysis of E2, caspase-9 and caspase-8 (left), bar diagram showing relative band intensity of cleaved caspase-9 and cleaved caspase-8 (right). Data represent mean $\pm$ SEM of at least three independent experiments. $p < 0.05$ was considered as statistically significant difference between the groups. (* $p < 0.05$; ** $p \leq 0.01$; *** $p \leq 0.001$).

To investigate the in-depth pathway of cleaved caspase-3 induction by CHIKV in macrophages, expression of both cleaved caspase-9 and cleaved caspase-8 was assessed by Western blot analysis. It was observed that both cleaved caspase-9 and -8 were induced during CHIKV infection in the macrophages (Figure 2E). This suggests that CHIKV might induce apoptosis in macrophages through intrinsic as well as extrinsic pathways. Hence, investigation was carried out to assess the importance of apoptosis in CHIKV infected macrophages using a pan caspase inhibitor, Z-VAD-FMK. First, the cytotoxicity of Z-VAD-FMK was assessed by MTT assay with different concentrations. As shown in Supplementary Figure S3A, around 100% and 90% cells were viable after 24 h with 40 μM and 80 μM concentrations of Z-VAD-FMK respectively. Thus, 40 μM concentration of Z-VAD-FMK was used for further studies. The CHIKV infection was assessed in Raw cells in the presence of Z-VAD-FMK by FC analysis. It was observed that, Z-VAD-FMK did not suppress the CHIKV E2 percent positive cells significantly (Figure S3B). Surprisingly, Z-VAD-FMK did reduce the newly synthesized CHIKV progenies around 2.5 fold (Figure S3C). This suggests that Z-VAD-FMK might not interfere in the CHIKV protein expression but it can reduce CHIKV infection by reducing the release of new virus particles in the host macrophages. Since CHIKV induced phosphatidyl serine in the host macrophages (Figure 2B), the effect of Z-VAD-FMK on Annexin V binding was also assessed. It was found that, Z-VAD-FMK reduced the Annexin V binding by 35% (Figure S3D) with 40 μM concentration. Moreover, the treatment of Z-VAD-FMK was found to reduce the inductions of cleaved caspase-3, cleaved caspase-9 and cleaved caspase-8 (Anisomycin treatment was used as positive control for cleaved caspase-9 and -8 [65]) during CHIKV infection in the host macrophages (Figure S3E,F). Taken together, the result suggests that inhibition of apoptosis by Z-VAD-FMK significantly affects CHIKV infection in macrophages, without altering virus protein expression.

3.3. CHIKV Infection Upregulates Pro-Inflammatory Response in Mouse Macrophages

To determine the physiological and functional relevance of macrophages during CHIKV infection, cell culture supernatants were collected at different time points (8, 12, and 24 hpi) and quantified for the secreted cytokines (e.g., IL-10, IL-12, TNF and IL-6) and chemokine, MCP-1 by ELISA. No significant changes were observed in both IL-10 and IL-12 cytokines in CHIKV infected cells as compared to mock in all the time points (Figure 3A,B). It was observed that TNF was upregulated at 8 hpi 405 ± 33 pg/mL (mock 211 ± 20 pg/mL, $p \leq 0.01$) and followed by a maximum peak at 24 hpi 1016 ± 30 pg/mL (mock 367 ± 15 pg/mL, $p \leq 0.001$) (Figure 3C). The production of IL-6 was found to be upregulated as early as 8 hpi (mock 20 ± 2 pg/mL, CHIKV 71 ± 8 pg/mL, $p \leq 0.001$) and followed by a maximum peak at 24 hpi (mock 34 ± 6 pg/mL, CHIKV 480 ± 17 pg/mL, $p \leq 0.001$) (Figure 3D). Together, it was found that both TNF and IL-6 were increased during CHIKV infection in Raw cells in a time dependent manner.

The role of MCP-1 has been shown as an important mediator of inflammation in a variety of diseases, which recruit other immune cells to the site of impact to induce inflammatory responses [66,67]. Monocytes and macrophages are one of the major sources of MCP-1. Accordingly, it has been assessed whether CHIKV infection induces MCP-1 production in macrophages. It was observed that there was no significant difference in MCP-1 production during CHIKV infection at early time points (8 and 12 hpi) as compared to mock. However, CHIKV infection positively modulated MCP-1 secretion around 24 hpi (1354 ± 19 pg/mL) as compared to mock (1114 ± 64 pg/mL, $p < 0.05$). This result indicates that CHIKV infection in macrophages may upregulate TNF, IL-6 and MCP-1 over time, while no such significant changes were observed for IL-10 and IL-12.

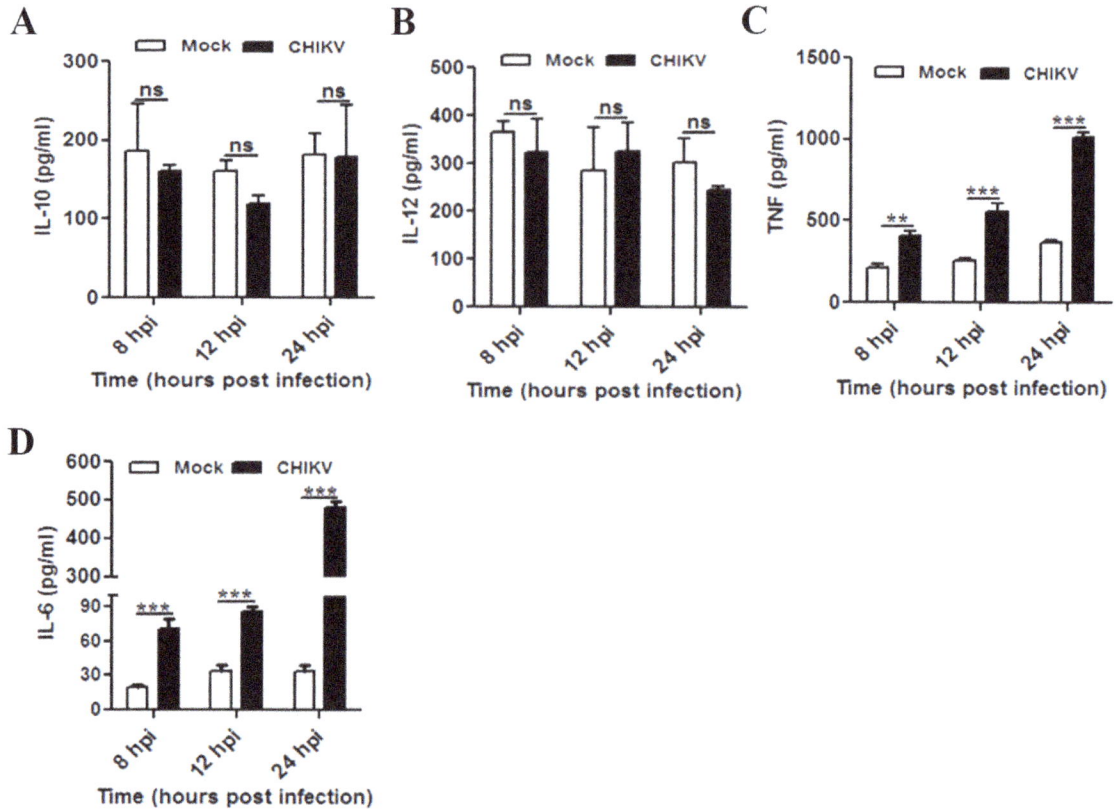

Figure 3. Modulation of macrophage derived cytokines during CHIKV infection. Raw264.7 cells were infected with CHIKV at MOI 5. The cell culture supernatants were collected at different time points and the levels of secreted cytokines in the samples were quantified. Graphical representation showing the amount of secreted: IL-10 (**A**); IL-12 (**B**); TNF (**C**); and IL-6 (**D**) in mock and CHIKV infected macrophage as quantified using sandwich ELISA. Data represent mean ± SEM of at least three independent experiments. $p < 0.05$ was considered as statistically significant difference between the groups. (ns, non-significant; ** $p \leq 0.01$; *** $p \leq 0.001$).

3.4. Induction of MHCs and Co-Stimulatory Molecules in CHIKV Infected Macrophages

APCs like macrophages are known to show altered expression of MHCs and co-stimulatory molecules (e.g., B7 molecules) during pathogenic encounter including viral infection [68–71]. Flow cytometry based investigations were carried out to detect the surface expression of MHC-I, MHC-II and inducible co-stimulatory molecule, CD86 (B7.2) in the CHIKV infected macrophages at different time points. It was found that the MHC-I surface expression was significantly up-regulated at 8, 12 and 24 hpi (Figure 4A,B). Unlike MHC-I, the surface expression of the MHC-II was increased significantly only at the later time point (24 hpi) after CHIKV infection (Figure 4A,B). Similar to MHC-I, the expression level of CD86 was found to be upregulated at different hpi (Figure 4C,D). This result indicates that CHIKV infection may significantly induce MHCs and co-stimulatory molecules in mouse macrophages.

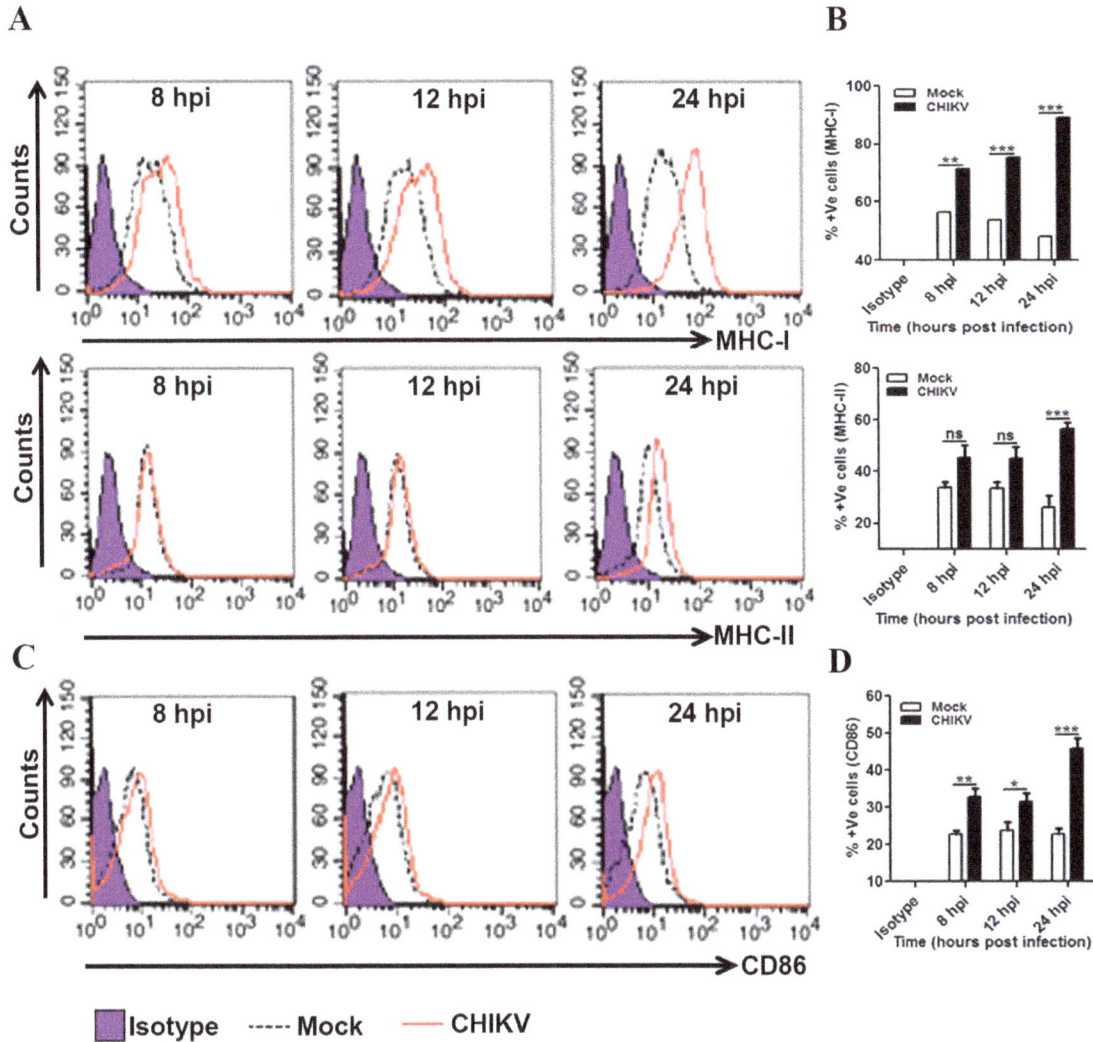

Figure 4. Expression pattern of MHC and co-stimulatory molecules during CHIKV infection in macrophages. CHIKV infected Raw264.7 cells were harvested at different time intervals followed by flow cytometry based analysis. (**A**) MFI representing MHC-I (upper panel) and MHC-II (lower panel) expressions in isotype (purple filled), mock (black dashed line) and CHIKV infected macrophage (red solid line). (**B**) Bar diagram showing percent positive cells for MHC-I (upper panel) and MHC-II (lower panel) in mock (white bar) and CHIKV infected samples (dark bar) at 8, 12 and 24 hpi. (**C**) MFI representing CD86 expression in isotype (purple filled), mock (black dashed line) and CHIKV infected macrophage (red solid line). (**D**) Bar diagram showing percent positive cells for CD86 in mock (white bar) and CHIKV infected samples (dark bar) at 8, 12 and 24 hpi. Data represent Mean $\pm$ SEM of at least three independent experiments. $p < 0.05$ was considered as statistically significant difference between the groups. (ns, non-significant; * $p \leq 0.05$; ** $p \leq 0.01$; *** $p \leq 0.001$).

3.5. Regulation of CHIKV Infection by 17-AAG in Macrophages

Our recent study showed that CHIKV nsP2 is stabilized by HSP90 during the active stage of replication in the Vero cells, which was abrogated by HSP90 inhibitor, GA [55]. Here, 17-AAG, a less toxic derivative of GA [72], was employed to study its effect on CHIKV infection in macrophages. The cytotoxicity of different concentrations of 17-AAG (0.0265, 0.125, 0.25, 0.5 and 1.0 μM) was tested on Raw cells by MTT assay for 24 h. It was observed that nearly 98% cells were viable up to 0.5 μM concentrations of 17-AAG (Figure 5A). However, with 1 μM of 17-AAG, approximately 50% cells were viable. Hence, 0.5 μM concentration of 17-AAG was selected as a non-cytotoxic dose for further study to work with the viable cells.

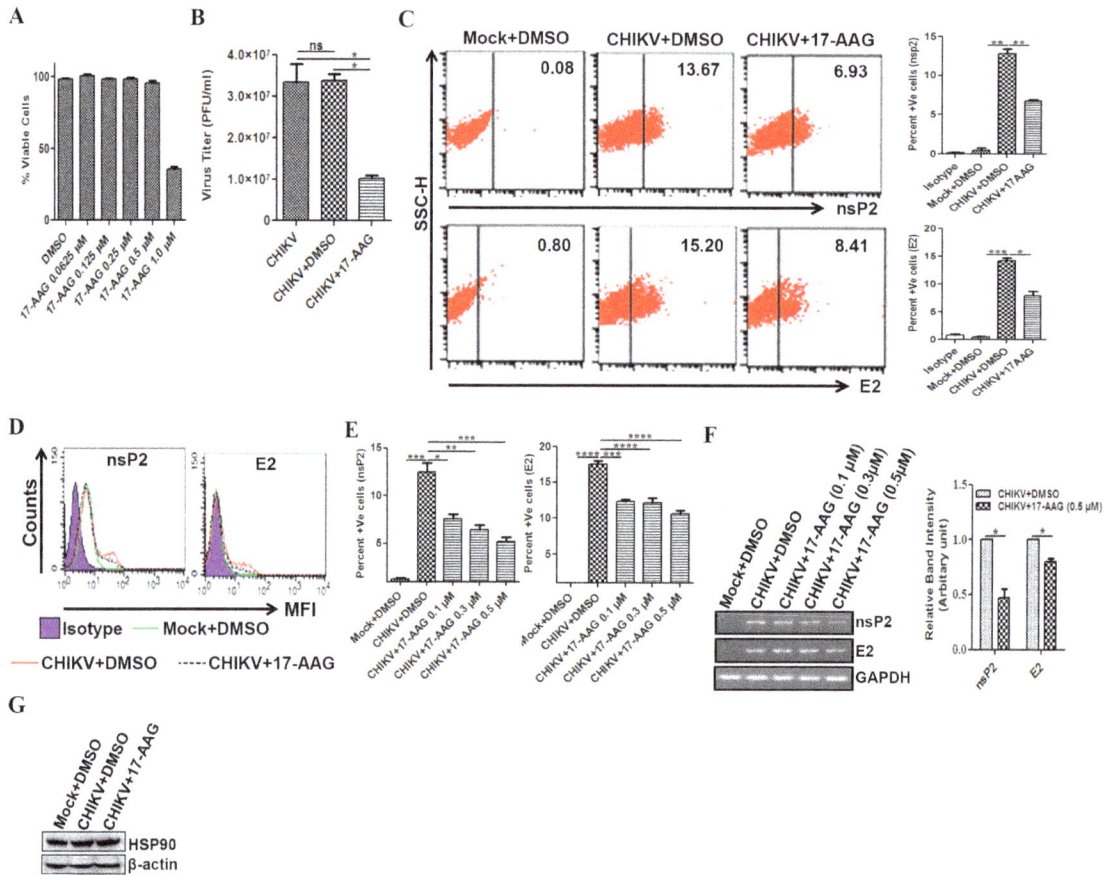

Figure 5. Modulation of CHIKV infection by 17-AAG in macrophages. Raw cells were infected with CHIKV at MOI 5. The cells were treated with either DMSO or 17-AAG as described earlier. (**A**) Percent viable cells treated with different concentrations of 17-AAG with respect to the solvent control (DMSO) as determined by MTT assay. (**B**) Line diagram showing CHIKV titer as PFU/mL in CHIKV, CHIKV + DMSO and CHIKV + 17-AAG (0.5 μM) at 12 hpi. (**C**) Representative dot plot analysis depicting percent positive cells for nsP2 and E2 at 8 hpi. Bar diagram showing percent positive cells for nsP2 (upper panel) and E2 (lower panel) in different samples at 8 hpi. (**D**) MFI representing expression of nsP2 and E2 in isotype (purple filled), mock + DMSO (green solid line), CHIKV + DMSO (red solid line) and CHIKV + 17AAG (black dashed line) at 8 hpi. (**E**) Graphical representation showing percent positive cells for nsP2 (left) and E2 (right) at 0.1 μM, 0.3 μM and 0.5 μM 17-AAG treatments at 8 hpi. (**F**) Agarose gel picture showing RT-PCR of nsP2 and E2 (left) and bar diagram (right) showing reduction in the CHIKV RNA synthesis in Raw cells after 0.5 μM 17-AAG treatment at 8 hpi. (**G**) Representative Western blot showing HSP90 level for mock + DMSO, CHIKV + DMSO and CHIKV + 17-AAG at 8 hpi. Data represent mean $\pm$ SEM of at least three independent experiments. $p < 0.05$ was considered as statistically significant difference between the groups. (ns, non-significant; * $p < 0.05$; ** $p \leq 0.01$; *** $p \leq 0.001$).

The Raw264.7 cells were infected with CHIKV in the presence of either DMSO or 0.5 μM concentration of 17-AAG and the cells were maintained along with the drug until harvesting. Then, the cells and supernatants were collected at 8, 12, 24 hpi and plaque assay was performed. It was observed that the treatment of 17-AAG reduced the number of new viral progeny production by 3.3 fold ($p < 0.05$) as compared to DMSO control (Figure 5B). Subsequently, the cells were collected to estimate the nsP2 and E2 protein levels. FC analysis showed that 17-AAG treatment (0.5 μM) was found to inhibit the level of both the viral proteins by 50% (Figure 5C,D). Furthermore, the dose kinetics depicts that 17-AAG inhibits nsP2 expression around 39% at 0.1 μM concentration, whereas it was further reduced to 58% with 0.5 μM concentration (Figure 5E). In case of the E2 protein, the expression was reduced to 30% at 0.1 μM concentration, however, it was further reduced to 40% with 0.5 μM of 17-AAG. This observation confirms that 17-AAG inhibits CHIKV protein synthesis and new viral progeny production in macrophages. Moreover, in this study it was explored whether 17-AAG treatment can reduce the level of CHIKV RNA. It was observed that the RNA levels were reduced by 2 fold ($p < 0.05$) for nsP2 and 1.25 fold ($p < 0.05$) for E2 only in the presence of 0.5 μM of 17-AAG (Figure 5F). This observation suggests that 17-AAG may reduce the production of viral progeny by inhibiting both the levels of nsP2 and E2 proteins as well as RNA. The result also indicates that all the concentrations of 17-AAG were able to suppress the viral protein levels remarkably, whereas the RNA levels were reduced significantly only with 0.5 μM concentration.

Next, the effect of 17-AAG treatment towards HSP90 expression in mock and CHIKV infected Raw cells were investigated. The Western blot analysis showed that expression of HSP90 remained unchanged in CHIKV + DMSO and CHIKV + 17-AAG treated macrophages as compared to Mock + DMSO. This suggests that 17-AAG may regulate CHIKV infection by abrogating HSP90 activity without modulating its expression (Figure 5G).

3.6. 17-AAG Regulates Apoptosis and Cellular Immune Responses during CHIKV Infection in Macrophages

Since apoptosis was detected by Annexin V binding during CHIKV infection (Figure 2A), experiments were performed to test whether 17-AAG can regulate CHIKV induced apoptosis and cellular immune responses in macrophages. It was observed that the Annexin V positive cells were 33.68% ± 4.15% after CHIKV infection, which was reduced to 18.43% ± 1.52% with 17-AAG at 24 hpi (Figure 6A). Furthermore, quantitative Western blot analysis showed that CHIKV induced cleaved caspase-3 upregulation was found to be reduced by around 30% with 17-AAG treatment as compared to the DMSO control at 12 hpi (Figure 6B, left and right panel). Together, it appears that 17-AAG treatment might regulate CHIKV induced apoptosis of host macrophages.

Since pro-inflammatory responses such as IL-6, TNF and MCP-1 were significantly induced in CHIKV infected macrophages (Figure 3), experiments were carried out to assess the efficacy of 17-AAG to suppress the cytokine and chemokine induction. The data showed that TNF level was 683 ± 26 pg/mL and 838 ± 18 pg/mL in CHIKV + DMSO at 12 and 24 hpi respectively, whereas, the levels were reduced to 304 ± 12 pg/mL and 385 ± 10 pg/mL in the presence of 17-AAG (Figure 6C). Similarly, the IL-6 levels were 85 ± 6 pg/mL and 476 ± 15 pg/mL in CHIKV + DMSO at 12 and 24 hpi respectively, whereas, the levels were reduced to 44 ± 3 pg/mL and 98 ± 2 pg/mL in the presence of 17-AAG (Figure 6D). The present data showed that TNF was reduced significantly (around 50%) upon 17-AAG treatment as compared to DMSO control. Similarly, the reduction of IL-6 after 17-AAG treatment was found to be around 50% at 8 and 12 hpi, whereas at 24 hpi it reached up to 80% as compared to DMSO control. In addition, the modulation of MCP-1 during CHIKV infection was also reduced after 17-AAG treatment (CHIKV+DMSO 1399 ± 15 pg/mL, CHIKV+17-AAG 641 ± 7 pg/mL, $p \leq 0.001$).

Unlike pro-inflammatory cytokine and chemokine production, inductions of MHC-I, MHC-II and CD86 by CHIKV were not suppressed upon 17-AAG treatment in all the time points (Figure S4). Together, the data indicate that 17-AAG might down regulate pro-inflammatory cytokine/chemokine production of host macrophages, without altering the induced immune activation markers like MHCs and CD86 during infection.

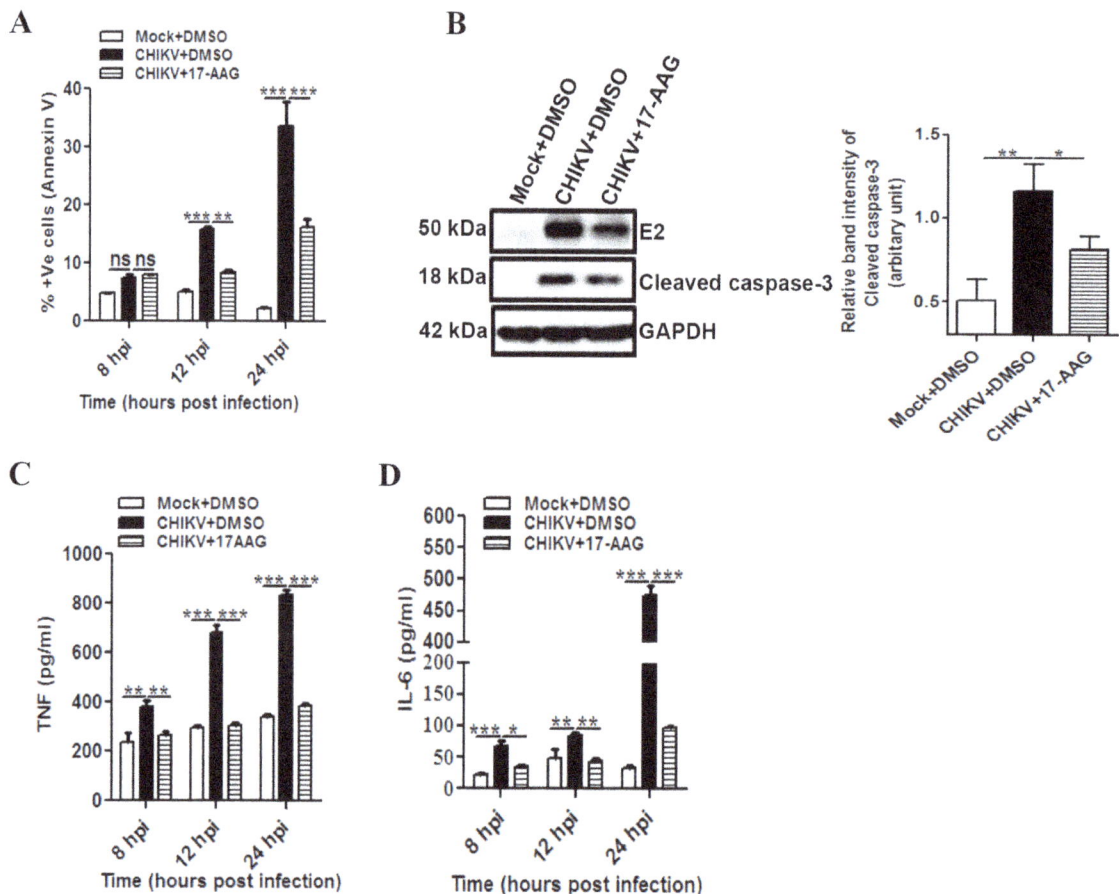

Figure 6. Suppression of CHIKV induced apoptosis and pro-inflammatory responses in macrophages by 17-AAG. Raw cells were infected with CHIKV at MOI 5. The cells were treated with either DMSO or 17-AAG as described earlier. (**A**) Percent Annexin V positive cells for mock + DMSO, CHIKV + DMSO and CHIKV + 17-AAG at 8, 12 and 24 hpi. (**B**) Western blot analysis showing the expression of the cleaved caspase-3, E2 and GAPDH in the presence of 17-AAG during CHIKV infection in macrophage at 12 hpi (left). Bar diagram showing relative band intensity of cleaved caspase-3 of mock + DMSO, CHIKV + DMSO and CHIKV + 17-AAG analyzed by the Quantity One 1-D analysis software (right). Cell culture supernatants were collected and were assessed for TNF (**C**) and IL-6 (**D**) secretion for mock + DMSO, CHIKV + DMSO and CHIKV + 17-AAG by sandwich ELISA at 8, 12 and 24 hpi. Data represent mean ± SEM of at least three independent experiments. $p < 0.05$ was considered as statistically significant difference between the groups. (ns, non-significant; * $p < 0.05$; ** $p \leq 0.01$; *** $p \leq 0.001$).

4. Discussion

CHIKV has been known to infect a wide range of cells including monocytes/ macrophages [22,36,61,73] and elicits strong immune response which involves the production of anti-viral IFNs, pro-inflammatory cytokines, chemokines and other growth factors [22–26,28–30]. The current study provides evidence that there is an alteration in immune responses of mouse macrophages (Raw264.7 cell line) comprising MHCs, co-stimulatory molecule (CD86), major pro-inflammatory cytokines/chemokine and host cell apoptosis during CHIKV infection in vitro. Furthermore, we report that 17-AAG, a potential HSP90 inhibitor, may effectively regulate CHIKV infection, apoptosis and associated inflammatory responses of host macrophages.

Acute phase CHIKV infection in human blood monocytes was found to induce robust and rapid innate immune responses [36]. In the current study, the CHIKV infected macrophages showed an increased level of viral proteins (both nsP2 and E2) at 8 hpi and maximum release of new viral progenies at 12 hpi. Like other Alphaviruses, it was reported that CHIKV infection induces apoptosis and CPE in other cell types [23,55,74–77]. However, induction of apoptosis in CHIKV infected macrophages is yet to be defined. Here, CHIKV infected macrophages showed characteristic features of an apoptotic cell in addition to CPE, membrane blebbings, rounding off and detachment. The results showed increased surface binding of Annexin V and increased cleaved caspase-3, affirming the induction of apoptosis in the CHIKV infected macrophages with MOI 5. Furthermore, it was demonstrated for the first time that CHIKV induces apoptosis in macrophages by both intrinsic and extrinsic pathways, as induced expression of the cleaved caspase-9 and -8 were observed during infection. Earlier, apoptosis was not detected in mouse macrophages infected with CHIKV at MOI 1 [22]. The differences in the observations could be due to the differences in MOIs being used in their experiments, different strain of viruses as well as different experimental set up. Apoptosis is an important anti-viral mechanism and is also considered to have a protective role in macrophages to ward off the viruses, thereby impairing viral propagation [78,79]. Interestingly, the treatment of Z-VAD-FMK was found to reduce new viral progeny release of CHIKV and infection, without altering the frequency of E2 percent positive host macrophages. The importance of Z-VAD-FMK on CHIKV infection was also studied earlier in HeLa cells, suggesting the reduction of both CHIKV replication and infection in vitro [64]. Moreover, Z-VAD-FMK was also found to differentially regulate other viral replication and infection [80–83]. Arguably, our study in macrophages may highlight the importance of CHIKV-induced apoptosis in increasing the viremia and propagation of infection to the nearby host cells. Further investigations are required to understand the mechanism in details.

CHIKV infection associated inflammatory response involves production and secretion of several pro-inflammatory cytokines. However, predominance of anti-inflammatory responses might also prevail under such conditions, which are against a common description and the notion of CHIKV infection [30]. IL-12 and TNF possess potent anti-viral property and also promote macrophage activation [84–86], while IL-10 plays an important cross-regulatory role during infection associated inflammation [87]. Moreover, IL-6 and MCP-1 induction during CHIKV infection have been shown to promote inflammatory and heightened cellular immune responses [30,66]. In this study, we assessed the expression of macrophage derived pro-inflammatory cytokines and found that there was a time-dependent increase in the secreted TNF, IL-6 and MCP-1 during CHIKV infection, while the production of IL-10 and IL-12 remain unchanged.

Viral infection in APCs like macrophages is known to enhance antigen processing and presentation via MHC to elicit adaptive immune responses. However, some viruses are known to manipulate the expression of MHC and co-stimulatory molecules to evade host cell immunity [32,68–71]. Moreover, our recent in silico analysis has identified several highly conserved CHIKV specific immunodominant MHC-I restricted peptide epitopes which may elicit strong anti-CHIKV CD8[+] T cell responses [88]. The present study demonstrated that the levels of MHC-I, MHC-II and CD86 activation markers were elevated in the CHIKV infected macrophages at various time intervals. It is plausible that the CHIKV induced MHCs may present specific immunodominant peptides and also induce CD86 expression in macrophages towards concomitant anti-CHIKV specific T cell immunity, which needs further investigations.

Host derived endogenous HSP90 has been identified to play a critical role in the pathogenesis of CHIKV [55,56]. In this study, we used 17-AAG on CHIKV-infected macrophages and our results demonstrated that the treatment of 17-AAG has reduced the production of live infectious virus particles significantly. Concurrently, there was reduction in the levels of viral proteins (nsP2 and E2) and synthesis of infectious virus particles in the 17-AAG treated-cells indicating its regulatory role during CHIKV infection in macrophages.

Subsequently, the efficacy of 17-AAG was tested towards the regulation of apoptosis in CHIKV-infected macrophages, if any. Although 17-AAG is known to induce cleaved caspase-3 in some cancerous cells [89,90], interestingly, in the current investigation, it was noticed that non-cytotoxic dose of 17-AAG abrogated CHIKV induced apoptosis (measured by Annexin V binding and the expression of cleaved caspase-3). This observation might be explained by the fact that the reduction of CHIKV induced Annexin V and cleaved caspase-3 by 17-AAG might be due to reduction of CHIKV infection in the host macrophages. Our current observation also suggests that the upregulated level of pro-inflammatory cytokines (TNF and IL-6) and chemokine (MCP-1) in CHIKV infected macrophages could be down regulated by 17-AAG treatment, while preserving the expression of MHC and co-stimulatory molecules in macrophages, probably to strengthen the subsequent anti-viral immune responses.

In conclusion, for the first time, we showed that CHIKV infection induces apoptosis, enhances MHCs and co-stimulatory molecule expressions along with IL-6 and MCP-1 production in mouse macrophages, in vitro. We have also identified an important role of 17-AAG to regulate CHIKV infection, pro-inflammatory cytokine/chemokine production and apoptosis of macrophages during viral infection as schematically summarized in Figure 7. Further studies are required to substantiate the mechanism of CHIKV infection in macrophage associated cellular pathways towards apoptosis, induction of host cell immunity and pathogenesis for designing rationale therapeutic drugs against CHIKV infection.

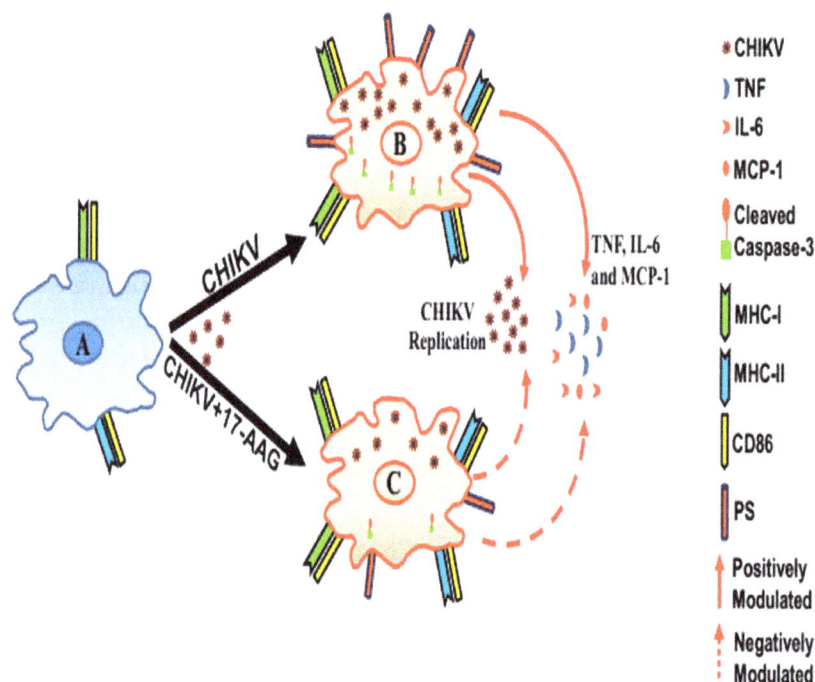

Figure 7. Proposed working model depicting CHIKV infection in macrophages and 17-AAG mediated possible regulation of its altered immune responses by inhibition of HSP90. Uninfected macrophage (**A**); and CHIKV infected macrophage (**B**) showing induction of MHC I/II, CD86 molecules as immune activation markers along with inflammatory cytokines/chemokine (TNF, IL-6 and MCP-1) production and apoptosis by phosphatidylserine (PS) and cleaved caspase-3 expression. (**C**) 17-AAG has been found to regulate the viral infection, apoptosis and inflammatory responses (TNF, IL-6 and MCP-1), suggesting its therapeutic implication in CHIKV infection.

Acknowledgments: We are thankful to Dr. Manmohan Parida, DRDE, Gwalior, India for kindly providing the virus strains (DRDE-06 and S 27), Vero cell line and E2 antibody. This study has been funded by Council of Scientific and Industrial Research (CSIR), New Delhi, India, vide grant no 37 (1542)/12/EMR-II. It was supported by Institute of life sciences, Bhubaneswar, under Department of Biotechnology and National Institute of Science, Education and Research, Bhubaneswar, under Department of Atomic Energy, Government of India. We wish to acknowledge the University Grant Commission (UGC), New Delhi, India for the fellowship of TKN.

Author Contributions: Conceived and designed the experiment: So.C., Su.C., T.K.N. and P.M. Contributed reagents/materials/analysis: So.C. and Su.C. Performed the experiments: T.K.N., P.M., A.K., L.P.K.S., and S.S.S. Wrote the paper: So.C., Su.C., T.K.N., P.M. and L.P.K.S.

References

1. Khatun, S.; Chakraborty, A.; Rahman, M.; Nasreen Banu, N.; Rahman, M.M.; Hasan, S.M.; Luby, S.P.; Gurley, E.S. An outbreak of chikungunya in rural bangladesh, 2011. *PLoS Negl. Trop. Dis.* **2015**, *9*, e0003907. [CrossRef] [PubMed]

2. Kosasih, H.; de Mast, Q.; Widjaja, S.; Sudjana, P.; Antonjaya, U.; Ma'roef, C.; Riswari, S.F.; Porter, K.R.; Burgess, T.H.; Alisjahbana, B.; et al. Evidence for endemic chikungunya virus infections in bandung, indonesia. *PLoS Negl. Trop. Dis.* **2013**, *7*, e2483. [CrossRef] [PubMed]

3. Powers, A.M.; Logue, C.H. Changing patterns of chikungunya virus: Re-emergence of a zoonotic arbovirus. *J. Gen. Virol.* **2007**, *88*, 2363–2377. [CrossRef] [PubMed]

4. Rodriguez-Barraquer, I.; Solomon, S.S.; Kuganantham, P.; Srikrishnan, A.K.; Vasudevan, C.K.; Iqbal, S.H.; Balakrishnan, P.; Solomon, S.; Mehta, S.H.; Cummings, D.A. The hidden burden of dengue and chikungunya in chennai, India. *PLoS Negl. Trop. Dis.* **2015**, *9*, e0003906. [CrossRef] [PubMed]

5. Robinson, M.C. An epidemic of virus disease in southern province, tanganyika territory, in 1952–53. I. Clinical features. *Trans. R. Soc. Trop. Med. Hyg.* **1955**, *49*, 28–32. [CrossRef]

6. Schwartz, O.; Albert, M.L. Biology and pathogenesis of chikungunya virus. *Nat. Rev. Microbiol.* **2010**, *8*, 491–500. [CrossRef] [PubMed]

7. Thiberville, S.D.; Boisson, V.; Gaudart, J.; Simon, F.; Flahault, A.; de Lamballerie, X. Chikungunya fever: A clinical and virological investigation of outpatients on Reunion Island, South-West Indian Ocean. *PLoS Negl. Trop. Dis.* **2013**, *7*, e2004. [CrossRef] [PubMed]

8. Mohd Zim, M.A.; Sam, I.C.; Omar, S.F.; Chan, Y.F.; AbuBakar, S.; Kamarulzaman, A. Chikungunya infection in Malaysia: Comparison with dengue infection in adults and predictors of persistent arthralgia. *J. Clin. Virol.* **2013**, *56*, 141–145. [CrossRef] [PubMed]

9. Borgherini, G.; Poubeau, P.; Staikowsky, F.; Lory, M.; le Moullec, N.; Becquart, J.P.; Wengling, C.; Michault, A.; Paganin, F. Outbreak of chikungunya on reunion island: Early clinical and laboratory features in 157 adult patients. *Clin. Infect. Dis.* **2007**, *44*, 1401–1407. [CrossRef] [PubMed]

10. Maiti, C.R.; Mukherjee, A.K.; Bose, B.; Saha, G.L. Myopericarditis following Chikungunya virus infection. *J. Indian Med. Assoc.* **1978**, *70*, 256–258. [PubMed]

11. Mittal, A.; Mittal, S.; Bharati, M.J.; Ramakrishnan, R.; Saravanan, S.; Sathe, P.S. Optic neuritis associated with Chikungunya virus infection in South India. *Arch. Ophthalmol.* **2007**, *125*, 1381–1386. [CrossRef] [PubMed]

12. Lalitha, P.; Rathinam, S.; Banushree, K.; Maheshkumar, S.; Vijayakumar, R.; Sathe, P. Ocular involvement associated with an epidemic outbreak of chikungunya virus infection. *Am. J. Ophthalmol.* **2007**, *144*, 552–556. [CrossRef] [PubMed]

13. Solanki, B.S.; Arya, S.C.; Maheshwari, P. Chikungunya disease with nephritic presentation. *Int. J. Clin. Pract.* **2007**, *61*. [CrossRef] [PubMed]

14. Simon, F.; Paule, P.; Oliver, M. Chikungunya virus-induced myopericarditis: Toward an increase of dilated cardiomyopathy in countries with epidemics? *Am. J. Trop. Med. Hyg.* **2008**, *78*, 212–213. [PubMed]

15. Chusri, S.; Siripaitoon, P.; Hirunpat, S.; Silpapojakul, K. Case reports of neuro-Chikungunya in Southern Thailand. *Am. J. Trop. Med. Hyg.* **2011**, *85*, 386–389. [CrossRef] [PubMed]

16. Renault, P.; Solet, J.L.; Sissoko, D.; Balleydier, E.; Larrieu, S.; Filleul, L.; Lassalle, C.; Thiria, J.; Rachou, E.; de Valk, H.; et al. A major epidemic of Chikungunya virus infection on reunion island, France, 2005–2006. *Am. J. Trop. Med. Hyg.* **2007**, *77*, 727–731. [PubMed]

17. Thiboutot, M.M.; Kannan, S.; Kawalekar, O.U.; Shedlock, D.J.; Khan, A.S.; Sarangan, G.; Srikanth, P.; Weiner, D.B.; Muthumani, K. Chikungunya: A potentially emerging epidemic? *PLoS Negl. Trop. Dis.* **2010**, *4*, e623. [CrossRef] [PubMed]

18. Strauss, J.H.; Strauss, E.G. The alphaviruses: Gene expression, replication, and evolution. *Microbiol. Rev.* **1994**, *58*, 491–562. [PubMed]

19. Couderc, T.; Chretien, F.; Schilte, C.; Disson, O.; Brigitte, M.; Guivel-Benhassine, F.; Touret, Y.; Barau, G.; Cayet, N.; Schuffenecker, I.; et al. A mouse model for Chikungunya: Young age and inefficient type-I interferon signaling are risk factors for severe disease. *PLoS Pathog.* **2008**, *4*, e29. [CrossRef] [PubMed]

20. Fros, J.J.; Liu, W.J.; Prow, N.A.; Geertsema, C.; Ligtenberg, M.; Vanlandingham, D.L.; Schnettler, E.; Vlak, J.M.; Suhrbier, A.; Khromykh, A.A.; et al. Chikungunya virus nonstructural protein 2 inhibits type I/II interferon-stimulated JAK-STAT signaling. *J. Virol.* **2010**, *84*, 10877–10887. [CrossRef] [PubMed]

21. Kelvin, A.A.; Banner, D.; Silvi, G.; Moro, M.L.; Spataro, N.; Gaibani, P.; Cavrini, F.; Pierro, A.; Rossini, G.; Cameron, M.J.; et al. Inflammatory cytokine expression is associated with Chikungunya virus resolution and symptom severity. *PLoS Negl. Trop. Dis.* **2011**, *5*, e1279. [CrossRef] [PubMed]

22. Kumar, S.; Jaffar-Bandjee, M.C.; Giry, C.; Connen de Kerillis, L.; Merits, A.; Gasque, P.; Hoarau, J.J. Mouse macrophage innate immune response to Chikungunya virus infection. *Virol. J.* **2012**, *9*. [CrossRef] [PubMed]

23. Priya, R.; Dhanwani, R.; Patro, I.K.; Rao, P.V.; Parida, M.M. Differential regulation of TLR mediated innate immune response of mouse neuronal cells following infection with novel ECSA genotype of Chikungunya virus with and without E1:A226V mutation. *Infect. Genet. Evol.* **2013**, *20*, 396–406. [CrossRef] [PubMed]

24. Priya, R.; Patro, I.K.; Parida, M.M. TLR3 mediated innate immune response in mice brain following infection with Chikungunya virus. *Virus Res.* **2014**, *189*, 194–205. [CrossRef] [PubMed]

25. Chirathaworn, C.; Poovorawan, Y.; Lertmaharit, S.; Wuttirattanakowit, N. Cytokine levels in patients with Chikungunya virus infection. *Asian Pac. J. Trop. Med.* **2013**, *6*, 631–634. [CrossRef]

26. Chow, A.; Her, Z.; Ong, E.K.; Chen, J.M.; Dimatatac, F.; Kwek, D.J.; Barkham, T.; Yang, H.; Renia, L.; Leo, Y.S.; et al. Persistent arthralgia induced by Chikungunya virus infection is associated with interleukin-6 and granulocyte macrophage colony-stimulating factor. *J. Infect. Dis.* **2011**, *203*, 149–157. [CrossRef] [PubMed]

27. Lohachanakul, J.; Phuklia, W.; Thannagith, M.; Thongsakulprasert, T.; Smith, D.R.; Ubol, S. Differences in response of primary human myoblasts to infection with recent epidemic strains of Chikungunya virus isolated from patients with and without myalgia. *J. Med. Virol.* **2015**, *87*, 733–739. [CrossRef] [PubMed]

28. Phuklia, W.; Kasisith, J.; Modhiran, N.; Rodpai, E.; Thannagith, M.; Thongsakulprasert, T.; Smith, D.R.; Ubol, S. Osteoclastogenesis induced by CHIKV-infected fibroblast-like synoviocytes: A possible interplay between synoviocytes and monocytes/macrophages in CHIKV-induced arthralgia/arthritis. *Virus Res.* **2014**, *177*, 179–188. [CrossRef] [PubMed]

29. Reddy, V.; Mani, R.S.; Desai, A.; Ravi, V. Correlation of plasma viral loads and presence of Chikungunya igm antibodies with cytokine/chemokine levels during acute Chikungunya virus infection. *J. Med. Virol.* **2014**, *86*, 1393–1401. [CrossRef] [PubMed]

30. Ng, L.F.; Chow, A.; Sun, Y.J.; Kwek, D.J.; Lim, P.L.; Dimatatac, F.; Ng, L.C.; Ooi, E.E.; Choo, K.H.; Her, Z.; et al. Il-1beta, IL-6, and rantes as biomarkers of Chikungunya severity. *PLoS ONE* **2009**, *4*, e4261. [CrossRef] [PubMed]

31. Daley-Bauer, L.P.; Wynn, G.M.; Mocarski, E.S. Cytomegalovirus impairs antiviral CD8+ T cell immunity by recruiting inflammatory monocytes. *Immunity* **2012**, *37*, 122–133. [CrossRef] [PubMed]

32. Pratheek, B.M.; Saha, S.; Maiti, P.K.; Chattopadhyay, S.; Chattopadhyay, S. Immune regulation and evasion of mammalian host cell immunity during viral infection. *Indian J. Virol.* **2013**, *24*, 1–15. [CrossRef] [PubMed]

33. Qin, X.; Yao, J.; Yang, F.; Nie, J.; Wang, Y.; Liu, P.C. Human immunodeficiency virus type 1 Nef in human monocyte-like cell line THP-1 expands treg cells via toll-like receptor 2. *J. Cell. Biochem.* **2011**, *112*, 3515–3524. [CrossRef] [PubMed]

34. Shi, C.; Pamer, E.G. Monocyte recruitment during infection and inflammation. *Nat. Rev. Immunol.* **2011**, *11*, 762–774. [CrossRef] [PubMed]

35. Veiga-Parga, T.; Sehrawat, S.; Rouse, B.T. Role of regulatory T cells during virus infection. *Immunol. Rev.* **2013**, *255*, 182–196. [CrossRef] [PubMed]

36. Her, Z.; Malleret, B.; Chan, M.; Ong, E.K.; Wong, S.C.; Kwek, D.J.; Tolou, H.; Lin, R.T.; Tambyah, P.A.; Renia, L.; et al. Active infection of human blood monocytes by chikungunya virus triggers an innate immune response. *J. Immunol.* **2010**, *184*, 5903–5913. [CrossRef] [PubMed]

37. Labadie, K.; Larcher, T.; Joubert, C.; Mannioui, A.; Delache, B.; Brochard, P.; Guigand, L.; Dubreil, L.;

Lebon, P.; Verrier, B.; et al. Chikungunya disease in nonhuman primates involves long-term viral persistence in macrophages. *J. Clin. Invest.* **2010**, *120*, 894–906. [CrossRef] [PubMed]

38. Gardner, J.; Anraku, I.; Le, T.T.; Larcher, T.; Major, L.; Roques, P.; Schroder, W.A.; Higgs, S.; Suhrbier, A. Chikungunya virus arthritis in adult wild-type mice. *J. Virol.* **2010**, *84*, 8021–8032. [CrossRef] [PubMed]

39. Hoarau, J.J.; Jaffar Bandjee, M.C.; Krejbich Trotot, P.; Das, T.; Li-Pat-Yuen, G.; Dassa, B.; Denizot, M.; Guichard, E.; Ribera, A.; Henni, T.; et al. Persistent chronic inflammation and infection by Chikungunya arthritogenic alphavirus in spite of a robust host immune response. *J. Immunol.* **2010**, *184*, 5914–5927. [CrossRef] [PubMed]

40. Arndt, V.; Rogon, C.; Hohfeld, J. To be, or not to be—Molecular chaperones in protein degradation. *Cell. Mol. Life Sci.* **2007**, *64*, 2525–2541. [CrossRef] [PubMed]

41. Ma, Y.; Hendershot, L.M. ER chaperone functions during normal and stress conditions. *J. Chem. Neuroanat.* **2004**, *28*, 51–65. [CrossRef] [PubMed]

42. McClellan, A.J.; Tam, S.; Kaganovich, D.; Frydman, J. Protein quality control: Chaperones culling corrupt conformations. *Nat. Cell Biol.* **2005**, *7*, 736–741. [CrossRef] [PubMed]

43. Welch, W.J. How cells respond to stress. *Sci. Am.* **1993**, *268*, 56–64. [CrossRef] [PubMed]

44. Sedger, L.; Ruby, J. Heat shock response to vaccinia virus infection. *J. Virol.* **1994**, *68*, 4685–4689. [PubMed]

45. Sullivan, C.S.; Pipas, J.M. The virus-chaperone connection. *Virology* **2001**, *287*, 1–8. [CrossRef] [PubMed]

46. Basha, W.; Kitagawa, R.; Uhara, M.; Imazu, H.; Uechi, K.; Tanaka, J. Geldanamycin, a potent and specific inhibitor of Hsp90, inhibits gene expression and replication of human cytomegalovirus. *Antivir. Chem. Chemother.* **2005**, *16*, 135–146. [CrossRef] [PubMed]

47. Dutta, D.; Bagchi, P.; Chatterjee, A.; Nayak, M.K.; Mukherjee, A.; Chattopadhyay, S.; Nagashima, S.; Kobayashi, N.; Komoto, S.; Taniguchi, K.; et al. The molecular chaperone heat shock protein-90 positively regulates rotavirus infectionx. *Virology* **2009**, *391*, 325–333. [CrossRef] [PubMed]

48. Geller, R.; Vignuzzi, M.; Andino, R.; Frydman, J. Evolutionary constraints on chaperone-mediated folding provide an antiviral approach refractory to development of drug resistance. *Genes Dev.* **2007**, *21*, 195–205. [CrossRef] [PubMed]

49. Hu, J.; Seeger, C. Hsp90 is required for the activity of a hepatitis B virus reverse transcriptase. *Proc. Natl. Acad. Sci. USA* **1996**, *93*, 1060–1064. [CrossRef] [PubMed]

50. Hung, J.J.; Chung, C.S.; Chang, W. Molecular chaperone Hsp90 is important for vaccinia virus growth in cells. *J. Virol.* **2002**, *76*, 1379–1390. [CrossRef] [PubMed]

51. Okamoto, T.; Nishimura, Y.; Ichimura, T.; Suzuki, K.; Miyamura, T.; Suzuki, T.; Moriishi, K.; Matsuura, Y. Hepatitis C virus RNA replication is regulated by FKBP8 and Hsp90. *EMBO J.* **2006**, *25*, 5015–5025. [CrossRef] [PubMed]

52. Sun, X.; Barlow, E.A.; Ma, S.; Hagemeier, S.R.; Duellman, S.J.; Burgess, R.R.; Tellam, J.; Khanna, R.; Kenney, S.C. Hsp90 inhibitors block outgrowth of EBV-infected malignant cells in vitro and in vivo through an EBNA1-dependent mechanism. *Proc. Natl. Acad. Sci. USA* **2010**, *107*, 3146–3151. [CrossRef] [PubMed]

53. Zheng, Z.Z.; Miao, J.; Zhao, M.; Tang, M.; Yeo, A.E.; Yu, H.; Zhang, J.; Xia, N.S. Role of heat-shock protein 90 in hepatitis E virus capsid trafficking. *J. Gen. Virol.* **2010**, *91*, 1728–1736. [CrossRef] [PubMed]

54. Kowalczyk, A.; Guzik, K.; Slezak, K.; Dziedzic, J.; Rokita, H. Heat shock protein and heat shock factor 1 expression and localization in vaccinia virus infected human monocyte derived macrophages. *J. Inflamm.* **2005**, *2*. [CrossRef] [PubMed]

55. Das, I.; Basantray, I.; Mamidi, P.; Nayak, T.K.; Pratheek, B.M.; Chattopadhyay, S.; Chattopadhyay, S. Heat shock protein 90 positively regulates Chikungunya virus replication by stabilizing viral non-structural protein nsP2 during infection. *PLoS ONE* **2014**, *9*, e100531. [CrossRef] [PubMed]

56. Rathore, A.P.; Haystead, T.; Das, P.K.; Merits, A.; Ng, M.L.; Vasudevan, S.G. Chikungunya virus nsP3 and nsP4 interacts with Hsp-90 to promote virus replication: Hsp-90 inhibitors reduce CHIKV infection and inflammation in vivo. *Antivir. Res.* **2014**, *103*, 7–16. [CrossRef] [PubMed]

57. Agnew, E.B.; Wilson, R.H.; Grem, J.L.; Neckers, L.; Bi, D.; Takimoto, C.H. Measurement of the novel antitumor agent 17-(allylamino)-17-demethoxygeldanamycin in human plasma by high-performance liquid chromatography. *J. Chromatogr. B Biomed. Sci. Appl.* **2001**, *755*, 237–243. [CrossRef]

58. Chattopadhyay, S.; Kumar, A.; Mamidi, P.; Nayak, T.K.; Das, I.; Chhatai, J.; Basantray, I.; Bramha, U.; Maiti, P.K.; Singh, S.; et al. Development and characterization of monoclonal antibody against non-structural protein-2 of Chikungunya virus and its application. *J. Virol. Methods* **2014**, *199*, 86–94. [CrossRef] [PubMed]

59. Kumar, A.; Mamidi, P.; Das, I.; Nayak, T.K.; Kumar, S.; Chhatai, J.; Chattopadhyay, S.; Suryawanshi, A.R. A novel 2006 Indian outbreak strain of Chikungunya virus exhibits different pattern of infection as compared to prototype strain. *PLoS ONE* **2014**, *9*, e85714. [CrossRef] [PubMed]

60. Mishra, P.; Kumar, A.; Mamidi, P.; Kumar, S.; Basantray, I.; Saswat, T.; Das, I.; Nayak, T.K.; Chattopadhyay, S.; Subudhi, B.B.; et al. Inhibition of chikungunya virus replication by 1-[(2-methylbenzimidazol-1-yl) methyl]-2-oxo-indolin-3-ylidene] amino] thiourea(mbzm-n-ibt). *Sci. Rep.* **2016**, *6*. [CrossRef] [PubMed]

61. Sourisseau, M.; Schilte, C.; Casartelli, N.; Trouillet, C.; Guivel-Benhassine, F.; Rudnicka, D.; Sol-Foulon, N.; le Roux, K.; Prevost, M.C.; Fsihi, H.; et al. Characterization of reemerging Chikungunya virus. *PLoS Pathog.* **2007**, *3*, e89. [CrossRef] [PubMed]

62. Barber, G.N. Host defense, viruses and apoptosis. *Cell Death Differ.* **2001**, *8*, 113–126. [CrossRef] [PubMed]

63. Coulombe, F.; Jaworska, J.; Verway, M.; Tzelepis, F.; Massoud, A.; Gillard, J.; Wong, G.; Kobinger, G.; Xing, Z.; Couture, C.; et al. Targeted prostaglandin E2 inhibition enhances antiviral immunity through induction of type I interferon and apoptosis in macrophages. *Immunity* **2014**, *40*, 554–568. [CrossRef] [PubMed]

64. Krejbich-Trotot, P.; Denizot, M.; Hoarau, J.J.; Jaffar-Bandjee, M.C.; Das, T.; Gasque, P. Chikungunya virus mobilizes the apoptotic machinery to invade host cell defenses. *FASEB J.* **2011**, *25*, 314–325. [CrossRef] [PubMed]

65. He, K.; Zhou, H.R.; Pestka, J.J. Mechanisms for ribotoxin-induced ribosomal RNA cleavage. *Toxicol. Appl. Pharmacol.* **2012**, *265*, 10–18. [CrossRef] [PubMed]

66. Chen, W.; Foo, S.S.; Taylor, A.; Lulla, A.; Merits, A.; Hueston, L.; Forwood, M.R.; Walsh, N.C.; Sims, N.A.; Herrero, L.J.; et al. Bindarit, an inhibitor of monocyte chemotactic protein synthesis, protects against bone loss induced by Chikungunya virus infection. *J. Virol.* **2015**, *89*, 581–593. [CrossRef] [PubMed]

67. Deshmane, S.L.; Kremlev, S.; Amini, S.; Sawaya, B.E. Monocyte chemoattractant protein-1 (MCP-1): An overview. *J. Interf. Cytokine Res.* **2009**, *29*, 313–326. [CrossRef] [PubMed]

68. Aleyas, A.G.; George, J.A.; Han, Y.W.; Rahman, M.M.; Kim, S.J.; Han, S.B.; Kim, B.S.; Kim, K.; Eo, S.K. Functional modulation of dendritic cells and macrophages by Japanese encephalitis virus through MyD88 adaptor molecule-dependent and -independent pathways. *J. Immunol.* **2009**, *183*, 2462–2474. [CrossRef] [PubMed]

69. Hansen, T.H.; Bouvier, M. MHC class I antigen presentation: Learning from viral evasion strategies. *Nat. Rev. Immunol.* **2009**, *9*, 503–513. [CrossRef] [PubMed]

70. Hegde, N.R.; Chevalier, M.S.; Johnson, D.C. Viral inhibition of MHC class II antigen presentation. *Trends Immunol.* **2003**, *24*, 278–285. [CrossRef]

71. Khan, N.; Gowthaman, U.; Pahari, S.; Agrewala, J.N. Manipulation of costimulatory molecules by intracellular pathogens: Veni, vidi, vici!! *PLoS Pathog.* **2012**, *8*, e1002676. [CrossRef] [PubMed]

72. Schulte, T.W.; Neckers, L.M. The benzoquinone ansamycin 17-allylamino-17-demethoxygeldanamycin binds to Hsp90 and shares important biologic activities with geldanamycin. *Cancer Chemother. Pharmacol.* **1998**, *42*, 273–279. [CrossRef] [PubMed]

73. Wikan, N.; Sakoonwatanyoo, P.; Ubol, S.; Yoksan, S.; Smith, D.R. Chikungunya virus infection of cell lines: Analysis of the east, central and south african lineage. *PLoS ONE* **2012**, *7*, e31102. [CrossRef] [PubMed]

74. Abraham, R.; Mudaliar, P.; Padmanabhan, A.; Sreekumar, E. Induction of cytopathogenicity in human glioblastoma cells by Chikungunya virus. *PLoS ONE* **2013**, *8*, e75854. [CrossRef] [PubMed]

75. Baer, A.; Lundberg, L.; Swales, D.; Waybright, N.; Pinkham, C.; Dinman, J.D.; Jacobs, J.L.; Kehn-Hall, K. Venezuelan equine encephalitis virus induces apoptosis through the unfolded protein response activation of EGR1. *J. Virol.* **2016**, *90*, 3558–3572. [CrossRef] [PubMed]

76. Dhanwani, R.; Khan, M.; Bhaskar, A.S.; Singh, R.; Patro, I.K.; Rao, P.V.; Parida, M.M. Characterization of chikungunya virus infection in human neuroblastoma SH-SY5Y cells: Role of apoptosis in neuronal cell death. *Virus Res.* **2012**, *163*, 563–572. [CrossRef] [PubMed]

77. Way, S.J.; Lidbury, B.A.; Banyer, J.L. Persistent ross river virus infection of murine macrophages: An in vitro model for the study of viral relapse and immune modulation during long-term infection. *Virology* **2002**, *301*, 281–292. [CrossRef] [PubMed]

78. Fujimoto, I.; Pan, J.; Takizawa, T.; Nakanishi, Y. Virus clearance through apoptosis-dependent phagocytosis of influenza a virus-infected cells by macrophages. *J. Virol.* **2000**, *74*, 3399–3403. [CrossRef] [PubMed]

79. Van den Berg, E.; van Woensel, J.B.; Bem, R.A. Apoptosis in pneumovirus infection. *Viruses* **2013**, *5*, 406–422. [CrossRef] [PubMed]

80. Deszcz, L.; Cencic, R.; Sousa, C.; Kuechler, E.; Skern, T. An antiviral peptide inhibitor that is active against picornavirus 2A proteinases but not cellular caspases. *J. Virol.* **2006**, *80*, 9619–9627. [CrossRef] [PubMed]

81. Deszcz, L.; Seipelt, J.; Vassilieva, E.; Roetzer, A.; Kuechler, E. Antiviral activity of caspase inhibitors: Effect on picornaviral 2A proteinase. *FEBS Lett.* **2004**, *560*, 51–55. [CrossRef]

82. Kim, M.S.; Lee, J.A.; Kim, K.H. Effects of a broad-spectrum caspase inhibitor, z-vad(ome)-fmk, on viral hemorrhagic septicemia virus (VHSV) infection-mediated apoptosis and viral replication. *Fish Shellfish Immunol.* **2016**, *51*, 41–45. [CrossRef] [PubMed]

83. Martin, U.; Jarasch, N.; Nestler, M.; Rassmann, A.; Munder, T.; Seitz, S.; Zell, R.; Wutzler, P.; Henke, A. Antiviral effects of pan-caspase inhibitors on the replication of coxsackievirus B3. *Apoptosi* **2007**, *12*, 525–533. [CrossRef] [PubMed]

84. Benedict, C.A. Viruses and the TNF-related cytokines, an evolving battle. *Cytokine Growth Factor Rev.* **2003**, *14*, 349–357. [CrossRef]

85. Mogensen, T.H.; Paludan, S.R. Molecular pathways in virus-induced cytokine production. *Microbiol. Mol. Biol. Rev.* **2001**, *65*, 131–150. [CrossRef] [PubMed]

86. Trinchieri, G. Cytokines acting on or secreted by macrophages during intracellular infection (IL-10, IL-12, IFN-gamma). *Curr. Opin. Immunol.* **1997**, *9*, 17–23. [CrossRef]

87. Redpath, S.; Ghazal, P.; Gascoigne, N.R. Hijacking and exploitation of IL-10 by intracellular pathogens. *Trends Microbiol.* **2001**, *9*, 86–92. [CrossRef]

88. Pratheek, B.M.; Suryawanshi, A.R.; Chattopadhyay, S.; Chattopadhyay, S. In silico analysis of MHC-I restricted epitopes of Chikungunya virus proteins: Implication in understanding anti-CHIKV CD8[+] T cell response and advancement of epitope based immunotherapy for CHIKV infection. *Infect. Genet. Evol.* **2015**, *31*, 118–126. [CrossRef] [PubMed]

89. Georgakis, G.V.; Li, Y.; Rassidakis, G.Z.; Medeiros, L.J.; Younes, A. The Hsp90 inhibitor 17-AAG synergizes with doxorubicin and U0126 in anaplastic large cell lymphoma irrespective of Alk expression. *Exp. Hematol.* **2006**, *34*, 1670–1679. [CrossRef] [PubMed]

90. Nimmanapalli, R.; O'Bryan, E.; Kuhn, D.; Yamaguchi, H.; Wang, H.G.; Bhalla, K.N. Regulation of 17-AAG-induced apoptosis: Role of Bcl-2, Bcl-XL, and bax downstream of 17-AAG-mediated down-regulation of Akt, Raf-1, and Src kinases. *Blood* **2003**, *102*, 269–275. [CrossRef] [PubMed]

Feline Panleucopenia Virus NS2 Suppresses the Host IFN-β Induction by Disrupting the Interaction between TBK1 and STING

Hongtao Kang [1,†], Dafei Liu [1,†], Jin Tian [1], Xiaoliang Hu [1], Xiaozhan Zhang [1], Hang Yin [2], Hongxia Wu [1], Chunguo Liu [1], Dongchun Guo [1], Zhijie Li [1], Qian Jiang [1], Jiasen Liu [1] and Liandong Qu [1,*]

[1] Division of Zoonosis of Natural Foci, State Key Laboratory of Veterinary Biotechnology, Harbin Veterinary Research Institute, Chinese Academy of Agricultural Sciences, 678 Haping road, Xiangfang District, Harbin 150000, China; kang1989462@sina.com (H.K.); hattman324@163.com (D.L.); tj6049345@126.com (J.T.); Liang679@163.com (X.H.); xiaozhan063@163.com (X.Z.); whx450650@163.com (H.W.); liuchunguo@caas.cn (C.L.); gdongchun@126.com (D.G.); lizhijie.1982@126.com (Z.L.); jiangqian623@sina.com (Q.J.); neauljs@163.com (J.L.)

[2] College of Veterinary Medicine, Northeast Agricultural University, Harbin 150000, China; yinhang_marine@126.com

* Correspondence: qld@hvri.ac.cn

† These authors contributed equally to this work.

Academic Editor: Joanna Parish

Abstract: Feline panleucopenia virus (FPV) is a highly infectious pathogen that causes severe diseases in pets, economically important animals and wildlife in China. Although FPV was identified several years ago, little is known about how it overcomes the host innate immunity. In the present study, we demonstrated that infection with the FPV strain Philips-Roxane failed to activate the interferon β (IFN-β) pathway but could antagonize the induction of IFN stimulated by Sendai virus (SeV) in F81 cells. Subsequently, by screening FPV nonstructural and structural proteins, we found that only nonstructural protein 2 (NS2) significantly suppressed IFN expression. We demonstrated that the inhibition of SeV-induced IFN-β production by FPV NS2 depended on the obstruction of the IFN regulatory factor 3 (IRF3) signaling pathway. Further, we verified that NS2 was able to target the serine/threonine-protein kinase TBK1 and prevent it from being recruited by stimulator of interferon genes (STING) protein, which disrupted the phosphorylation of the downstream protein IRF3. Finally, we identified that the C-terminus plus the coiled coil domain are the key domains of NS2 that are required for inhibiting the IFN pathway. Our study has yielded strong evidence for the FPV mechanisms that counteract the host innate immunity.

Keywords: Feline panleucopenia virus (FPV); Nonstructural protein 2 (NS2); interferon β (IFN-β)

1. Introduction

Feline panleukopenia is an acute, highly contagious, and fatal infectious disease. The causative pathogen, feline panleucopenia virus (FPV), is a small (18–25 nm) negative-sense single stranded DNA virus, which has the widest host range and highest pathogenicity of the viruses in the carnivore parvovirus subgroup. Many carnivore parvoviruses, including canine parvovirus and mink enteritis virus, evolved from FPV to adapt into new hosts. Its genome includes two major open reading frames that express the nonstructural (NS) proteins (NS1 and NS2) and capsid protein (VP), including VP1 and VP2 by alternative splicing. The viruses replicate using the host cell polymerases and other DNA replication machinery, which causes the virus to infect the rapidly dividing cells [1,2]. In 2014, in Zhengzhou, Henan Province, and in 2015 in Guangyuan, Sichuan Province of China, it was reported that giant pandas were infected with FPV. Furthermore, in recent years, FPV has been epidemic in pets, economically important animals, and wildlife in China. Clinically, the disease has a high rate of infection, atypical symptoms, long-term infectious carriers and a significantly higher prevalence of co-infection with feline calicivirus (FCV) and feline herpesvirus 1 (FHV-1), which may be due to an insufficient host immune response against FPV infection, suggesting that FPV has the ability to antagonize the host antiviral response.

The innate immune system is the first line of host defenses against viral infection, and the induction of interferon (IFN)-α/β is a crucial antiviral mechanism of the innate immune system, which plays an important role in the defense against invading viruses, the termination of early viral replication and the development of an adaptive immune response [3,4]. The initiation of IFN expression is triggered by pathogen-associated molecular patterns (PAMPs) through host pattern recognition receptors (PRRs) [5,6]. When a host is invaded by a virus, PRRs transmit signals to different downstream adaptor molecules such as mitochondrial antiviral-signaling protein (MAVS), TIR-domain-containing adapter-inducing interferon-β (TRIF), and MyD88 (myeloid differentiation primary response gene 88) to recruit IκB kinase (IKK)-related kinases. With the help of the IKK-related kinases, transcription factors, including IFN regulatory factors (IRFs), subunits of the nuclear factor (NF)-κB/Rel family, and ATF-2/c-Jun (AP-1), are activated by phosphorylation and translocate into the nucleus. Once in the nucleus, the transcription factors bind to their specific binding elements, termed positive regulatory domains (PRDs), within the IFN-β promoter region to initiate IFN transcription [7,8].

Due to the powerful antiviral effect of type I interferon, the virus must encode some components that inhibit the host interferon signaling pathway for survival during the co-evolution of the virus and host. In the present study, we found that FPV NS2 could inhibit Sendai virus (SeV) mediated IFN-β induction by disrupting the TBK1–STING interaction and identified that the C-terminus plus the coiled coil domain are the regulatory elements of NS2 that inhibit the IRF3 signaling pathway. To the best of our knowledge, there is no related research about the inhibition of type I interferon induction by FPV. Our work reveals a novel mechanism that explains how FPV NS2 efficiently inhibits host innate immunity.

2. Materials and Methods

2.1. Viruses and Cells

F81 cells were obtained from the Cell Resource Center of the Shanghai Institutes for Biological Science, Chinese Academy of Science, and maintained in Roswell Park Memorial Institute (RPMI) 1640 medium with 10% fetal bovine serum (Life Technologies, San Diego, CA, USA) at 37 °C with 5% CO_2. The FPV strain Philips-Roxane was obtained from American Type Culture Collection (ATCC).

2.2. Plasmids

The plasmids pIFN-β-Luc, p3×PRDII-Luc, PRDIII/I-Luc, and p6×PRDIV-Luc contain a luciferase (Luc) expression cassette driven by the feline IFN-β promoter, three copies of the NF-κB binding region, three copies of the IRF3 binding region and six copies of the AP-1 binding region, respectively, which have been described previously [9]. The pRL-TK plasmid (Promega, Madison, WI, USA), a vector that encodes Renilla luciferase, was used as an internal control for the normalization of gene transfection efficiency. The p3×Flag-MAVS and p3×Flag-STING vectors encode the Flag-tagged feline MAVS and STING proteins, respectively [10,11]. The p3×Flag-VP1, p3×Flag-VP2, p3×Flag-NS1, and p3×Flag-NS2 plasmids express Flag-tagged FPV VP1, VP2, NS1, and NS2, respectively. The pMyc-TBK1 plasmid produces the Myc-tagged feline TBK1. The pEGFP-IRF3-5D plasmid expresses the feline IRF3 with an EGFP tag. The feline TBK1 sequence was amplified from the plasmid pMyc-TBK1 and inserted into the pDsRed2-C1 plasmid (pDsred-TBK1). The FPV NS2 sequence was amplified from the plasmid p3×Flag-NS2 and inserted into the pEGFP-C1 plasmid (pEGFP-NS2) and pcDNA™3.1/V5-Hi plasmid (pV5-NS2).

2.3. Transfections and Luciferase Reporter Assays

F81 cells were seeded in 48-well plates prior to transfection and then co-transfected with luciferase reporter plasmids, the internal control pRL-TK plasmid and various expression plasmids or an empty control plasmid. After 24 h of co-transfection, the cells were inoculated with SeV at an MOI of 10 as an interferon pathway activator. The cells were lysed with 65 μl of Passive Lysis buffer (Promega) 12 h after stimulation, and the supernatants were used to measure firefly and Renilla luciferase activities using the Dual Luciferase Assay System according to the manufacturer's instructions (Promega). The data are presented as relative firefly luciferase activities normalized to Renilla luciferase activities (means ± SD) and are representative of three independent experiments.

2.4. Western Blotting

The cell lysates were separated using 12% SDS-PAGE and then transferred onto nitrocellulose membranes (Millipore, Bedford, MA, USA). The membranes were blocked with tris-buffered saline Tween 20 (TBST) containing 5% skim milk for 2 h at room temperature (RT) and then incubated with the indicated primary antibodies for 1 h at RT or overnight at 4 °C. After being washed with TBST, the membranes were incubated with an IRDye 800-conjugated secondary antibody (Rockland Immunochemicals, Limerick, PA, USA) for 1 h at RT. The membranes were washed three times in TBST and then visualized and analyzed with an Odyssey Infrared Imaging System (LI-COR Biosciences, Lincoln, NE, USA). The intensities of bands were analyzed with Image J1.49 software.

2.5. Co-Immunoprecipitation (Co-IP) Analysis

F81 cells were co-transfected with pMyc-TBK1 and p3×Flag-NS2 or with pV5-NS2, pMyc-TBK1, and p3×Flag-STING using Lipofectamine 2000 (Invitrogen, Grand Island, NY, USA). Thirty-six hours later, the cells were lysed in radioimmunoprecipitation assay (RIPA) lysis buffer (50 mM Tris–HCl, pH 7.4, 150 mM NaCl, 1% NP-40, 0.5% sodium deoxycholate, 0.1% SDS, 1 mM phenylmethylsulfonyl fluoride (PMSF)) and incubated on ice with shaking. Next, the cell lysates were centrifuged for 10 min at 12,000 g. The supernatants were incubated with equilibrated ANTI-FLAG M2 magnetic beads (Sigma-Aldrich, Saint Louis, MO, USA) for 1 h at RT with gentle mixing. After the binding step, the magnetic beads were collected by a magnetic separator, and the supernatant was removed. Then, the bead-protein complexes were washed with TBS buffer to remove all of the nonspecifically bound proteins. The proteins were eluted from the beads in 40 μL of 2×SDS-PAGE sample buffer and subjected to boiling for 10 min. The proteins were separated by SDS-PAGE and transferred to nitrocellulose membranes for western blot analyses. Each experiment was repeated at least three times.

2.6. Separation of Cytoplasmic and Nuclear Extracts

The cytoplasmic and nuclear extracts of F81 cells were separated using NE-PER® Nuclear and Cytoplasmic Extraction Reagents according to the manufacturer's instructions (Thermo Fisher Scientific, Rockford, IL, USA). Briefly, after being transfected with p3×Flag-NS2 and stimulated with SeV the F81 cells were harvested with trypsin-EDTA and then centrifuged at 500× g for 5 min. After centrifugation, the cell pellets were washed with phosphate buffered saline (PBS), and the supernatant was discarded. Next, the cell pellets were homogenized in Cytoplasmic Extraction Reagent I (CER I) (Thermo Fisher Scientific) and vortexed to fully suspend the cell pellet. The cell pellets were then incubated on ice. Cytoplasmic Extraction Reagent II (CER II) was added to the cell pellet. After vortexing and centrifugation, the supernatants (cytoplasmic extract) were transferred to a clean tube. The insoluble (pellet) fraction was suspended in Nuclear Extraction Reagent (NER). After vortexing and centrifugation, the supernatants (nuclear extract) were transferred to a clean tube. The cytoplasmic and nuclear extracts were analyzed by western blotting.

2.7. Subcellular Localization

To further investigate the interaction between the feline TBK1 protein and FPV NS2, F81 cells were plated in Nunc® glass-bottom dishes (Sigma-Aldrich) and co-transfected with the pDsred-TBK1 plasmid with red fluorescence and pEGFP-NS2 plasmid with green fluorescence. Twenty-four hours after transfection, the cells were washed with PBS, fixed with 4% paraformaldehyde at 4 °C for 30 min and then permeabilized with 0.2% Triton X-100 for 15 min at RT, followed by staining with 1 μM DAPI (Sigma-Aldrich). Fluorescent images were obtained with a confocal laser scanning microscope (Leica Microsystems, Heidelberg GmbH, Mannheim, Germany).

2.8. Statistical Analysis

The data were analyzed using a one-way analysis of variance (ANOVA) followed by the Tukey-Kramer test. All analyses were performed using the GraphPad Prism (version 6.03) software. Significant differences between the experimental groups were established as p-values less than 0.001(***), 0.01(**), or 0.05(*).

3. Results

3.1. FPV Infection Fails to Activate IFN-β and Interrupts SeV-Mediated IFN-β Induction

To explore if FPV affects IFN-β induction, the IFN-β promoter luciferase reporter system was used to analyze IFN-β expression after FPV infection. F81 cells were co-transfected with the luciferase reporter plasmid IFN-β-Luc and the internal control plasmid pRL-TK, followed by a mock infection or infection with FPV. After 12 h of co-transfection, the cells were stimulated with SeV (SeV+) or were left untreated (SeV-). The cells were lysed 8–12 h after stimulation, and both the firefly and Renilla luciferase activities were evaluated. As shown in Figure 1, without SeV stimulation, no significant differences in luciferase activity were detected between cells infected with FPV and mock-infected cells, suggesting that FPV infection could not activate the IFN-β promoter. Next, with SeV stimulation, the cells infected with FPV produced a significantly lower level of luciferase activity compared with the mock-infected cells, which indicated that FPV infection inhibits SeV-induced IFN-β promoter activity and interrupts SeV-mediated IFN-β production.

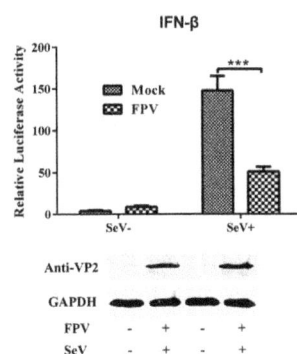

Figure 1. Effect of feline panleucopenia virus (FPV) on interferon (IFN)-β production and Sendai virus (SeV)-mediated IFN-β production as indicated by luciferase activity in F81 cells. The cells in this experiment were co-transfected with IFN-β-Luc and the Renilla luciferase construct pRL-TK, followed by a mock infection or infection with FPV. After 12 h of co-transfection, the cells were stimulated with SeV (SeV+) or left untreated (SeV-). The data represent the relative firefly luciferase activity normalized to the Renilla luciferase activity. The data represent the mean values of three independent experiments. The error bars represent standard deviations, and the asterisks indicate significant differences (*: $p < 0.05$; **: $p < 0.01$; ***: $p < 0.001$) between groups. The FPV infection was monitored by immunoblotting using a mouse anti-capsid protein 2 (VP2) antibody, and glyceraldehyde 3-phosphate dehydrogenase (GAPDH) was used as a loading control.

3.2. FPV NS2 as a Negative Regulator Impedes SeV-Mediated IFN-β Induction

We next evaluated which viral protein(s) could modulate the IFN-β induction. F81 cells were co-transfected with IFN-β-Luc, pRL-TK, and a plasmid expressing one of the FPV viral proteins (VP1, VP2, NS1, and NS2). As shown in Figure 2A, we found that NS2 could significantly inhibit the activation of the IFN-β promoter. We then validated the NS2-mediated inhibition of the IFN-β induction by measuring the IFN-β expression in cells stimulated with SeV after transfection with NS2 (Figure 2B). In accordance with the IFN-β promoter activity, in the presence of NS2, the expression of IFN-β was decreased after the cells were infected with SeV for 12 h compared with the mock group without NS2 transfection. These results indicated that FPV NS2 is a negative regulator of type I IFN induction and that the inhibitory effect occurred in a dose-dependent manner (Figure 2C). These results consistently supported the suppression of IFN-β induction by NS2.

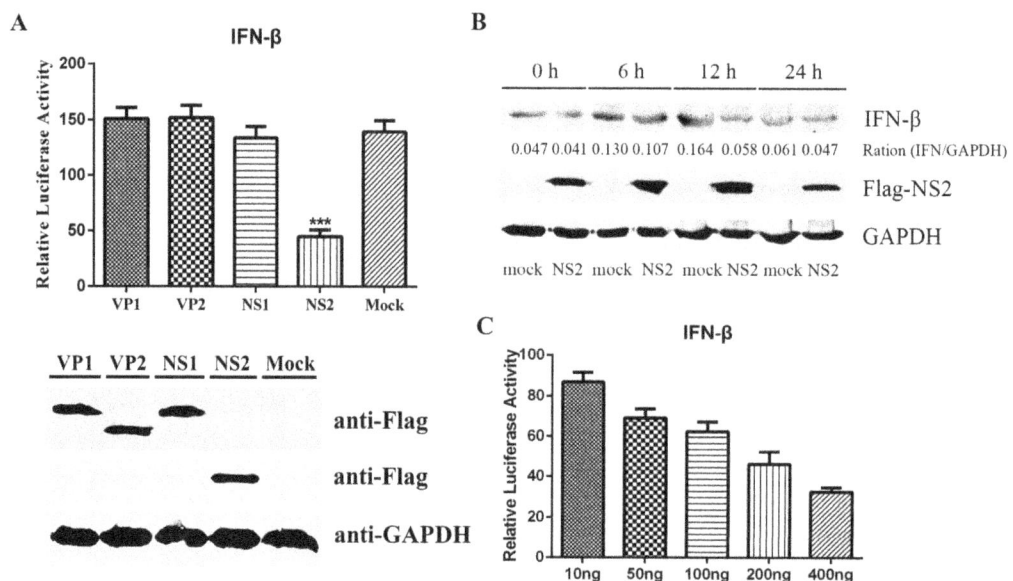

Figure 2. FPV NS2 as a negative regulator impedes SeV-mediated IFN-β induction. (**A**) Effects of protein-coding genes of FPV on the SeV-induced IFN-β promoter activation in F81 cells. The cells in this experiment were co-transfected with IFN-β-Luc, the Renilla luciferase construct pRL-TK and one of the recombinant plasmids pFlag-vp1, pFlag-vp2, pFlag-ns1, or pFlag-ns2. Twenty-four hours later, the cells were stimulated with SeV. The luciferase activity was measured at 12 h after simulation. The values were normalized to the Renilla activity. The data represent the mean values of three independent experiments. The error bars represent standard deviations, and "*" indicates significant differences ($p < 0.05$) between groups. The expression of VP1, VP2, NS1, or NS2 was monitored by immunoblotting using a mouse anti-Flag antibody; GAPDH was used as a loading control. (**B**) The SeV-mediated IFN-β expression is disrupted by NS2. F81 cells were transfected with p3×Flag-NS2. At 12 h post transfection, the cells were inoculated with SeV. The cell lysates at 0, 6, 12 and 24 h after SeV infection were analyzed by immunoblotting (IB) using anti-Flag and anti-IFN-β antibodies. (**C**). NS2 inhibits IFN promoter activity in a dose-dependent manner. F81 cells were co-transfected with IFN-β-Luc, pRL-TK and different amounts of p3×Flag-NS2 (10, 50, 100, 200 or 400 ng). At 12 h post transfection, the cells were inoculated with SeV. Twelve hours after infection, the cells were harvested, and the luciferase activities were measured. The values were normalized to the Renilla activity. The data represent the mean values of three independent experiments. The error bars represent standard deviations, and the asterisks indicate significant differences (*: $p < 0.05$; **: $p < 0.01$; ***: $p < 0.001$) between groups.

3.3. NS2 Interrupts the SeV-Mediated Activation of IFN-β by Blocking the IRF3 Pathway

To investigate which signal pathway(s) of type I IFN induction were inhibited by NS2, F81 cells were co-transfected with the luciferase reporter plasmids pNF-κB-Luc, pIRF3-Luc or pAP-1-Luc, pRL-TK, and p3×Flag-NS2 (p3×Flag as control). After 12 h of co-transfection, the cells were stimulated with SeV. The cells were lysed 12 h after infection, and the cell supernatant was used to evaluate the firefly and Renilla luciferase activities. Compared to the cells transfected with pIRF3-Luc and empty vector, the luciferase activity of the cells transfected with pIRF3-Luc and p3×Flag-NS2 were significantly reduced, as shown in Figure 3B. However, the same results were not observed in cells transfected with pNF-κB-Luc and p3×Flag-NS2 or pAP-1-Luc and p3×Flag-NS2 (Figure 3A,C). Furthermore, we examined the phosphorylation level of IRF3 in the cells stimulated with SeV after transfection with NS2, as shown in Figure 3D. From 0 h to 9 h, the level of phosphorylated IRF3 increased after SeV inoculation, but in the presence of NS2, the level of phosphorylated IRF3 decreased after SeV infection. These results indicated that NS2 could impede SeV-mediated activation of the transcription factor IRF3 but does not suppress the activation of the transcription factors NF-κB and AP-1. Hence, we hypothesized that NS2 interrupts the SeV-mediated activation of IFN-β by blocking the IRF3 pathway.

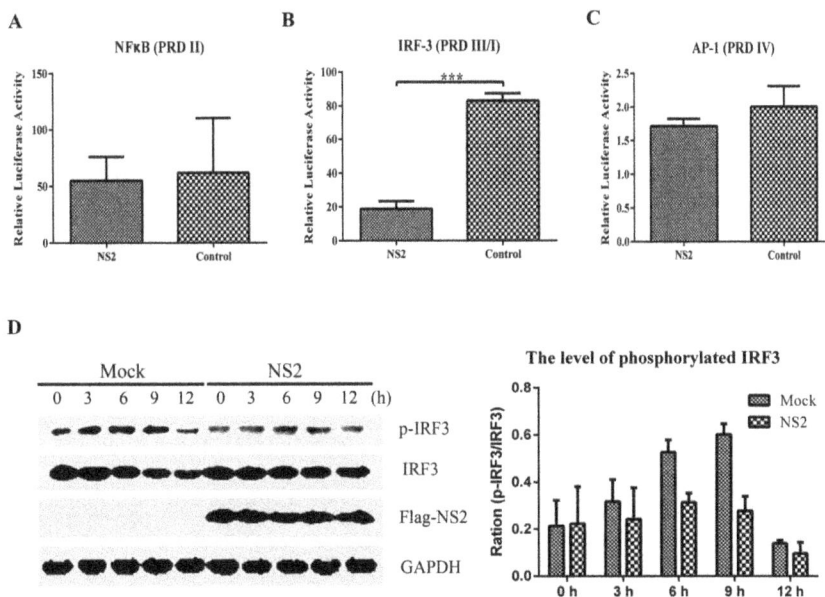

Figure 3. Inhibition of the activity of transcription factors on specific PRDs of the IFN-β promoter by NS2. F81 cells were co-transfected with p3×Flag-NS2, pRL-TK and either p3×PRDII-Luc (**A**), PRDIII/I-Luc (**B**), or p6×PRDIV-Luc (**C**). At 12 h post-transfection, the cells were inoculated with SeV. Twelve hours after infection, the cells were harvested and the luciferase activities were measured. The values were normalized to the Renilla activity. The error bars represent standard deviations, and the asterisks indicate significant differences (*: $p < 0.05$; **: $p < 0.01$; ***: $p < 0.001$) between groups. (**D**) The phosphorylation levels of IRF3 in the cells stimulated with SeV after transfection with NS2. F81 cells were transfected with p3×Flag-NS2. At 12 h post-transfection, the cells were inoculated with SeV. The cell lysates at 0, 3, 6, 9 and 12 h after SeV infection were analyzed by immunoblotting using antibodies against IRF3 or phosphorylated IRF3 and the quantification of relative p-IRF3 band intensities to IRF3 was showed with histogram.

Next, we screened which adaptor molecules in the IRF3 pathway were inhibited by NS2. F81 cells were co-transfected with IRF3-Luc, p3×Flag-NS2 and a plasmid expressing one of several adaptors in the IRF3 signaling pathway, including MAVS, STING, TBK1, and IRF3-5D. The cell lysates were used to measure the firefly and Renilla luciferase activities at 24 h post-transfection. As shown in Figure 4, the transfection of FPV NS2 significantly suppressed the activation of the IFN-β promoter stimulated by molecules upstream of IRF3 such as MAVS, STING, and TBK1 but did not counteract the IRF3-5D-mediated IRF3 activation. These results provide evidence supporting the hypothesis that

FPV NS2 interrupts SeV-mediated IFN-β induction by acting on the proteins upstream of IRF3 and that TBK1 appears to be a target protein of NS2 suppression.

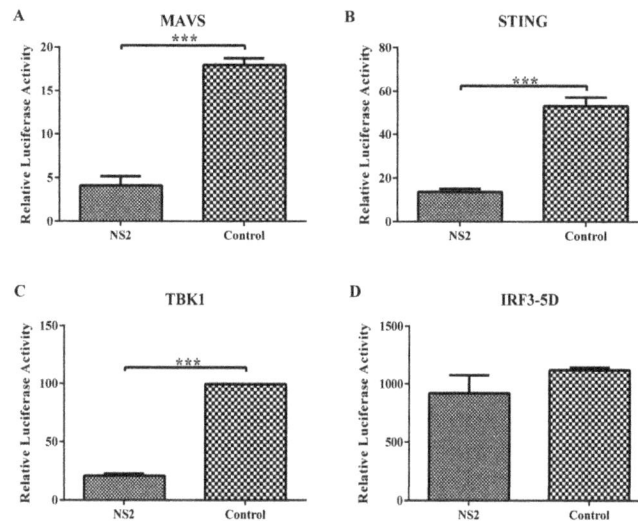

Figure 4. FPV NS2 blocks interferon induction at a step upstream of IRF3. F81 cells were co-transfected with IRF3-Luc, pRL-TK, p3×Flag-NS2 and a plasmid expressing one of the molecules in the IRF3 signaling pathway: MAVS (**A**), STING (**B**), TBK1 (**C**), and IRF3-5D (**D**), respectively. The cell lysates were used to measure the firefly and Renilla luciferase activities at 24 h post-transfection. The values were normalized to the Renilla activity. The data represent the mean values of three independent experiments. The error bars represent standard deviations, and the asterisks indicate significant differences (*: $p < 0.05$; **: $p < 0.01$; ***: $p < 0.001$) between groups.

3.4. NS2 Can Interact Directly with TBK1 and Disrupt the TBK1–STING Interaction

To confirm whether FPV NS2 interacted with TBK1, co-immunoprecipitation and western blotting assays were used, and cells were co-transfected with p3×Flag-NS2 and pMyc-TBK1. As shown in Figure 5A, Flag-tagged NS2 interacted with Myc-tagged TBK1, demonstrating the association between NS2 and TBK1. Furthermore, this result was validated by the confocal assay. The results showed the co-localization of NS2 and TBK1 in the cytoplasm, with a co-localization coefficient of 0.934 based on the digital analysis of cell images. Taken together, our data demonstrate that TBK1 interacts directly with FPV NS2.

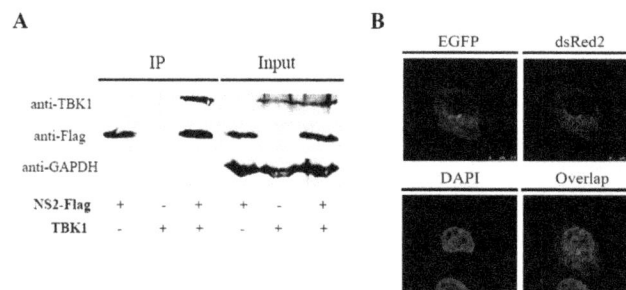

Figure 5. Analysis of the interaction between NS2 and TBK1. (**A**) Flag-tagged NS2 was co-transfected with Myc-tagged TBK1 into F81 cells for 36 h. The cell lysates were immunoprecipitated using ANTI-FLAG M2 magnetic beads. The whole-cell lysates and immunoprecipitation complexes were analyzed by immunoblotting using anti-Flag or anti-Myc antibodies. (**B**) F81 cells were co-transfected with the pDsred-TBK1 plasmid and pEGFP-NS2 plasmid. Twenty-four hours after transfection, the nuclei were stained with DAPI. The fluorescent images of the cells were obtained with a confocal laser scanning microscope.

To determine whether the interaction between NS2 and TBK1 affects the function of TBK1, we assessed the impact of NS2 on the assembly of the STING-TBK1 complex. p3×Flag-STING and pMyc-TBK1 were co-transfected into F81 cells in the absence or presence of pV5-NS2. The cell lysates were then co-immunoprecipitated. We observed that the co-immunoprecipitation of STING with TBK1 was disrupted in the presence of NS2. Moreover, the phosphorylation of STING was also reduced when the cells were co-transfected with p3×Flag-STING and pV5-NS2, as determined by western blot analysis (Figure 6). Furthermore, the phosphorylation of IRF3 in the cytoplasm and nucleus was also attenuated in the presence of NS2. These results indicated that the interaction between NS2 and TBK1 inhibited the recruitment of TBK1 to STING, which reduced the phosphorylation of STING and the phosphorylation of downstream IRF3, leading to the inhibition of the IRF3 signaling pathway. Thus, NS2 inhibits the function of the IRF3 signaling pathway by interacting with TBK1 to disrupt the TBK1–STING interaction.

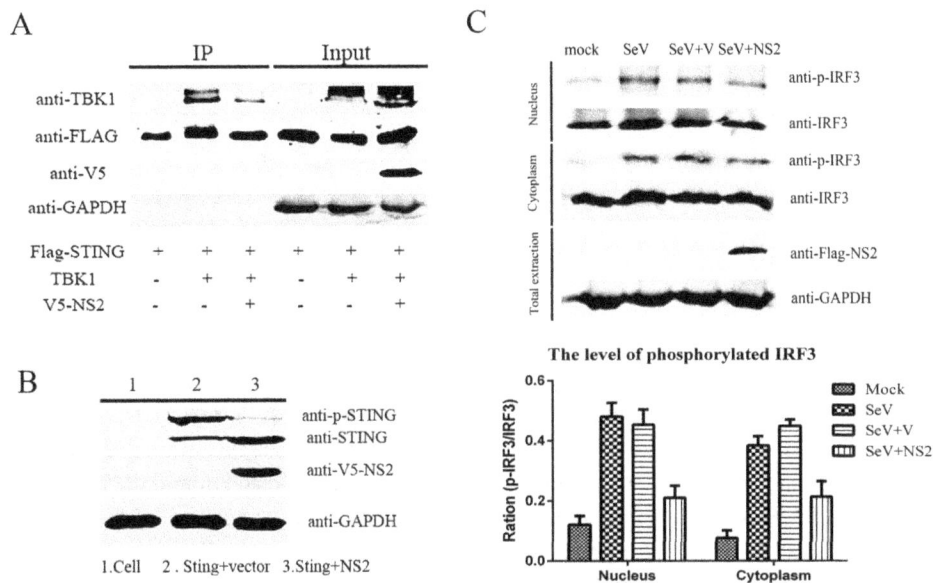

Figure 6. NS2 disrupts the TBK1–STING interaction and reduces the phosphorylation of STING and downstream IRF3. (**A**) F81 cells were transfected with Flag-tagged STING, Myc-tagged TBK1 and V5-tagged NS2 for 36 h. The cell lysates were subjected to immunoprecipitation with ANTI-FLAG M2 magnetic beads. The whole-cell lysates and IP complexes were analyzed by western blot using anti-Flag or anti-Myc antibodies. (**B**) F81 cells were transfected with Flag-tagged STING and V5-tagged NS2 for 36 h. The cell lysates were resolved by SDS-PAGE, after which they were analyzed by western blot with antibodies against STING or phosphorylated STING. (**C**) F81 cells were transfected with Flag-tagged NS2. Twenty-four hours later, the cells were stimulated with SeV. The cytoplasmic and nuclear extracts of the F81 cells were separated at 6 h after simulation and analyzed by western blot with antibodies against IRF3 or phosphorylated IRF3 and the quantification of relative p-IRF3 band intensities to IRF3 was showed with histogram.

3.5. The C-terminus Plus the Coiled Coil Domain of NS2 can Inhibit the IRF3 Signaling Pathway to the Same Extent as Full-Length NS2

To define the regulatory elements within NS2, we analyzed the sequence and structure of the NS2 protein and created truncated forms NS2, including the N-terminal domain (1–87aa), which is the same sequence as the NS1 N-terminal domain, a C-terminal domain (88–165aa) and the C-terminal domain plus the coiled coil domain (53–165aa). The effect of each truncated form of NS2 on the expression of the luciferase gene under the control of the IRF3 binding region PRDIII/I of the feline IFN-β promoter was determined using a luciferase assay. As shown in Figure 7, the N-terminal domain could not affect the activity of the feline IFN-β promoter through the PRDIII/I region. Compared with the full-length

NS2, the C-terminal domain of NS2 exhibited reduced ability to inhibit the reporter gene expression, with about 50% of inhibitory ability of the full-length NS2. Interestingly and importantly, the construct containing the C-terminus plus the coiled coil domain of NS2 produced an inhibitory effect similar to that of the full-length NS2. These findings suggest that the C-terminus plus the coiled coil domain of NS2 are the functional determinants of the NS2-mediated inhibition of the IRF3 signaling pathway.

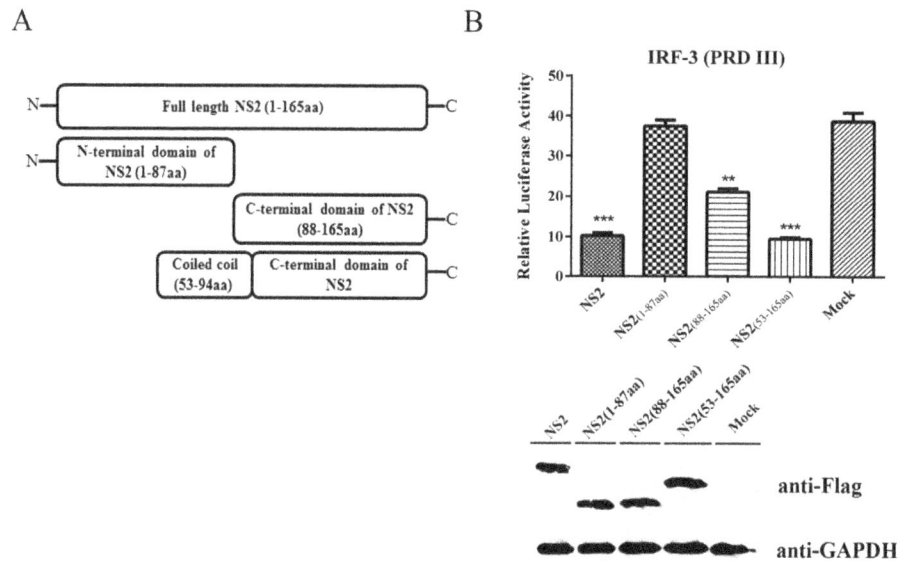

Figure 7. The C-terminus plus the coiled coil domain of NS2 can inhibit the IRF3 signaling pathway. (**A**) Schematic diagram of the truncated NS2 constructs. (**B**) F81 cells were co-transfected with IRF3-Luc, pRL-TK and plasmids expressing either the intact NS2, the N-terminal domain of NS2, the C-terminal domain of NS2 or the C-terminal domain plus the coiled coil domain of NS2. The effect of each truncated NS2 on the expression of the luciferase gene under the control of the IRF3 binding region PRDIII/I of the feline IFN-β promoter was determined using a luciferase assay. The values were normalized to the Renilla activity. The data represent the mean values of three independent experiments. The error bars represent standard deviations, and the asterisks indicate significant differences (*: $p < 0.05$; **: $p < 0.01$; ***: $p < 0.001$) between groups. The expression of the intact NS2, the N-terminal domain of NS2, the C-terminal domain of NS2 or the C-terminal domain plus the coiled coil domain of NS2 was monitored by immunoblotting using a mouse anti-Flag antibody; GAPDH was used as a loading control.

4. Discussion

Viruses of the carnivore parvovirus group infect a wide range of hosts and cause severe diseases. However, the mechanism of their evasion of the innate immune system has rarely been reported. In the *Parvoviridae* family, only a few viruses, such as minute virus of mice (MVM), porcine parvovirus (PPV), and porcine bocavirus (PBoV), have been reported to block IFN-β production to attenuate innate immune responses [12–14]. In the present study, we focused on the immune evasion of FPV and showed the antagonistic function of the FPV NS2 protein against the IRF3 signaling pathway to inhibit IFN-β induction.

To combat the host antiviral effects, viruses have evolved elaborate mechanisms to antagonize the innate immune response [15,16]. The inhibition of interferon transcription is a common method viruses use to escape the innate immune response [3]. In the present study, we first found that infection with the FPV strain Philips-Roxane failed to activate IFN-β transcription but antagonized the type I IFN response stimulated by SeV in F81 cells. Next, we overexpressed the structural and nonstructural proteins of FPV and found that NS2 could inhibit the IFN-β induction. In this experiment, the expression of luciferase gene induced by IFN-β promoter can be accumulated, but the induction of

IFN-α/β during the early phase of viral infection is detectable, then the negative feedback pathway against IFN-α/β is activated and begins to degrade the IFN-α/β. So, the different expression of IFN-β was significant only at some time points, but the significantly different expression of luciferase gene presented during the whole infection stage, which contributed that the suppression result in luciferase reporter assays were more significant than that in western blot. The initiation of IFN expression is tightly regulated by transcription factors consisting of IRFs, NF-κB, and AP-1. These transcription factors bind specific PRD motifs in the IFN-β promoter, and an association with the transcriptional coactivator CREB-binding protein (CBP) and p300 leads to the initiation of IFN transcription in the nucleus. In the current study, transfection with NS2 only inhibited the luciferase activity of the PRDIII motif, which has specific affinity to IRF3, and did not interrupt the luciferase activity of PRDII, which has specific affinity to NF-κB and PRD IV, which has specific affinity to AP-1. These results indicated that NS2 impedes the IRF3 signaling pathway to interrupt the SeV-mediated activation of IFN production.

To pinpoint the key step and the target protein of the NS2-mediated suppression of IFN-β induction, effectors in the IRF3 signaling pathway were investigated by epistatic analysis. The induction of type I IFN production is governed by signal transduction pathways in which an activation signal is relayed by a cascade of signaling proteins to IRF3 transcription factors which then turn on IFN promoters [17]. Theoretically, NS2 should suppress the effect of all upstream inducers and have no influence on the effect of its downstream effectors. Therefore, the action point of NS2 can be determined by assessing its suppressive effects on a series of transducer proteins. In this study, our findings demonstrate that FPV NS2 inhibited MAVS-, STING-, and TBK1-directed IFN-transcription (>80%) but failed to inhibit IFN-induction directed by constitutively active IRF3-5D. We therefore reasoned that NS2 might act at the TBK1 step. As a phosphokinase, TBK1 plays critical roles in the IRF3 signaling pathway. TBK1 and IRF3 can be recruited by STING to form a complex that facilitates TBK1 phosphorylation and activation followed by STING and IRF3 phosphorylation by TBK1, thus leading to the activation of the IFN-β response and triggering the host immune response [18,19]. In the present study, we found that NS2 could interact with TBK1 and significantly inhibit both the binding of TBK1 to STING and the subsequent phosphorylation of STING and IRF3. These results indicate that NS2 inhibits the function of the IRF3 signaling pathway by preventing TBK1 from binding to STING. Thus, we report that NS2 can block the interaction between TBK1 and STING to disturb the TBK1-mediated IRF3 phosphorylation, which is the key step in the NS2-mediated suppression of IFN-β induction.

The functions of NS2 in the viral life cycle or replication are not well understood [20]. FPV NS2 is formed from a spliced transcribed message, with its 87 N-terminal amino acids, which it shares with nonstructural protein 1 (NS1), being joined to 78 amino acids from an alternative open reading frame, and has been shown to have little effect on efficient viral DNA replication and the assembly of viral capsid proteins. In the current study, we demonstrated that FPV NS2 could significantly suppress IFN expression. To define the important regulatory elements within NS2, the effects of several truncated forms of NS2 on the inhibition of SeV-induced IFN-β induction were examined. In accordance with the function of NS1 in anti-IFN activity, the N-terminal domain of NS2 could not inhibit reporter gene expression. However, the C-terminal domain of NS2 had 50% of the inhibitory ability of full-length NS2. By analyzing the sequence and structure of NS2, we found a coiled coil domain between the N-terminal domain and C-terminal domain. Coiled coil-mediated protein-protein interactions have been frequently detected [21]. A truncated form of NS2 in which the coiled coil domain was joined to the C-terminal domain produced a similar inhibitory effect as full-length NS2. These results indicated that the C-terminus plus the coiled coil domain of NS2 are the key regulatory elements required for FPV to suppress type I IFN induction. Although the specific mechanism needs further investigation, due to the high conservation of the NS2 sequence, the function of the NS2 protein in inhibiting host innate immunity may be widespread in the carnivore parvovirus subgroup.

In summary, this study is the first report identifying the viral proteins encoded by FPV that are responsible for inhibiting the induction of IFN-β. We found that NS2 inhibits the induction of IFN-β by targeting TBK1 to prevent it from interacting with STING, thereby inhibiting the downstream IRF3 phosphorylation and resulting in an inhibition of the IRF3 signaling pathway. The present study has shed light on a novel function of the FPV NS2 protein as a negative regulator of IFN-β production. Moreover, these findings contribute to our understanding of the molecular mechanisms of innate immunity evasion strategies utilized by FPV and the outcomes of infection such as the establishment of a persistent FPV infection.

Acknowledgments: This work was funded by the Open Fund of the State Key Laboratory of Veterinary Biotechnology of the Harbin Veterinary Research Institute (grant No. SKLVBF20160) and the National Natural Science Foundation of China (grant no. 31402201).

Author Contributions: J.T. and L.Q. conceived and designed the experiments; H.K., X.H., and X.Z. performed the experiments; H.K., D.L., C.L., and H.W. analyzed the data; D.L., D.G., J.L., and Q.J. contributed reagents/materials/analysis tools; H.K., H.Y., and J.L. wrote the paper. D.L., H.K., and X.H. revised the paper.

References

1. Truyen, U.; Parrish, C.R. Feline panleukopenia virus: Its interesting evolution and current problems in immunoprophylaxis against a serious pathogen. *Vet. Microbiol.* **2013**, *165*, 29–32. [CrossRef] [PubMed]
2. Poncelet, L.; Garigliany, M.; Ando, K.; Franssen, M.; Desmecht, D.; Brion, J.P. Cell cycle S phase markers are expressed in cerebral neuron nuclei of cats infected by the feline panleukopenia virus. *Cell Cycle* **2016**, *15*, 3482–3489. [CrossRef] [PubMed]
3. Bowie, A.G.; Unterholzner, L. Viral evasion and subversion of pattern-recognition receptor signalling. *Nat. Rev. Immunol.* **2008**, *8*, 911–922. [CrossRef] [PubMed]
4. Stetson, D.B.; Medzhitov, R. Type I Interferons in host defense. *Immunity* **2006**, *25*, 373–381. [CrossRef] [PubMed]
5. Kawai, T.; Akira, S. Innate immune recognition of viral infection. *Nat. Immunol.* **2006**, *7*, 131–137. [CrossRef] [PubMed]
6. Kawai, T.; Akira, S. Antiviral signaling through pattern recognition receptors. *J. Biochem.* **2007**, *141*, 137–145. [CrossRef] [PubMed]
7. Dempsey, A.; Bowie, A.G. Innate immune recognition of DNA: A recent history. *Virology* **2015**, *479–480*, 146–152. [CrossRef] [PubMed]
8. Yoneyama, M.; Onomoto, K.; Jogi, M.; Akaboshi, T.; Fujita, T. Viral RNA detection by RIG-I-like receptors. *Curr. Opin. Immunol.* **2015**, *32*, 48–53. [CrossRef] [PubMed]
9. Tian, J.; Zhang, X.; Wu, H.; Liu, C.; Liu, J.; Hu, X.; Qu, L. Assessment of the IFN-β response to four feline caliciviruses: Infection in CRFK cells. *Infect. Genet. Evol.* **2015**, *34*, 352–360. [CrossRef] [PubMed]
10. Wu, H.; Zhang, X.; Liu, C.; Liu, D.; Liu, J.; Wang, G.; Tian, J.; Qu, L. Molecular cloning and functional characterization of feline MAVS. *Immunol. Res.* **2016**, *64*, 82–92. [CrossRef] [PubMed]
11. Zhang, X.; Wu, H.; Liu, C.; Sun, X.; Zu, S.; Tian, J.; Qu, L.; Li, S. The function of feline stimulator of interferon gene (STING) is evolutionarily conserved. *Vet. Immunol. Immunopathol.* **2016**, *169*, 54–62. [CrossRef] [PubMed]
12. Lin, W.; Qiu, Z.; Liu, Q.; Cui, S. Interferon induction and suppression in swine testicle cells by porcine parvovirus and its proteins. *Vet. Microbiol.* **2013**, *163*, 157–161. [CrossRef] [PubMed]
13. Mattei, L.M.; Cotmore, S.F.; Tattersall, P.; Iwasaki, A. Parvovirus evades interferon-dependent viral control in primary mouse embryonic fibroblasts. *Virology* **2013**, *442*, 20–27. [CrossRef] [PubMed]
14. Zhang, R.; Fang, L.; Wang, D.; Cai, K.; Zhang, H.; Xie, L.; Li, Y.; Chen, H.; Xiao, S. Porcine bocavirus NP1 negatively regulates interferon signaling pathway by targeting the DNA-binding domain of IRF9. *Virology* **2015**, *485*, 414–421. [CrossRef] [PubMed]
15. Randall, R.E.; Goodbourn, S. Interferons and viruses: An interplay between induction, signalling, antiviral responses and virus countermeasures. *J. Gen. Virology* **2008**, *89*, 1–47. [CrossRef] [PubMed]
16. Versteeg, G.A.; Garcia-Sastre, A. Viral tricks to grid-lock the type I Interferon system. *Curr. Opin. Microbiol.* **2010**, *13*, 508–516. [CrossRef] [PubMed]

17. Kagan, J.C.; Barton, G.M. Emerging principles governing signal transduction by pattern-recognition receptors. *Cold Spring Harb. Perspect. Biol.* **2014**, *7*, a016253. [CrossRef] [PubMed]

18. Liu, S.; Cai, X.; Wu, J.; Cong, Q.; Chen, X.; Li, T.; Du, F.; Ren, J.; Wu, Y.T.; Grishin, N.V. Phosphorylation of innate immune adaptor proteins MAVS, STING, and TRIF induces IRF3 activation. *Science* **2015**, *347*. [CrossRef] [PubMed]

19. Tanaka, Y.; Chen, Z.J. STING specifies IRF3 phosphorylation by TBK1 in the cytosolic DNA signaling pathway. *Sci. Signal.* **2012**, *5*, ra20. [CrossRef] [PubMed]

20. Wang, D.; Yuan, W.; Davis, I.; Parrish, C.R. Nonstructural protein-2 and the replication of canine parvovirus. *Virology* **1998**, *240*, 273–281. [CrossRef] [PubMed]

21. Lupas, A. Coiled coils: New structures and new functions. *Trends Biochem. Sci.* **1996**, *21*, 375–382. [CrossRef]

PreC and C Regions of Woodchuck Hepatitis Virus Facilitate Persistent Expression of Surface Antigen of Chimeric WHV-HBV Virus in the Hydrodynamic Injection BALB/c Mouse Model

Weimin Wu [1], Yan Liu [1], Yong Lin [1], Danzhen Pan [1], Dongliang Yang [2], Mengji Lu [1,3] and Yang Xu [1,*]

[1] Department of Pathogen Biology, School of Basic Medicine, Tongji Medical College, Huazhong University of Science and Technology, Wuhan 430030, China; wuweimin2012@hotmail.com (W.W.); liuyan_54321@126.com (Y.L.); linyong1027@163.com (Y.L.); pande20080901@163.com (D.P.); mengji.lu@uni-due.de (M.L.)

[2] Department of Infectious Diseases, Union Hospital, Tongji Medical College, Huazhong University of Science and Technology, Wuhan 430022, China; dlyang@hust.edu.cn

[3] Institute of Virology, University Hospital of Essen, 45147 Essen, Germany

* Correspondence: orangexuyang@hotmail.com

Academic Editor: Joanna Parish

Abstract: In the hydrodynamic injection (HI) BALB/c mouse model with the overlength viral genome, we have found that woodchuck hepatitis virus (WHV) could persist for a prolonged period of time (up to 45 weeks), while hepatitis B virus (HBV) was mostly cleared at week four. In this study, we constructed a series of chimeric genomes based on HBV and WHV, in which the individual sequences of a 1.3-fold overlength HBV genome in pBS-HBV1.3 were replaced by their counterparts from WHV. After HI with the WHV-HBV chimeric constructs in BALB/c mice, serum viral antigen, viral DNA (vDNA), and intrahepatic viral antigen expression were analyzed to evaluate the persistence of the chimeric genomes. Interestingly, we found that HI with three chimeric WHV-HBV genomes resulted in persistent antigenemia in mice. All of the persistent chimeric genomes contained the preC region and the part of the C region encoding the N-terminal 1–145 amino acids of the WHV genome. These results indicated that the preC region and the N-terminal part of the C region of the WHV genome may play a role in the persistent antigenemia. The chimeric WHV-HBV genomes were able to stably express viral antigens in the liver and could be further used to express hepadnaviral antigens to study their pathogenic potential.

Keywords: woodchuck hepatitis virus; hepatitis B virus; chimeric genome; mouse model

1. Introduction

The hydrodynamic injection (HI) mouse model has been explored in different studies in hepatitis B virus (HBV) research [1]. It has been proven that the HBV mouse model based on HI could be used to study viral replication and persistence in order to analyze the immunological factors required for HBV clearance and to evaluate novel antiviral therapy strategies [1–10]. HI of plasmid pAAV/HBV1.2 (containing a 1.2-fold overlength HBV genome) was found in a previous study to cause HBV replication and HBV surface antigen (HBsAg) expression that persisted for more than six months in approximately 40% of the injected mice [3]. It was speculated that the outcome after HI was also determined by the backbone of plasmids and the genetic background of injected mice. In our previous studies, we explored pAAV/HBV1.3 (containing a 1.3-fold overlength HBV genome) in C57BL/6 mice.

We found that high levels of serum HBsAg and HBV DNA were detected at seven days post injection (dpi), and they declined to undetectable levels at 28 dpi, which indicated temporary HBV replication and antigen expression [2,7,11]. Therefore, we speculated that in addition to the backbone of plasmids and the genetic background of the injected specimen, viral characteristics could also be an independent determinant for viral replication and persistence in the HI mouse model.

Woodchuck hepatitis virus (WHV), like HBV, is a member of the family *Hepadnaviridae*. WHV and HBV share a remarkable similarity in genome organization and replication strategy [12]. Their nucleotide (nt) sequences were found to have a homology of 62%–70% [13]. Early studies revealed that the HBV capsid could encapsidate WHV polymerase-epsilon complexes and vice versa [14]. Likewise, defective HBV polymerase (POL) could be complemented by WHV counterparts, and vice versa [15]. Furthermore, HBV core antigen (HBcAg) and WHV core antigen (WHcAg) could interact with each other to produce chimeric capsids [15]. In our previous studies, we constructed two plasmids, pBS-HBV1.3 (pHBV1.3, containing a 1.3-fold overlength HBV genome) and pBS-WHV1.3 (pWHV1.3, containing a 1.3-fold overlength WHV genome) in pBluescript II SK(+) vector. After HI of pHBV1.3 in BALB/c mice, serum HBV antigen and HBV DNA peaked at seven dpi and normally disappeared at 28 dpi. However, we found that WHV viral DNA (vDNA) and antigens could persist up to 45 weeks after HI with pWHV1.3. The reason for WHV persistence in BALB/c mice was unknown and therefore, we wished to explore it.

In the present study, we constructed a series of chimeric WHV-HBV genomes based on pHBV1.3, in which different fragments of the HBV genome were substituted by the counterparts from WHV. We studied whether the chimeric WHV-HBV genomes would persist or be cleared in HI mice model and attempted to determine the viral genomic fragments that assisted in the persistence of the chimeric WHV-HBV genomes.

2. Materials and Methods

2.1. Ethics Statement

Female BALB/c mice (six to eight weeks old) used in this study were purchased from SJA Co., Ltd. (Changsha, China). Mice were kept in specific-pathogen-free (SPF) conditions with free access to water and foods. Guidelines for laboratory animal experiments were strictly followed. This study was conducted under the Permit Number 2010-361 from the Institutional Animal Care and Use Committee of Tongji Medical College (Wuhan, China).

2.2. Chimeric Woodchuck Hepatitis Virus and Hepatitis B Virus (WHV-HBV) Genomes

pWHV1.3 contained 1.3 copies of WHV genome (nt 1050–2190, GenBank J04514). pHBV1.3 contained 1.3 copies of the HBV genome (nt 1040–1986, GenBank AY220698). Both pWHV1.3 and pHBV1.3 were constructed based on the pBluescript II SK(+) vector. We constructed a series of chimeric genomes based on pHBV1.3 and pWHV1.3, in which the individual HBV sequences of pHBV1.3 were substituted by the corresponding WHV sequences.

First, we constructed the chimeric plasmid pWHBV3, in which an HBV fragment (nt 1040-2817) was substituted by its counterpart WHV fragment (nt 1050–2950) (Figure 1, Figure S1 and Table S1). The WHV fragment (nt 1050–2950) was amplified from pWHV1.3 using high-fidelity polymerase chain reaction (PCR) with primers WHBV3F and WHBV3R (Table S2). Both pHBV1.3 and the cloned WHV fragment (nt 1050–2950) were double-digested with the *Kpn*I and *Bst*EII and then fused together. pWHBV3 contains the intact HBV *S* gene and WHV *C* gene, the chimeric *P* gene and two *X* genes.

Next, we constructed the chimeric plasmids of pWHBV5 and pWHBV5C, in which the inserted WHV fragment (nt 1050–2950) in pWHBV3 was subdivided. For pWHBV5, the WHV fragment (nt 1050–1933) replaced the corresponding HBV fragment (nt 1040–1818), based on pHBV1.3 (Figure 1, Figure S1 and Table S1). The WHV fragment (nt 1050–1933) was amplified from pWHV1.3 with primers WHBV4F and WHBV5FR, and the HBV fragment (nt 1819–2817) was amplified from pHBV1.3 with primers WHBV4R and WHBV5RF (Table S2). Then, the cloned WHV fragment (nt 1050–1933) and HBV fragment (nt 1819–2817) were fused together by overlapping PCR. The fused WHV-HBV fragment and pHBV1.3 were both digested with the restriction enzymes *Xho*I and *Bst*EII and then fused together, resulting in pWHBV5. pWHBV5C was constructed by the same strategy. For pWHBV5C, the HBV fragment (nt 1820–2817) was replaced by the corresponding WHV fragment (nt 1935–2950), based on pHBV1.3 (Figure 1, Figure S1 and Table S1). The HBV fragment (nt 1040–1819) was amplified from pHBV1.3 with primers WHBV5CF and WHBV5CFR, and the WHV fragment (nt 1935–2950) was amplified from pWHV1.3 using primers WHBV5CRF and WHBV5CR (Table S2). The cloned HBV fragment (nt 1040–1819) and WHV fragment (nt 1935–2950) were fused together by overlapping PCR. The fused HBV-WHV fragment and pHBV1.3 were both digested by the restriction enzymes *Pst*I and *Bst*EII and then fused together, resulting in pWHBV5C.

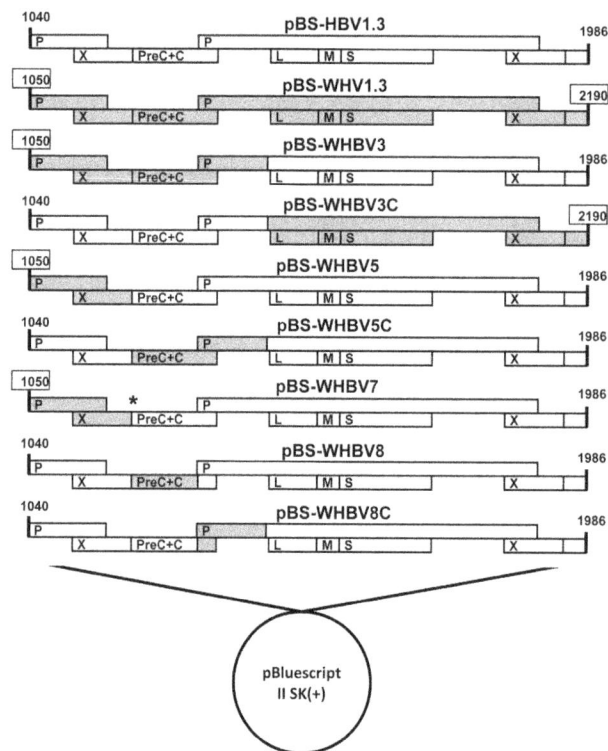

Figure 1. Recombinant woodchuck hepatitis virus-hepatitis B virus (WHV-HBV) genomes used in this study. pHBV1.3 (white) and pWHV1.3 (gray), which contained the 1.3-fold overlength genome of HBV and WHV, respectively, were both based on the pBluescript II SK(+) vector. The respective HBV genome sequences were substituted by the corresponding WHV sequences (gray bars), resulting in a series of chimeric WHV-HBV plasmids. *, the point mutation C1819T in the HBV preC region.

Finally, the chimeric plasmids of pWHBV8 and pWHBV8C were constructed based on pWHBV5C, in which the inserted WHV fragment (nt 1935–2950) in pWHBV5C was further subdivided (Figure 1, Figure S1, and Table S1). For pWHBV8, the HBV fragment (nt 1820–2331) was substituted by the corresponding WHV fragment (nt 1935–2449) based on pHBV1.3. The fragment (HBV nt 1040–1819, WHV nt 1935–2449) was digested from pWHBV5C by the restriction enzymes *Pst*Iand *Bsp*EI and then fused with the digested pHBV1.3 fragment using the same restriction enzymes, resulting in pWHBV8. For pWHBV8C, the HBV fragment (nt 2332–2817) was substituted by the corresponding WHV fragment (nt 2450–2950) based on pHBV1.3 (Figure 1, Figure S1 and Table S1). Two HBV fragments (nt 1040–2031 and nt 2035–2331) were amplified from pHBV1.3 using primers WHBV5CF/WHBV8CFR and WHBV8CRF/WHBV4R, respectively (Table S2), and were fused by overlapping PCR to change the redundant restriction enzyme *Bsp*EI site (TCT to AGC). The fused HBV fragment (HBV nt 1040–2031,

AGC, HBV nt 2035–2331) and pWHBV5C were both digested with *Pst*I and *Bsp*EI and then fused together, resulting in pWHBV8C.

All of the chimeric HBV-WHV plasmids were sequenced by a commercial service (Beijing Genomics Institute, Shenzhen, China). We found that pWHBV5 had a stop codon at the beginning of the HBV preC region, leading to the termination of HBeAg production. Therefore, pWHBV7 was generated by site-directed mutagenesis using primers WHBV7FR and WHBV7RF to correct the second encoding codon from TAA to CAA in the HBV preC region (Table S2).

2.3. Hydrodynamic Injection (HI) of BALB/c Mice

HI procedures were performed as previously described [6]. Briefly, for each mouse, 10 μg of plasmids were diluted in normal saline (10% of mouse body weight) and injected via the tail vein within 8 s.

2.4. Serological Assays

Mouse sera were diluted (1:10), then enzyme-linked immunosorbent assay (ELISA; Kehua Tech, Xiamen, China) was used to measure viral antigens and related antibodies in sera (HBsAg, anti-HBs, HBeAg, anti-HBe, and anti-HBc). The results were indicated as optical density (OD) 450 nm, as measured on a microplate reader (cutoff = 0.1).

2.5. Measurementof Serum Viral DNA (vDNA)

At first, residual input DNA in sera was eliminated by DNase I (TaKaRa Bio Inc., Kusatsu, Japan). vDNA kits (Omega Bio-tek Inc., Norcross, GA, USA) were used to isolate DNA from viruses according to the manufacturer's instructions. Primer QuantS and QuantAS (Table S3) were employed to measure vDNA via real-time PCR (qPCR) [16].

2.6. Measurementof WHV Core Antigen (WHcAg) and HBV Core Antigen (HBcAg) via Immunohistochemical Staining (IHC)

Mice were sacrificed and livers were collected. Then, livers were fixed with formalin and embedded in paraffin. WHcAg and HBcAg in paraffin sections were measured as previously described [17].

2.7. Measurement of Cytotoxic T Lymphocyte (CTL) Responses

Cytotoxic T lymphocyte (CTL) responses were measured using peptide stimulation and flow cytometry as previously described [18]. Briefly, at four weeks after HI with pHBV1.3 or pWHV-Sa (WHV) in C57BL/6 mice, the splenocytes were stimulated for six hours with the HBcAg peptide C93-100 (MGLKFRQL) or WHcAg peptide C13-21 (YQLLNFLPL; Sangon Biotech, Shanghai, China), respectively. For cell staining, anti-mIFN-γ-APC and anti-mCD8-PE (BD Biosciences, San Jose, CA, USA) were used. Dead cells were excluded using the LIVE/DEAD fixable dead cell stain kit (Invitrogen, Waltham, MA, USA). Cells were measured on FACSCalibur and analyzed using FlowJo software (Ashland, OR, USA).

2.8. Statistical Analysis

In this study, GraphPad Prism (GraphPad Software Inc., La Jolla, CA, USA) was applied to perform all of the statistical analyses. Student's t-test was utilized to analyze any differences between the two independent groups. A statistical significance was set at $p < 0.05$. Results were indicated as means $\pm$ SD.

3. Results

3.1. WHV Could Persist in BALB/c Mice after HI

We have explored HI of pHBV1.3 and pWHV1.3 in female BALB/c mice. After HI of pHBV1.3, HBsAg was detectable in the serum of all of the mice at one dpi. HBsAg expression levels peaked at seven dpi and began to decrease at 14 dpi. Then, serum HBsAg disappeared in nearly all mice at five weeks post infection (wpi). Serum HBeAg was detectable at one dpi, declined slowly and disappeared at nine wpi. HBV DNA in serum peaked at seven dpi and then decreased, finally disappearing at nine wpi (Figure 2a). Specific humoral immune responses were induced, as measured by specific viral antibodies (anti-HBs, anti-HBc, and anti-HBe), which was consistent with the kinetics of viral clearance (Figure S3).

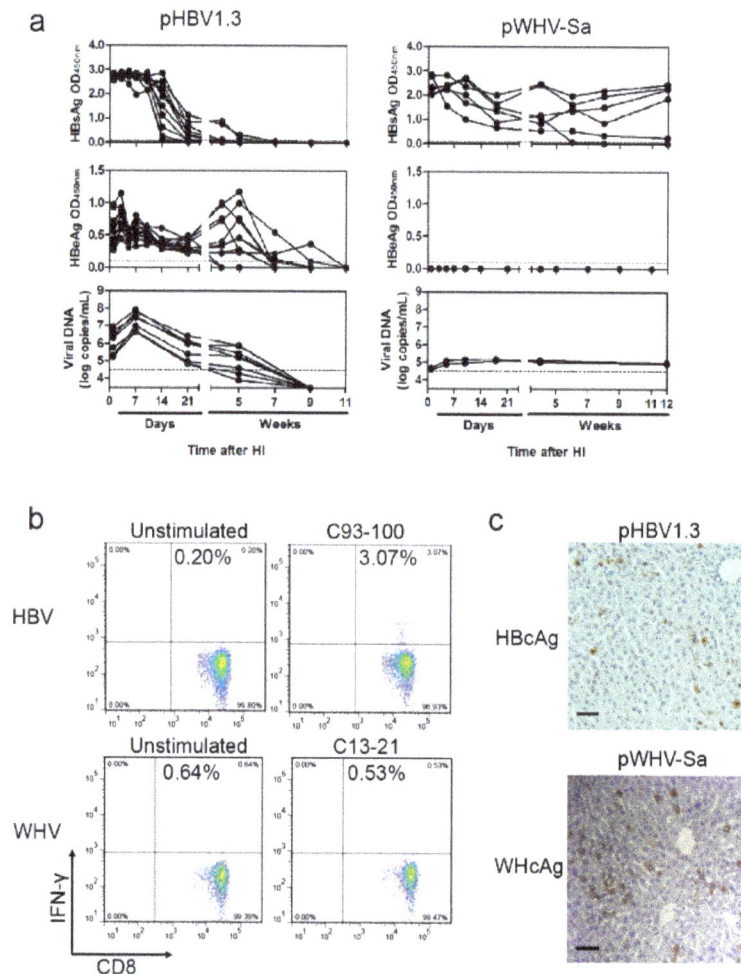

Figure 2. The kinetics of viral antigens and viral DNA (vDNA) in sera, the specific immune response, and hepatic antigen expression after HI with pHBV1.3 or pWHV-Sa in BALB/c mice. (**a**) In mouse sera, HBsAg and HBeAg were detected via ELISA, and the encapsidated vDNA was measured via real-time polymerase chain reaction (qPCR) at the indicated time points with pHBV1.3 or pWHV-Sa. The vector pBluescript II was used as negative control. The cutoff values for ELISA and qPCR were set as 0.1 and 3×10^4 copies/mL, respectively, which were indicated by the dotted line; (**b**) At four weeks after HI with pHBV1.3 (HBV) or pWHV-Sa (WHV) in C57BL/6 mice, the splenocytes were stimulated with HBV core antigen (HBcAg) peptide C93-100 or WHV core antigen (WHcAg) peptide C13-21, respectively. IFN-γ-producing CD8+ T cells were measured via flow cytometry; (**c**) At 10 days post infection (dpi), HBcAg or WHcAg expression in mouse livers were detected by immunohistochemical staining (IHC) after HI with pHBV1.3, pWHV-Sa or pBluescript II. Magnification: 200×.

After HI with pWHV1.3, we found that WHV vDNA and antigens could persist up to 45 weeks [19]. We constructed a chimeric WHV-HBV genome of pWHV-Sa, which contained the HBsAg a-determinant (amino acids (aa) 121–147) in place of the corresponding WHV sequence based on pWHV1.3 [19]. Similarly, the serum HBsAg expression by pWHV-Sa was highly positive from one dpi to 12 wpi after HI, though HBeAg was not produced by pWHV-Sa. Furthermore, the vDNA was maintained at a persistently low level in mouse serum until 12 wpi (Figure 2a). HI with the vector pBluescript II was used as negative control. At 10 dpi with pHBV1.3 or pWHV-Sa, HBcAg or WHV core antigen (WHcAg), respectively, were detectable by immunohistochemical staining (IHC) in mouse liver (Figure 2c). We compared the virus-specific CD8+ T cell responses induced in C57BL/6 mice at week 4 after HI with pHBV1.3 or pWHV-Sa. It was shown that HI with pHBV1.3 induced a stronger T cell response than with pWHV-Sa, which barely induced a specific immune response (Figure 2b). In this study, we tried to explore the characteristics of chimeric viruses that could promote the persistence of viral antigen.

3.2. Persistence of Chimeric WHV-HBV Genome of pWHBV3 in BALB/c Mice

To determine which region of the WHV genome is important for facilitating the persistence of chimeric virus, we first divided the 1.3-fold overlength WHV genome into two parts (WHV nt 1050–2950 and nt 2951–3323/0–2190). Then, we inserted the two WHV sequences into pHBV1.3, replacing the corresponding HBV sequences, to construct two chimeric genomes of pWHBV3 and pWHBV3C, respectively (Figure 1). In pWHBV3, the X gene, preC and C gene, and part of the P gene of HBV (nt 1040–2817) were substituted by the corresponding WHV genome (nt 1050–2950). The substitutions were carefully designed and the translation frames of viral proteins were unaffected. The S gene of HBV was kept intact in pWHBV3. In pWHBV3C, the S gene, part of the P gene and the entire X gene of HBV (nt 2818–3215/0–1986) were substituted by the corresponding WHV genome (nt 2954–3323/0–2190), but the HBV C gene was kept intact. To examine whether pWHBV3 and pWHBV3C were able to replicate and express antigens, Huh7 cells were transfected with pWHBV3 and pWHBV3C, respectively. ELISA assay of supernatants showed that pWHBV3 and pWHBV3C could produce high levels of HBsAg and HBeAg, respectively, at the indicated time points (Figure S2). pWHBV3 and pWHBV3C could replicate in Huh7 cells and had comparable replication levels (data not shown).

We examined the replicative capacity of pWHBV3 and pWHBV3C in vivo by HI in mice. After HI with pWHBV3 in 13 BALB/c mice, high levels of HBsAg were produced in mouse sera from one dpi and remained positive in 92% of mice until 11 wpi (Figure 3a). However, only one mouse showed detectable serum vDNA at one dpi, which disappeared at seven dpi. The serum vDNA loads of other mice were very low and were under the detection limit of qPCR, which might be due to the fused POL encoded by the recombinant P gene. The POL of pWHBV3 was composed of WHV POL 1–175 aa and HBV POL 170–497 aa. However, HBeAg was detectable in 11 BALB/c mice after HI with pWHBV3C. At 11 wpi, approximately 50% of the mice showed persistent, low levels of HBeAg expression (Figure 3a). Similarly, pWHBV3C showed a weak replicative ability and serum vDNA was undetectable. At 10 dpi, HBcAg and WHcAg were distinctly detectable in mouse liver by IHC (Figure 3c). A specific humoral immune response, such as to anti-HBc, was undetectable in mice receiving pWHBV3C, though HBcAg was expressed in mouse livers (Figure S3). In mice that received HI with pWHBV3, the prolonged and high level of HBsAg antigenemia with undetectable levels of vDNA attracted our interest.

3.3. Persistent HBsAg Antigenemia after HI with Chimeric WHV-HBV Genomes of pWHBV5C and pWHBV8

We tried to define the shorter WHV region, which, replacing the homogenous HBV sequence, could support the persistent HBsAg antigenemia of the chimeric genomes in the mouse model. We divided the WHV sequence in pWHBV3 into two parts—nt 1050–1933 and nt 1935–2950—and then inserted them into pHBV1.3 to replace the corresponding HBV sequences, resulting in two chimeric plasmids. pWHBV5 contained the WHV fragment (nt 1050–1933), which primarily replaced the first X region (HBV nt 1040–1819) based on pHBV1.3, in which four open reading frames (ORFs) of the HBV genome were unaffected (Figure 1). pWHBV5C contained the inserted WHV fragment (nt 1935–2950), which replaced the corresponding HBV sequence (nt 1820–2817; Figure S1). The construction procedure of pWHBV5 caused a point mutation (C1819T) in the HBV genome, leading to a stop codon in the HBV preC region and to cessation of HBeAg expression. To eliminate the point mutation, pWHBV7 was generated by site-directed mutagenesis, in which the second encoding codon, TAA, was converted into CAA, so that HBeAg could be produced by pWHBV7. In vitro, both pWHBV7 and pWHBV5C could produce HBsAg in the supernatant of Huh7 cells after transfection. pWHBV7 also could produce HBeAg in the cellular supernatant (Figure S2). pWHBV5C and pWHBV7 showed comparable replication levels in Huh7 cells. We compared the characteristics of pWHBV7 and pWHBV5C after HI in BALB/c mice.

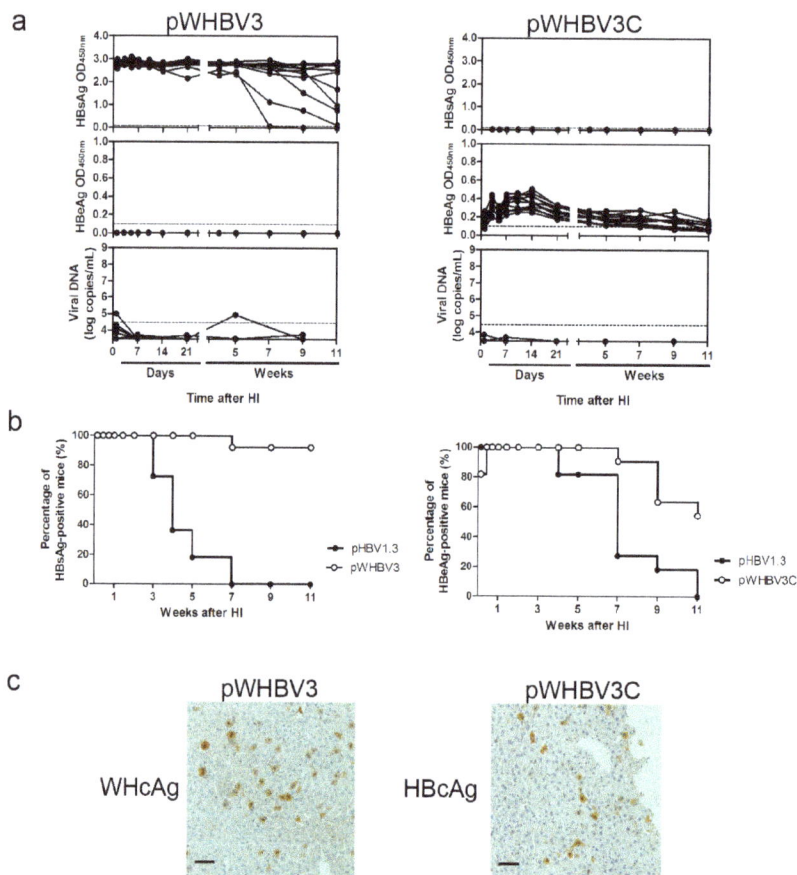

Figure 3. The kinetics of viral antigens and vDNA in sera, percentage of HBsAg and HBeAg antigenemia, and hepatic antigen expression after HI with pWHBV3 or pWHBV3C. (**a**) In mouse sera, viral antigens were measured via ELISA assay. The encapsidated vDNA was measured via qPCR. The cutoff values for ELISA and qPCR were set as 0.1 and 3×10^4 copies/mL, respectively, which were indicated by the dotted line; (**b**) The percentages of HBsAg antigenemia in pWHBV3-injected mice and the percentage of HBeAg antigenemia in pWHBV3C-injected mice were compared with pHBV1.3-challenged mice at the indicated time points; (**c**) At 10 dpi, HBcAg and WHcAg in mouse liver was measured via IHC. Magnification: 200×.

In mice receiving pWHBV5C, HBsAg expression remained persistent in 54.5% of mice at 11 wpi (Figure 4). However, HBeAg and vDNA were undetectable in mouse serum. The low level of viral replication might be due to the recombinant POL in pWHBV5C. In mice receiving pWHBV7, HBsAg expression levels decreased and almost disappeared at five wpi. Accordingly, HBeAg expression gradually dropped and essentially disappeared at nine wpi. Serum vDNA peaked at seven dpi and disappeared at nine wpi, consistent with the kinetics of viral antigen (Figure 4a). At 10 dpi, HBcAg was expressed in approximately 7% of hepatocytes in mice receiving pWHBV7, while WHcAg was produced in approximately 2% of hepatocytes in mice receiving pWHBV5C (Figure 4c). Due to persistent HBsAg antigenemia and weak viral replication in pWHBV5C challenged mice, undetectable levels of antibodies were produced (Figure S4). In pWHBV7-challenged mice, antibodies, including anti-HBs and anti-HBe, were induced following the disappearance of antigen expression (Figure S4).

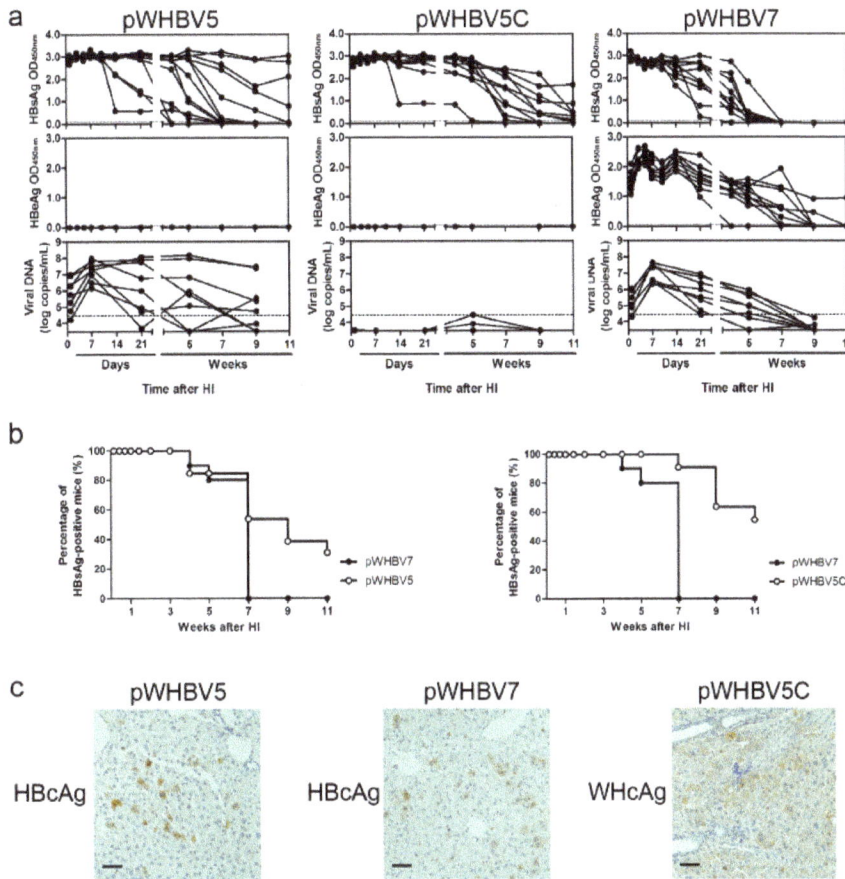

Figure 4. The kinetics of viral antigens and vDNA in sera, percentage of HBsAg antigenemia, and hepatic antigen expression after HI with pWHBV5, pWHBV5C or pWHBV7. (**a**) In mouse sera, viral antigens, and encapsidated vDNA were detected as described in Figure 3; (**b**) The percentages of HBsAg antigenemia in pWHBV5-, pWHBV5C-, and pWHBV7-challenged mice were calculated and compared at the indicated time points; (**c**) At 10 dpi, HBcAg and WHcAg expression in mouse liver was detected by IHC. Magnification: 200×.

We wondered whether the inserted WHV genome (nt 1935-2950) in pWHBV5C could be divided into smaller fragments that could be used to replace the corresponding HBV genome in pHBV1.3 to construct new chimeric plasmids leading to persistent HBsAg expression in mouse serum. Therefore, we further divided the inserted WHV genome in pWHBV5C into two parts—nt 1935–2449 and nt 2450–2950—to construct the chimeric plasmids pWHBV8 and pWHBV8C, using the start codon of HBV pre-genomic RNA (pgRNA; nt 2450). In pWHBV8, the HBV preC and C region fragments encoding the N-terminal 1–145 aa of HBcAg were replaced by the corresponding WHV sequences (nt 1935–2449), in which the HBV pgRNA was kept intact. In vitro, pWHBV8 could produce HBsAg, and pWHBV8C could produce high levels of HBsAg and HBeAg in the supernatant of Huh7 cells (Figure S2). We observed that HBsAg antigenemia persisted in 67% (8/12) of the mice receiving pWHBV8. In mice receiving pWHBV8C, HBsAg expression gradually reduced and then disappeared at 11 wpi. The serum HBeAg expression level was very low (Figure 5a). Replacing the HBV sequence encoding the C-terminal 150–183 aa of HBcAg with the corresponding WHV sequence in pWHBV8C might reduce HBeAg secretion. The serum vDNA loads were under the detection limit in mice receiving either pWHBV8 or pWHBV8C (Figure 5a). pWHBV8 persisted in 67% of mice at 11 wpi; however, pWHBV8C was cleared. At 10 dpi, WHcAg and HBcAg were expressed in the livers of mice that received pWHBV8 and pWHBV8C, respectively (Figure 5c). Anti-HBc antibody was induced in 18% (2/11) of mice receiving pWHBV8C. Finally, anti-HBs antibody was only produced in one mouse receiving pWHBV8 following HBsAg disappearance (Figure S4).

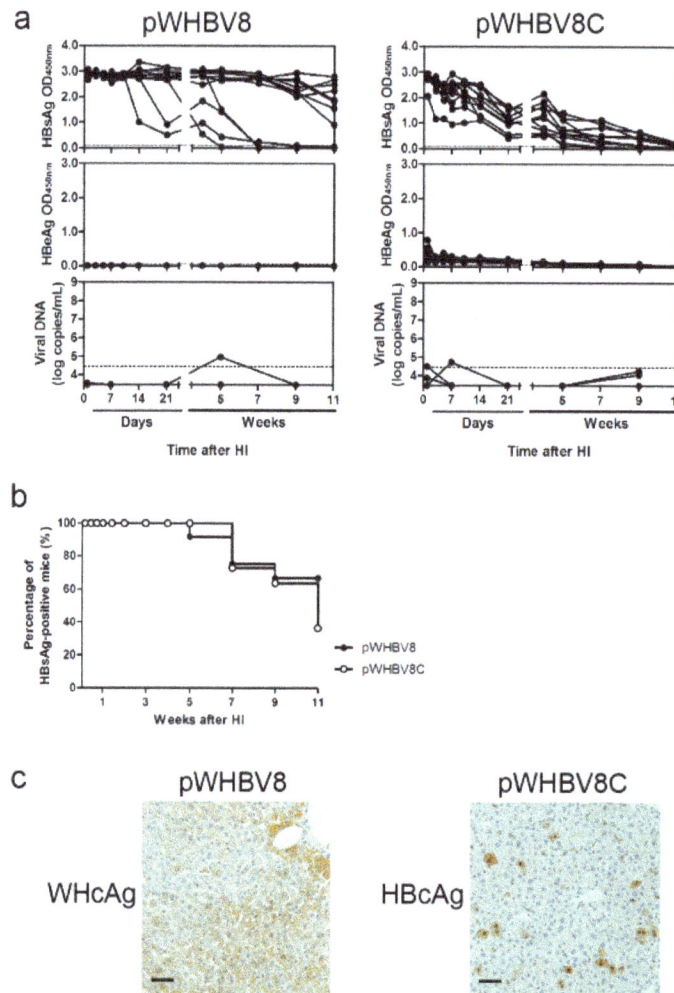

Figure 5. The kinetics of viral antigens and vDNA in sera, percentage of HBsAg antigenemia, and hepatic antigen expression after HI with pWHBV8 or pWHBV8C. (**a**) In mouse sera, viral antigens and encapsidated vDNA were detected as described in Figure 3; (**b**) The percentages of HBsAg antigenemia in pWHBV8- and pWHBV8C-injected mice were calculated at the indicated time points; (**c**) At 10 dpi, HBcAg and WHcAg expression in mouse liver was detected by ICS. Magnification: 200×.

3.4. C1819T Mutation in pWHBV5 Led to HBsAg Persistence

Plasmid pWHBV5 is identical to pWHBV7, except that it contains a point mutation of C1819T in the HBV preC region. After HI in BALB/c mice, both pWHBV5- and pWHBV7-challenged mice could express HBsAg in serum; however, only mice that received pWHBV7 could produce serum HBeAg. In pWHBV5-challenged mice, serum HBsAg was produced at a high level and remained positive in 40% of mice at 11 wpi. However, in pWHBV7-challenged mice, serum HBsAg, HBeAg and vDNA simultaneously disappeared at seven wpi, indicating clearance of pWHBV7 (Figure 4a). The C1819T mutation in pWHBV5 makes the HBV initiator element (Inr) sequence more similar to the optimal Inr sequence (Figure S5), which could result in more effective transcription of pgRNA, a higher level of HBcAg production, and stronger viral replication than pWHBV7. We observed that HBcAg expression in liver at 10 dpi was stronger in pWHBV5-injected mice than in pWHBV7-injected mice (Figure 4a). Meanwhile, a more robust anti-HBc antibody response was detected in pWHBV5-injected mice due to the higher level of HBcAg expression (Figure S4). HBsAg persisted with the high titer of anti-HBc antibody in pWHBV5-injected mice.

4. Discussion

In this study, we created a series of chimeric WHV-HBV genomes based on a commonly used vector and explored hydrodynamic injection of the chimeric plasmids in BALB/c mice to study genomic factors that determine HBsAg persistence. Consistent with our previous results [19], a part of such constructs (pWHBV3, pWHBV5C, and pWHBV8) was able to persist in mice and continuously produce HBsAg. For example, pWHBV3 containing a WHV fragment (nt 1050–2950) supported persistent expression of HBsAg in 92% of challenged mice at 11 wpi, while pWHBV5C containing the shorter WHV fragment (nt 1935–2950) supported HBsAg persistence in only 54.5% of mice. Furthermore, pWHBV8 containing the smallest WHV fragment (nt 1935–2449) showed persistent HBsAg antigenemia in 67% of mice. Thus, we assume that the vector backbone does not represent a determinant for the persistence of WHV or the chimeric WHV-HBV genome in the HI mouse model. Compared with pHBV1.3, HBsAg expression in pWHV-Sa challenged mouse sera persisted until week 12 (Figure 2). We speculated that the low-level replication and the weak immune responses induced by pWHV-Sa might lead to persistent HBsAg expression. It is likely that the persistence is because the expression of viral antigens was below a critical level and did not provoke an effective host immune response before tolerance was established. pWHBV3-, pWHBV5C-, and pWHBV8-challenged mice consistently expressed intrahepatic WHcAg. Moreover, the complementary chimeric plasmids of pWHBV3C, pWHBV7, and pWHBV8C producing intrahepatic HBcAg could not support HBsAg persistence in BALB/c mice. It was reported that HBcAg played an important role in the clearance of HBV infection in the liver. Intrahepatic antiviral response may be triggered by HBcAg. In the HI mouse model, the intrahepatic antiviral response triggered by HBcAg is helpful not only for the clearance of infected hepatocytes but also for the input HBV DNA [3,4,20]. Our results showed that the mice receiving pWHBV3, pWHBV5C, or pWHBV8 with intrahepatic WHcAg expression provoked weak immune responses and failed to clear the chimeric plasmids in the liver, leading to HBsAg persistence. We speculated that WHcAg might present weaker antigenicity than HBcAg in BALB/c mice.

The C-terminal of HBcAg is rich in arginine residues and the C-terminal 39 residues of HBcAg are not required for capsid assembly, but function in pgRNA binding and vDNA synthesis [21]. It was reported that the last 10 residues of HBcAg played a key role in the immune response induction in the hydrodynamic mouse model [4]. The chimeric plasmid pWHBV8 encoded the recombinant core protein of the N-terminal 1–145 aa of WHcAg fused with the C-terminal 38 residues of HBcAg. We aligned the C-terminal 38 residues of HBcAg with WHcAg and found that this region was highly conserved. In our study, pWHBV8 persisted in mice with weak immune responses, though it contained only the C-terminal 10 residues of HBcAg. We speculated that the N-terminal 1–145 aa of HBcAg might be equally important for inducing strong immune responses. The weak immune response provoked by pWHBV8 might be due to the weak antigenicity of the assembled WHcAg particle by the N-terminal 1–145 aa of WHcAg in BALB/c mouse. Using the same strategy, we have constructed pWHV-HBV-Sa and pWHV-HBV-SaC145 based on pWHV1.3, in which WHV sequences were substituted by the HBV counterparts. In detail, pWHV-HBV-Sa contained HBsAg a-determinant (aa 121–147). On the other hand, pWHV-HBV-SaC145 was based on pWHV-HBV-Sa and contained one additional HBV fragment, which included the HBV preC region and the part of the C region encoding the N-terminal 1–145 aa of HBcAg. Of the BALB/c mice injected with pWHV-HBV-Sa, 83% of mice presented serum HBsAg positive at 12 wpi. However, the persistence rate of HBsAg antigenemia at 12 wpi in mice receiving pWHV-HBV-SaC145 was only approximately 33% (Figure S6). These results confirmed that in the HI BALB/c mouse model, the WHV genome fragment (nt 1935–2449) encoding the preC and N-terminal 145 aa of WHcAg could facilitate serum HBsAg persistence for the weak antigenicity of WHcAg. On the other hand, the HBV genome fragment (nt 1814–2331) encoding the preC and N-terminal 145 aa of HBcAg might promote the clearance of chimeric genomes and serum HBsAg in BALB/c mice. In the future, our approach may be used to further study the requirement of viral persistence or clearance in the mouse model based on the HI method.

Interestingly, the chimeric HBV-WHV genomes were able to stably express viral antigens in the liver. These constructs could be further used to express hepadnaviral antigens to study their pathogenic potential; for example, whether the persistent expression of HBsAg or HBeAg may impair intrahepatic immune responses to other antigens. Constructs with mutated viral antigens, such as secretion-defective HBsAg, may also be created to examine their impact on cellular processes.

It is still not clear how the persistent HBsAg and HBeAg expression was achieved in the HI mouse model. We showed that the replication-defective pWHV-Sa genome was not able to persistently express viral antigens under the same conditions [19], suggesting that viral replication is a prerequisite of continuous expression of viral antigens in the mouse liver. However, it is not yet possible to detect covalently closed circle DNA (cccDNA) in the mouse liver after HI due to the large quantity of input plasmids and replication intermediates. The copy numbers of cccDNA may be extremely low if their existence can even be confirmed. Thus, this will be an issue to be clarified in the future.

We speculated that HBeAg expression might act as an immune activator in BALB/c mice, especially in the innate immune response phase. Therefore, pWHBV7 with HBeAg expression could be cleared at seven wpi, though the viral replication level and the titer of anti-HBc were higher. Without HBeAg expression, HBsAg persisted for a longer time in pWHBV5-injected mice.

Supplementary Materials: The following figures and tables are available online at www.mdpi.com/1999-4915/9/2/35/s1: Figure S1: The construction of recombinant WHV-HBV genomes; Figure S2: Viral antigens detected in Huh7 cells transfected with the chimeric WHV-HBV constructs; Figure S3: Antibody responses in pHBV1.3-, pWHBV3- and pWHBV3C-challenged mice; Figure S4: Antibody responses in pWHBV5-, pWHBV5C-, pWHBV7-, and pWHBV8C-challenged mice; Figure S5: Alignment of the basal core promoter (BCP) region and initiator (Inr) sequence of pHBV1.3 (HBV), pWHV1.3 (WHV), pWHBV3, pWHBV5C, pWHBV8, pWHBV5, and pWHBV7; Figure S6: HBsAg antigenemia in pWHV-HBV-Sa- and pWHV-HBV-SaC145-challenged mice; Table S1: The detailed composition of the chimeric WHV-HBV constructs; Table S2: Primers for chimeric WHV-HBV plasmids.

Acknowledgments: This work was supported by the National Natural Science Foundation of China (81461130019, 81201290, http://www.nsfc.gov.cn) to Y.X.; and grants of Deutsche Forschungsgemeinschaft (GRK1045/2 and Transregio TRR60) to M.L.

Author Contributions: M.L. and Y.X. conceived and designed the experiments; W.W. and Y.L. performed the experiments; W.W. and Y.X. analyzed the data; Y.L., D.P. and D.Y. contributed reagents/materials/analysis tools; Y.X. wrote the paper.

References

1.   Yang, P.L.; Althage, A.; Chung, J.; Chisari, F.V. Hydrodynamic injection of viral DNA: A mouse model of acute hepatitis b virus infection. *Proc. Natl. Acad. Sci. USA* **2002**, *99*, 13825–13830. [CrossRef] [PubMed]

2.   Cao, L.; Wu, C.; Shi, H.; Gong, Z.; Zhang, E.; Wang, H.; Zhao, K.; Liu, S.; Li, S.; Gao, X.; et al. Coexistence of hepatitis b virus quasispecies enhances viral replication and the ability to induce host antibody and cellular immune responses. *J. Virol.* **2014**, *88*, 8656–8666. [CrossRef] [PubMed]

3.   Huang, L.R.; Wu, H.L.; Chen, P.J.; Chen, D.S. An immunocompetent mouse model for the tolerance of human chronic hepatitis b virus infection. *Proc. Natl. Acad. Sci. USA* **2006**, *103*, 17862–17867. [CrossRef] [PubMed]

4.   Lin, Y.J.; Huang, L.R.; Yang, H.C.; Tzeng, H.T.; Hsu, P.N.; Wu, H.L.; Chen, P.J.; Chen, D.S. Hepatitis b virus core antigen determines viral persistence in a c57bl/6 mouse model. *Proc. Natl. Acad. Sci. USA* **2010**, *107*, 9340–9345. [CrossRef] [PubMed]

5. Tian, Y.; Chen, W.L.; Ou, J.H. Effects of interferon-alpha/beta on hbv replication determined by viral load. *PLoS Pathog.* **2011**, *7*, e1002159. [CrossRef] [PubMed]

6. Yin, Y.; Wu, C.; Song, J.; Wang, J.; Zhang, E.; Liu, H.; Yang, D.; Chen, X.; Lu, M.; Xu, Y. DNA immunization with fusion of ctla-4 to hepatitis b virus (hbv) core protein enhanced th2 type responses and cleared hbv with an accelerated kinetic. *PLoS ONE* **2011**, *6*, e22524. [CrossRef] [PubMed]

7. Wu, C.; Deng, W.; Deng, L.; Cao, L.; Qin, B.; Li, S.; Wang, Y.; Pei, R.; Yang, D.; Lu, M.; et al. Amino acid substitutions at positions 122 and 145 of hepatitis b virus surface antigen (hbsag) determine the antigenicity and immunogenicity of hbsag and influence in vivo hbsag clearance. *J. Virol.* **2012**, *86*, 4658–4669. [CrossRef] [PubMed]

8. Song, J.; Zhou, Y.; Li, S.; Wang, B.; Zheng, X.; Wu, J.; Gibbert, K.; Dittmer, U.; Lu, M.; Yang, D. Susceptibility of different hepatitis b virus isolates to interferon-alpha in a mouse model based on hydrodynamic injection. *PLoS ONE* **2014**, *9*, e90977. [CrossRef] [PubMed]

9. Yang, P.L.; Althage, A.; Chung, J.; Maier, H.; Wieland, S.; Isogawa, M.; Chisari, F.V. Immune effectors required for hepatitis b virus clearance. *Proc. Natl. Acad. Sci. USA* **2010**, *107*, 798–802. [CrossRef] [PubMed]

10. Wang, J.; Wang, B.; Huang, S.; Song, Z.; Wu, J.; Zhang, E.; Zhu, Z.; Zhu, B.; Yin, Y.; Lin, Y.; et al. Immunosuppressive drugs modulate the replication of hepatitis b virus (hbv) in a hydrodynamic injection mouse model. *PLoS ONE* **2014**, *9*, e85832. [CrossRef] [PubMed]

11. Qin, B.; Budeus, B.; Cao, L.; Wu, C.; Wang, Y.; Zhang, X.; Rayner, S.; Hoffmann, D.; Lu, M.; Chen, X. The amino acid substitutions rtp177g and rtf249a in the reverse transcriptase domain of hepatitis b virus polymerase reduce the susceptibility to tenofovir. *Antiviral Res.* **2013**, *97*, 93–100. [CrossRef] [PubMed]

12. Menne, S.; Cote, P.J. The woodchuck as an animal model for pathogenesis and therapy of chronic hepatitis b virus infection. *World J. Gastroenterol.* **2007**, *13*, 104–124. [CrossRef] [PubMed]

13. Galibert, F.; Chen, T.N.; Mandart, E. Nucleotide sequence of a cloned woodchuck hepatitis virus genome: Comparison with the hepatitis b virus sequence. *J. Virol.* **1982**, *41*, 51–65. [PubMed]

14. Ziermann, R.; Ganem, D. Homologous and heterologous complementation of hbv and whv capsid and polymerase functions in rna encapsidation. *Virology* **1996**, *219*, 350–356. [CrossRef] [PubMed]

15. Okamoto, H.; Omi, S.; Wang, Y.; Imai, M.; Mayumi, M. Trans-Complementation of the c gene of human and the p gene of woodchuck hepadnaviruses. *J. Gen. Virol.* **1990**, *71 (Pt. 4)*, 959–963. [CrossRef] [PubMed]

16. Schaefer, S.; Glebe, D.; Wend, U.C.; Oyunbileg, J.; Gerlich, W.H. Universal primers for real-time amplification of DNA from all known orthohepadnavirus species. *J. Clin. Virol.* **2003**, *27*, 30–37. [CrossRef]

17. Wang, B.J.; Tian, Y.J.; Meng, Z.J.; Jiang, M.; Wei, B.Q.; Tao, Y.Q.; Fan, W.; Li, A.Y.; Bao, J.J.; Li, X.Y.; et al. Establishing a new animal model for hepadnaviral infection: Susceptibility of chinese marmota-species to woodchuck hepatitis virus infection. *J. Gen. Virol.* **2011**, *92*, 681–691. [CrossRef] [PubMed]

18. Kosinska, A.D.; Johrden, L.; Zhang, E.; Fiedler, M.; Mayer, A.; Wildner, O.; Lu, M.; Roggendorf, M. DNA prime-adenovirus boost immunization induces a vigorous and multifunctional t-cell response against hepadnaviral proteins in the mouse and woodchuck model. *J. Virol.* **2012**, *86*, 9297–9310. [CrossRef] [PubMed]

19. Pan, D.; Lin, Y.; Wu, W.; Song, J.; Zhang, E.; Wu, C.; Chen, X.; Hu, K.; Yang, D.; Xu, Y.; et al. Persistence of the recombinant genomes of woodchuck hepatitis virus in the mouse model. *PLoS ONE* **2015**, *10*, e0125658. [CrossRef] [PubMed]

20. Lucifora, J.; Xia, Y.; Reisinger, F.; Zhang, K.; Stadler, D.; Cheng, X.; Sprinzl, M.F.; Koppensteiner, H.; Makowska, Z.; Volz, T.; et al. Specific and nonhepatotoxic degradation of nuclear hepatitis b virus cccdna. *Science* **2014**, *343*, 1221–1228. [CrossRef] [PubMed]

21. Nassal, M. The arginine-rich domain of the hepatitis b virus core protein is required for pregenome encapsidation and productive viral positive-strand DNA synthesis but not for virus assembly. *J. Virol.* **1992**, *66*, 4107–4116. [PubMed]

Virus Resistance Is Not Costly in a Marine Alga Evolving under Multiple Environmental Stressors

Sarah E. Heath [1,*], Kirsten Knox [2], Pedro F. Vale [1] and Sinead Collins [1]

[1] Institute of Evolutionary Biology, School of Biological Sciences, University of Edinburgh, Ashworth Laboratories, The King's Buildings, Charlotte Auerbach Road, Edinburgh EH9 3FL, UK; Pedro.Vale@ed.ac.uk (P.F.V.); s.collins@ed.ac.uk (S.C.)

[2] Institute of Molecular Plant Sciences, School of Biological Sciences, University of Edinburgh, Rutherford Building, Max Born Crescent, Edinburgh EH9 3BF, UK; kirsten.knox@ed.ac.uk

* Correspondence: s.heath-2@sms.ed.ac.uk

Academic Editors: Mathias Middelboe and Corina Brussaard

Abstract: Viruses are important evolutionary drivers of host ecology and evolution. The marine picoplankton *Ostreococcus tauri* has three known resistance types that arise in response to infection with the Phycodnavirus OtV5: susceptible cells (S) that lyse following viral entry and replication; resistant cells (R) that are refractory to viral entry; and resistant producers (RP) that do not all lyse but maintain some viruses within the population. To test for evolutionary costs of maintaining antiviral resistance, we examined whether *O. tauri* populations composed of each resistance type differed in their evolutionary responses to several environmental drivers (lower light, lower salt, lower phosphate and a changing environment) in the absence of viruses for approximately 200 generations. We did not detect a cost of resistance as measured by life-history traits (population growth rate, cell size and cell chlorophyll content) and competitive ability. Specifically, all R and RP populations remained resistant to OtV5 lysis for the entire 200-generation experiment, whereas lysis occurred in all S populations, suggesting that resistance is not costly to maintain even when direct selection for resistance was removed, or that there could be a genetic constraint preventing return to a susceptible resistance type. Following evolution, all S population densities dropped when inoculated with OtV5, but not to zero, indicating that lysis was incomplete, and that some cells may have gained a resistance mutation over the evolution experiment. These findings suggest that maintaining resistance in the absence of viruses was not costly.

Keywords: evolution; trade-off; cost of resistance; Phycodnavirus; Prasinovirus; environmental change; virus-host interactions; marine viral ecology; *Ostreococcus tauri*

1. Introduction

Viruses are the most abundant biological entities in the oceans, with an estimated 10^{30} particles globally [1]. Viruses play a key role in marine food webs, partially because viral infection of unicellular organisms often results in cell lysis, where the infected cell bursts to release the new viruses; products of lysis feed back into the microbial loop and provide organic matter to organisms at the base of the food web daily [2]. In addition to being a large cause of mortality to their hosts, viruses can exert strong selection on host immune defense, leading to the evolution of host resistance mechanisms. Strong immune defenses, in turn, impose strong selection on viruses to evade these resistance responses leading to an ongoing co-evolutionary process between hosts and viruses [3]. Experimental evidence of host-virus coevolution has come mainly from bacteria-phage systems [3,4]. Viruses evolve rapidly due to their small size and high mutation rates [5] which can strongly influence the evolution of their hosts. However, in addition to infection, hosts are also subject to other selection pressures, such as severe or stressful environmental changes. In the case of marine hosts, they will be subject to natural selection both from their viruses, and from, for example, the changes in nutrients, temperature and light associated with global change in the oceans [6], which opens up the possibility that the genetic and physiological changes associated with resistance may affect host evolution in response to challenges other than the virus itself. This in turn has the potential to affect how primary productivity at the base of marine food webs evolves in response to global change. Studies have examined environmental effects on interactions between microalgae and their viruses under a range of conditions including changes in temperature [7,8], nutrients [9–13], UV radiation [14], light intensity [11,15,16], and CO_2 levels [13,17,18]. Environmental change can have direct effects on marine viruses, for example by damaging and/or deactivating the particles through UV exposure or extreme temperatures [8,14]. However, viral abundance is thought to be mainly dependent on host availability and, therefore, the effects of environmental change on viruses are expected to be mainly indirect (e.g., [19]). Here we focus on host evolution rather than viral selection.

Hosts are capable of evolving resistance to their viruses, though resistance often entails a fitness cost, which can vary in form and magnitude [20]. Costs of resistance that have been reported in microorganisms include reduced competitive ability [20,21], reduced growth rate [22,23], reduced original function of a receptor protein [24,25], and increased susceptibility to other viruses [26–28]. If the cost of resistance is substantial and related to growth or competitive ability, resistance might be lost when the selection pressure for it is removed (i.e., when viruses are absent) [29]. For example, under conditions where viruses are present and able to interact with their host cells, resistant hosts should have a selective advantage over susceptible hosts by avoiding lysis. However, in the absence of viruses, the selection pressure for resistance is removed and costs of resistance, if present and substantial, should reduce host fitness, so that there is an advantage to losing resistance. Most studies have focused on costs of resistance in bacteria (e.g., [22,28,30,31]), however data for eukaryotic microalgae are lacking, which limits our ability to translate the literature on host-virus interactions to primary producers in the oceans. Because marine algae are the dominant primary producers in oceans [32], changes in the abundance, distribution and composition of microalgal assemblages in response to climate change are likely to have important implications for marine communities.

The marine picoeukaryote *Ostreococcus tauri* and its viruses, *Ostreococcus tauri* viruses (OtVs), are abundant in Mediterranean lagoons [33]. OtVs are lytic viruses belonging to the family Phycodnaviridae that cause susceptible (S) host *O. tauri* cells to burst following infection [34]. However, two resistant host types have been observed [35,36]. In the first type, viruses can attach to the resistant (R) host cells but are unable to replicate and cause lysis. In the second type, resistant producer (RP) populations consist mainly of resistant cells with a minority of susceptible cells (<0.5%) that maintains a population of viruses. These two resistance mechanisms have been observed repeatedly and remain resistant to lysis over many generation of sub-culturing [35,36]. Previous work found that there was no difference in growth rates between the three resistance types when they were maintained separately under standard laboratory culturing conditions, although long term competitions indicated a cost of resistance with susceptible cells outcompeting resistant cells and resistant cells outcompeting resistant producers after 100 and 200 days, respectively [35].

This study examined whether a cost of resistance could be detected in *O. tauri* in terms of the ability to adapt to different environmental conditions, and whether the evolutionary responses to environmental change were affected by resistance type. Populations of S, R and RP *O. tauri* were evolved under different environmental conditions in the absence of viruses for 200 generations to answer whether resistance type was maintained and how resistance type affected evolutionary responses, even in the absence of coevolutionary dynamics imposed by the presence of viruses. We found that all R and RP populations remained resistant to OtV5 inoculation across all environments, whereas S populations had a lower proportion of cell lysis at the end than at the start of the evolution experiment. Additionally, resistance type affected cell division rates, size and chlorophyll content, whereas selection environment affected cell division rates and competitive ability.

2. Materials and Methods

2.1. Susceptible and Resistant Lines

O. tauri lines were obtained from N. Grimsley, Observatoire Océanologique, Banyuls-sur-Mer, France. Three susceptible lines (NG'2, NG'3 and NG'4), three resistant lines (NG5, NG'13 and NG26) and three resistant producer lines (NG'10, NG'16 and NG27) were used. All lines were derived from a single clone of *O. tauri* (RCC4221) and therefore had the same starting genotype.

2.2. Culturing Conditions

For each of the nine lines described above, three biological replicates were evolved per environment (27 independent populations in total per environment). We refer to each independent replicate as a population. Populations were grown in batch culture. Culture medium was prepared using 0.22 μm filtered Instant Ocean artificial seawater (salinity 30 ppt) supplemented with Keller and f/2 vitamins [37]. Control cultures were maintained in a 14:10 light:dark cycle at 85 μmol photon m^{-2} s^{-1} at a constant temperature of 18 °C (Table 1). Each population was grown in 20 mL media and each week, 200 μL was transferred to fresh media to ensure populations were always growing exponentially. Cultures were resuspended by gentle shaking every 2–3 days to prevent cells sticking to the bottom of the flask. For the evolution experiment, *O. tauri* populations were grown either in the control environment as described above, in low light, low phosphate, low salinity or high temperature (Table 1), or a changing environment (random) in which one of the environments from those listed was chosen at random at each transfer. We refer to the environments where the populations evolved as "selection environments". Populations were grown in the absence of viruses for 32 weeks, corresponding to approximately 200 generations.

Table 1. A comparison of the control environment and the treatments used for each selection environment used in this study.

Selection Environment	Control	Treatment
Light (μmol m^{-2} s^{-1})	85	60
Phosphate (μM)	10	5
Salinity (ppt)	30	25
Temperature (°C)	18	20

For the low light environment, culture flasks were wrapped in 0.15 neutral density foil to reduce light intensity. For the low phosphate environment, phosphate was reduced by preparing Keller medium with half the amount of β-glycerophosphate present in the control media. For low salt, Instant Ocean was added to reach a salinity of 25 ppt. Cultures in the high temperature environment were maintained on a heat mat (Exo Terra Heat Wave substrate heat mat, Yorkshire, UK) set at 20 °C. These selection environments were chosen so that the populations responded to them by changing their growth rates relative to the control environment—in batch culture rapid growth is favored by natural selection, so any environment that decreases growth rates should then result in natural selection for traits that will allow cell division rates to recover in that environment. However, the selection environments were not extreme, so that populations were still able to grow at a measurable rate and survive the dilution rate of the experiment. This is in part so that a similar number of generations elapse in all environments over the course of the experiment.

2.3. Testing RP Lines for Viral Production

All resistant producer (RP) lines were tested for viral production prior to the start of the experiment. To check whether the three producing lines (NG'10, NG'16 and NG27) were releasing infectious viruses, we used the supernatant to infect susceptible *O. tauri* strain RCC4221. Two milliliters of each population were transferred to an Eppendorf tube and centrifuged at $8000 \times g$ for 15 min. Four hundred milliliters of the supernatant were removed carefully without drawing up any of the cells from the pellet at the bottom of the tube, and used to inoculate 1 mL of susceptible *O. tauri*. OtV5 was used as a positive control and Keller media was used as a negative control. Eight replicates were performed before the experiment was started. The test was performed every four weeks with three replicates per population. Samples were checked for lysis either by observing by eye whether they were green or clear, or by measuring cell densities using a BD FACSCanto II (BD Biosciences, Oxford, UK) flow cytometer.

In addition to liquid lysis tests, frozen stocks of RP supernatant were made by adding dimethyl sulfoxide (DMSO) (final concentration 10%) and storing at $-80\,^{\circ}$C. We tested these samples for viruses using the plaque assay technique [34]. A 1.5% agarose suspension was made and 5 mL aliquots were prepared in Falcon tubes and held at $70\,^{\circ}$C in a water bath. In a 50 mL Falcon tube, 30 mL exponentially growing *O. tauri* culture, 15 mL Keller media and 5 mL agarose were mixed rapidly but gently by inverting the tube (final agarose concentration 0.15%). The agarose was poured into a 12 cm square petri dish and left to set. Tenfold serial dilutions of the RP supernatant were made in 96-well plates using one row per sample. A Boekel Replicator was used to transfer all of the serial dilutions from one 96-well plate to one square petri dish. The replicator was sterilized between each use using ethanol and a flame. Petri dishes were checked daily for lysis plaques for a maximum of 10 days.

2.4. Testing Resistance Type Using OtV5 Inoculation

OtV5 inoculum was prepared prior to the start of the experiment and stored at $-80\,^{\circ}$C in 10% DMSO (final concentration) and inoculations were performed from the frozen stocks. The experiment did not include a co-evolving virus which allowed us to measure host evolution relative to the ancestral virus. After 32 weeks of evolution, each population was inoculated with a suspension of OtV5 particles to test whether it was susceptible or resistant to viral lysis. Samples were tested by inoculating 1 mL cell culture at a density of 10^5 with 10 μL OtV5 in 48-well plates with three replicates for each sample. Negative controls that were not inoculated with OtV5 were used as a comparison of cell growth. Cell density was measured using a FACSCanto flow cytometer 3 days after inoculation. Samples were run on 96-well plates by counting the total number of cells in 10 μL with a flow rate of 2.0 μL per second.

Data were analyzed with linear mixed effects models using the statistical packages lme4 [38] and lmerTest [39] in R (version 3.2.0, R Core Team, Vienna, Austria) to identify differences in cell densities after OtV5 inoculation compared to controls that were not inoculated. Selection environment, resistance type and treatment (inoculated or not inoculated) were set as fixed effects with population as a random effect. Post hoc Tukey tests were performed using lsmeans to confirm where significant differences occurred within the different effects.

2.5. Population Growth Rates, Cell Size and Cell Chlorophyll Content after Evolution

At the end of the evolution experiment, we quantified evolutionary responses by measuring average cell division rates and by measuring cell size and chlorophyll content for each population. All evolved populations were assayed in their selection environment and in the control environment, and all control populations were assayed in all selection environments except high temperature, since all populations in the high temperature environment went extinct and therefore there were no high temperature evolved strains. The populations that had evolved in a random environment for each transfer were only assayed in the control environment, which was not one of the environments they had been exposed to during the experiment, meaning only a correlated response (rather than a direct response) to selection could be obtained. Each population was assayed in triplicate. Due to the size of the experiment, assays were divided randomly into seven time blocks. This was factored into the statistical analysis.

Average cell division rates, which we refer to as "growth rates" are the average number of cell divisions per day over seven days, which corresponds to one transfer cycle. All populations were first maintained in their assay environment for an acclimation period of one week, which was one full transfer cycle, prior to measuring growth rates. After acclimation, cells were counted using a FACSCanto flow cytometer before the transfer into the assay environment (to calculate the number of cells transferred into fresh media) and again after seven days of growth. Each sample was counted in triplicate. The cell counts were converted to cells per milliliter and the number of divisions per day was calculated using Equation (1).

$$\mu \left(d^{-1} \right) = \frac{\log_2 \left(\frac{N_t}{N_0} \right)}{t - t_0} \tag{1}$$

where μ is population growth rate, and N_t and N_0 are the cell densities (cells mL^{-1}) at times t and t_0 (days), respectively. This measures the average number of cell divisions per ancestor over a single growth cycle and allows a comparison of offspring production between environments even if there are differences in the shape of the population growth curve, or in cases where r cannot be accurately estimated. To avoid biases of cell divisions being dependent on the time of the cell cycle, cells were always measured at the same time of day (at the beginning of the light period when cells are in G1 phase).

Cell size was inferred from FSC (forward scatter), which was calibrated using beads of known sizes (1 μm, 3 μm and 6.6 μm). Chlorophyll fluorescence was inferred by measuring PerCP-Cy5.5 emission with excitation at 488 nm. Relative chlorophyll was analyzed by taking the average chlorophyll fluorescence for all susceptible strains in the control environment and setting this to a value of 1, with chlorophyll measurements of all other strains relative to this value.

Data were analyzed with linear mixed effects models. To analyze differences in growth rate, cell size and chlorophyll under different environments, selection environment, assay environment and resistance type were fixed effects and population and block ware random effects that were treated as un-nested. An additional model was fitted to examine whether there was a difference in growth rate when populations were assayed in their selection environment or when they were assayed in a different environment, with assay as the only fixed effect and population and block set as random effects.

2.6. Competition Assay

To measure competitive fitness, all evolved populations were competed against a green fluorescent protein (GFP) line of *O. tauri*. A Gateway enabled entry clone containing roGFP2 was obtained by linearizing pH2GW7-roGFP2 [40] with EcoRV. The linearized vector was recombined with pDONR207, creating a pDONR207-roGFP2 clone. A pOtOX binary vector [41] was adapted to become a Gateway® destination vector and pDON207-roGFP2 was recombined into the vector, downstream of the high-affinity phosphate transporter (HAPT) promoter [41]. The pOtOx-roGFP2 vector was subsequently transformed into *O. tauri* using the procedure previously described [42].

All evolved populations competed in the selection environment that they evolved in, and all control populations competed in the control environment as well as in each selection environment to measure plastic response. All of the random populations competed in the control environment. All populations, including the roGFP line, were acclimated for one week in the corresponding assay environment prior to the assay. Equal starting densities of 5×10^5 of each evolved population and the roGFP line were grown in 20 mL media for one week, after which cells were counted using a FACSCanto flow cytometer. GFP and non-GFP populations were distinguished by measuring fluorescein isothiocyanate A (FITC-A) emission at 519 nm with excitation at 495 nm. Competitiveness of the evolved populations was measured relative to the roGFP line as fold change in cell density. Data were analyzed with a linear mixed effects model, with selection environment, assay environment and resistance type as fixed effects and population and assay replicate as random effects.

3. Results

3.1. Susceptibility to OtV5 after Evolution

3.1.1. Host Resistance Type Was Maintained during Evolution

After 200 generations of evolution in the selection environments, all surviving R and RP populations remained resistant to OtV5 lysis and all S populations remained susceptible to viral lysis in those environments (Figure 1). A significant interaction between selection environment, resistance type and treatment (OtV5 inoculation) affected susceptibility of *O. tauri* to OtV5 (ANOVA environment × resistance type × treatment, $F_{8,238} = 15.22$, $p < 0.0001$). A post hoc Tukey test showed that this was due to cell lysis of S populations ($t_{8,238} = 10.66$, $p < 0.001$), whereas cell density of R and RP lines did not decrease compared to controls that were not inoculated. The highest cell densities were observed in the low salt (post hoc Tukey test, $t_{8,238} = -29.90$, $p < 0.0001$) and random (post hoc Tukey test, $t_{8,238} = -7.54$, $p < 0.0001$) environments. The OtV5-inoculated S populations in low phosphate were the only populations where cell density fell below the starting cell density across all populations, indicating almost complete cell lysis and no cell growth for this combination of resistance type and selection environment. R and RP lines did not show decreases in cell density after inoculation with OtV5 compared to controls that were not inoculated, whereas S lines did.

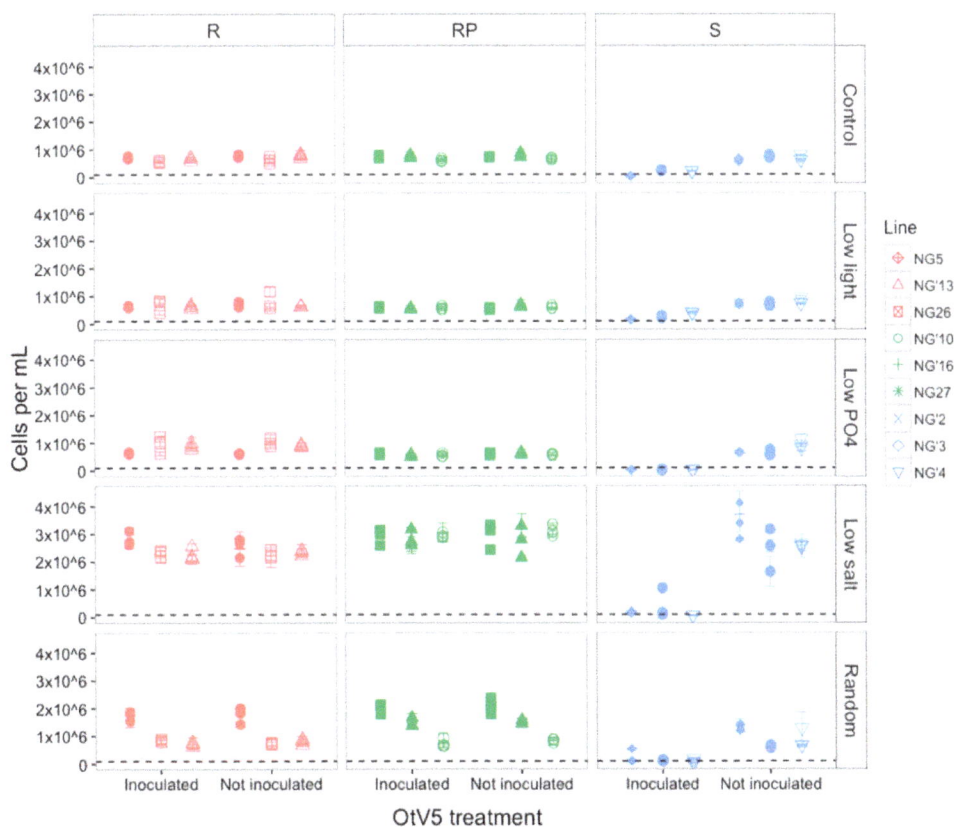

Figure 1. Mean (± SE) cell density mL^{-1} of resistant (R), resistant producer (RP) and susceptible (S) *O. tauri* lines three days after OtV5 inoculation in five environments. Points represent the average of the three assay replicates for each evolved population. Inoculated = populations inoculated with OtV5, Not inoculated = negative control populations that were grown for the same period without OtV5 inoculation. There were three evolved populations of each line. The dashed line represents the starting cell density at 100,000 cell mL^{-1}.

R and RP populations did not show a significant difference in cell density between populations that had been inoculated with OtV5 and populations that had not (Figure S1). In contrast, all S populations inoculated with OtV5 showed a change in cell density relative to non-inoculated S populations in the same environments (ANOVA effect of resistance type on difference $F_{2,125} = 66.51$, $p < 0.0001$). The largest differences in cell densities between inoculated and non-inoculated populations were observed in S populations evolved in the low salt environment, showing that whilst all populations in this environment were able to reach high densities in the absence of viruses, they were unable to grow in the presence of OtV5 (Figure 1). The large difference in S populations in low salt was due to the high growth rate of populations that had not been inoculated, since inoculated populations did not fall to lower densities than inoculated S populations in any other environments.

3.1.2. OtV5-Mediated Lysis Decreased in Susceptible Populations

Although S populations remained sensitive to viral lysis at the end of the evolution experiment, complete lysis was not observed in all populations, with a small proportion of populations able to reach numbers above the starting density of 100,000 cells mL^{-1} (Figure 1). This was in contrast to the beginning of the evolution experiment, when all susceptible populations fell below 100,000 cells mL^{-1} after inoculation with OtV5, indicating near-complete lysis (ANOVA effect of time point on cell density, $F_{1,65} = 21.87$, $p < 0.0001$) (Figure 2). The highest proportion of S cells that did not lyse was found in low salt evolved populations, suggesting that resistance mutations had been maintained in this environment, despite no selection by OtV5. To eliminate the possibility that the infection dynamics had changed and that the population decline was still in process, we measured the population density seven days after inoculation and did not observe any further decrease in population density (Figure S2).

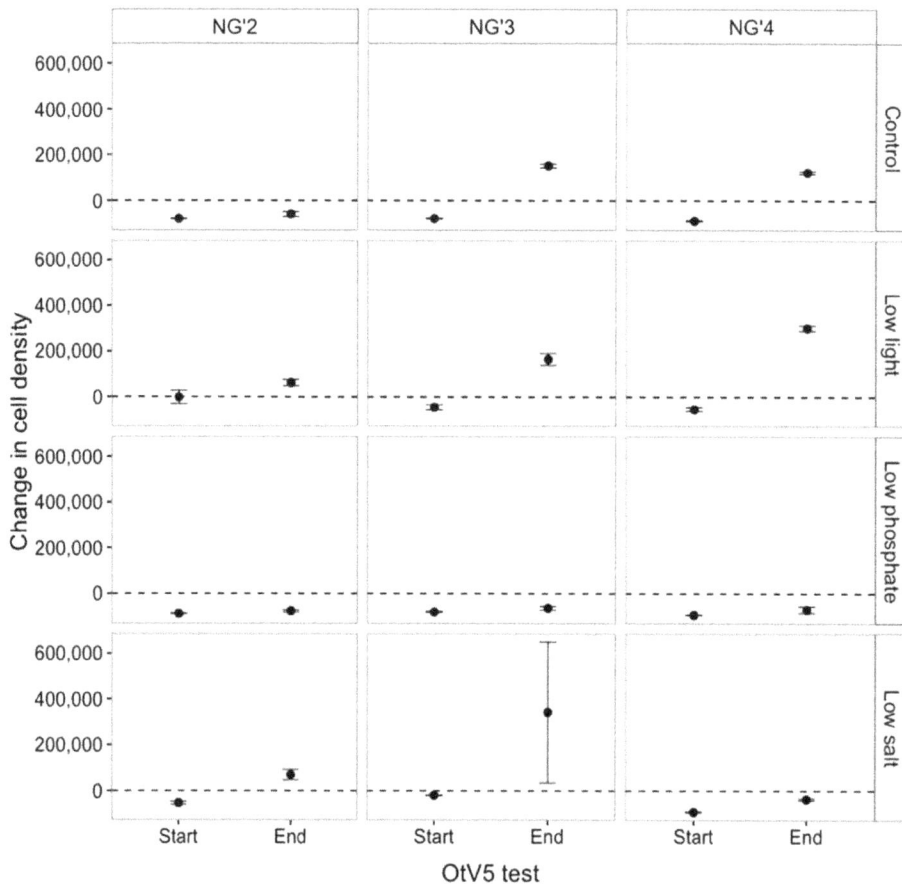

Figure 2. Change in cell density of the susceptible lines NG'2, NG'3 and NG'4 after OtV5 inoculation one week into the selection experiment (Start) and after 32 transfer cycles of evolution (End). The dashed line represents no change.

3.1.3. RPs Stopped Producing Viruses Early in the Evolution Experiment

During the evolution experiment, RP populations (NG27, NG'10 and NG'16) were tested to check that they were still producing viruses. Seven transfer cycles into the evolution experiment, all NG27 populations in all environments were still producing infectious viruses, as observed by cell lysis when their supernatant was used to inoculate the susceptible *O. tauri* strain RCC4221. In contrast, RCC4221 cultures that were inoculated with the supernatant of all populations of NG'10 and NG'16 continued growing, showing that no observable lysis had occurred. After 17 transfers in the selection environments, all RP populations in all environments had stopped producing infectious viruses (Figure S3), as observed by flow cytometric cell counts of RCC4221 populations inoculated with the supernatant of RP populations. When it was clear that all RP populations had stopped producing infectious viruses, frozen supernatant samples collected at transfers 9, 12, 14 and 15 were tested using the plaque assay method. No plaques were observed in any samples tested, thus we concluded that all RP populations in all environments had stopped producing viruses within nine weeks of the selection experiment.

3.2. Changes in Trait Values after Evolution

3.2.1. Changes in Cell Division Rate and Population Persistence during the Selection Experiment

Here, we focus on how growth rates vary with resistance type, selection environment and the number of transfer cycles in the selection environment. Growth rates of all populations were measured as the number of cell divisions per day, at four time points during the experiment (including at the beginning and end) (Figure 3). When comparing these time points, growth was significantly affected by environment, resistance type and time point ($p < 0.0001$ for all effects). In the first transfer cycle, which measured the population growth rates at the very start of the experiment following one week of acclimation, two out of the three RP lines (NG'10 and NG'16) had increased growth rates across all environments except for low phosphate (ANOVA effect of growth rate on cell divisions, $F_{3,5} = 17.19$, $p = 0.046$). These results are reported in [43].

After 14 transfer cycles, growth rates of all populations were approximately one division per day in the high salt, low phosphate, low light and random environments (Figure 3). In the control environment, growth rate varied across all S lines, even between populations of the same starting line, ranging from 0.18 to 0.87 divisions per day. The increased growth of all lines evolving in low phosphate to one division per day, which is the normal growth rate reported for *O. tauri* in phosphate-replete media, is consistent with adaptation to low phosphate in less than 100 generations. Additionally, RP lines that had been dividing more rapidly at transfer 1 were dividing at the same rate as other lines within each environment (Figure 3). This may be because the RP populations had stopped producing viruses and shifted to the R resistance type (see Section 3.1.3), thereby losing the growth advantage associated with the RP resistance type early on in this experiment. By transfer 24, all populations in the high temperature environment had gone extinct. RP populations went extinct more quickly than S and R populations, with 66% of RP lines extinct by T14 compared to 33% and 22% of S and R, respectively (Figure 3). At transfer 20, only three high temperature populations remained: one S (NG'4) and two R (NG'13 and NG26).

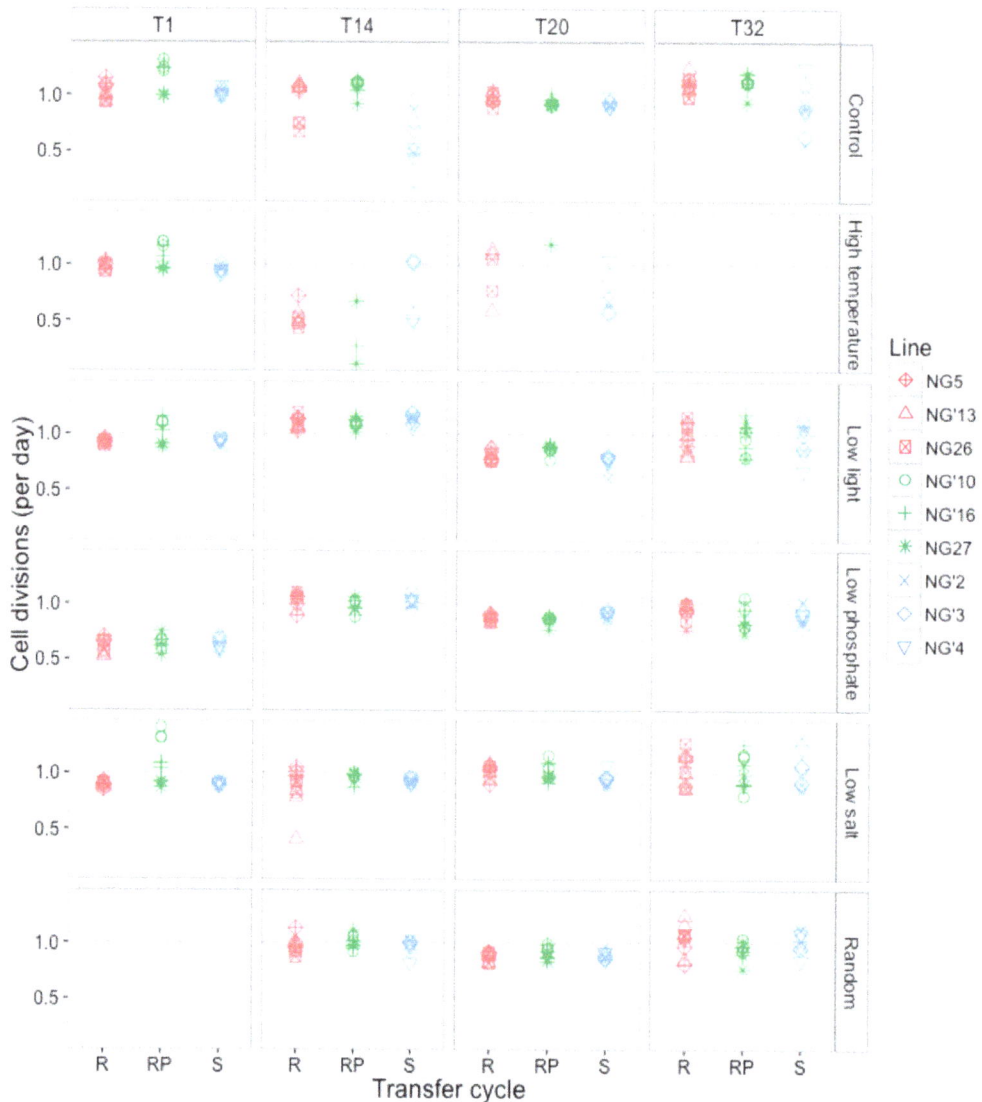

Figure 3. Growth rates as measured by mean cell divisions per day for each evolving population over four time points (1, 14, 20 and 32 transfer cycles). The dashed line represents one cell division per day. T1 is the growth rate following acclimation at the beginning of the experiment. There are no growth measurements for the randomized environment at T1 because lines had only been growing for one transfer cycle.

3.2.2. Growth Rates Varied with Selection Environment and Assay Environment after Evolution

After approximately 200 generations of evolution in each environment, a transplant assay was performed to quantify environmental effects on population growth rate, cell size and cell chlorophyll content for each evolved population. Here we define the selection environment as the environment that the population evolved in, and the assay environment as the environment in which measurements were taken. The direct response to selection compares the growth rate of a population evolved in a given selection environment with the growth rate of a population evolved in the control environment when both are grown (separately) in that given selection environment. The effect of selection environment on the direct response to evolution was large, and driven by the direct response to selection in the low phosphate environment (ANOVA effect selection environment on direct response, $F_{2,228} = 9.26$, $p = 0.0001$), whereas the effect of resistance type was smaller (ANOVA effect of resistance type on direct response, $F_{2,228}$ 2.87, $p = 0.06$).

Selection environment alone and assay environment alone both had a significant effect on population growth rate (ANOVA effect of selection environment on growth, $F_{4,200} = 19.92$, $p < 0.0001$; ANOVA effect of assay environment on growth, $F_{3,758} = 32.43$, $p < 0.0001$), which shows that environment affected growth rates. Resistance type also had an effect on growth rate (ANOVA effect of resistance type on growth, $F_{2,195} = 4.21$, $p = 0.02$), with R populations having the fastest cell division rates and S populations having the slowest cell division rates. Additionally, an interaction between selection environment and assay environment affected growth rate, indicating that the way in which selection environment affected growth differed between assay environments (ANOVA selection environment × assay environment, $F_{3,757} = 2.89$, $p = 0.03$). The fastest growth rates were seen in the evolved control populations that were assayed in low salt (Figure 4). Better performance was not due to being assayed in the same selection environment that the populations had evolved in (ANOVA effect of being assayed in selection environment on growth, $F_{1,831} = 1.70$, $p = 0.19$).

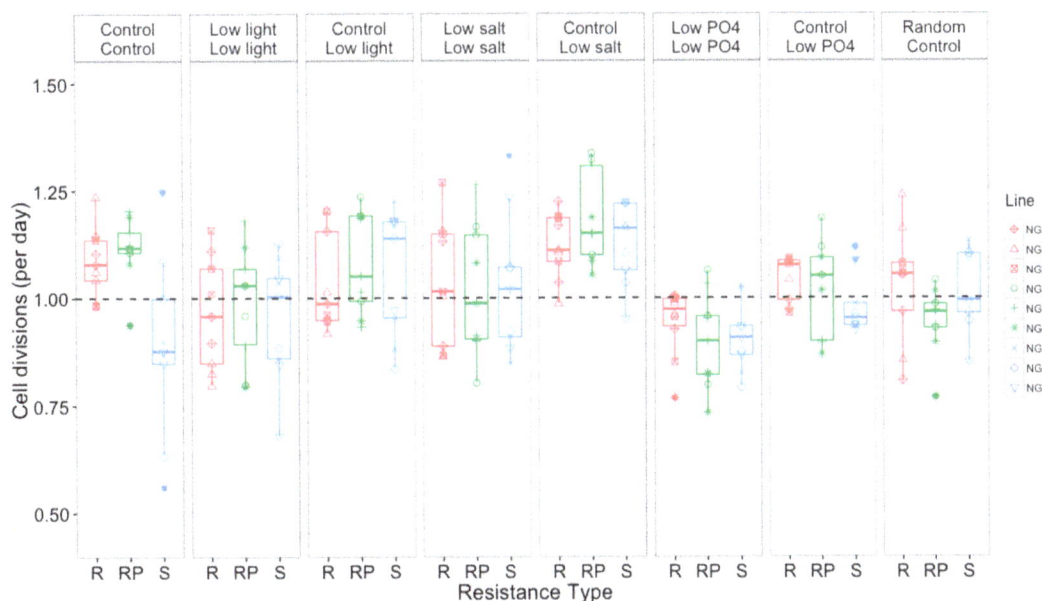

Figure 4. Mean cell divisions per day (±SEM). R = resistant, RP = resistant producer, S = susceptible. Each panel represents a growth assay, with cells evolved in the selection environment (top label) and growth rates measured in the assay environment (bottom label). The dashed line indicates, for reference, one cell division per day.

3.2.3. Resistance Type Affected Cell Size and Chlorophyll Content

Cells from different resistance types had different cell sizes (ANOVA effect of resistance type on size, $F_{2,140} = 9.49$, $p = 0.0001$) (Figure S4) and this was not affected during evolution in any of the environments (ANOVA effect of selection environment on size, $F_{4,155} = 0.66$, $p = 0.62$; ANOVA effect of assay environment on size, $F_{3,735} = 1.60$, $p = 0.19$). The greatest variation in cell size between populations was observed when control-evolved cells were assayed in low salt (0.86–0.97 μm) across all resistance types. Less variation was found in the control-evolved cells assayed in low phosphate (0.82–0.97 μm).

The environment in which populations were assayed had a significant effect on the relative chlorophyll content per cell volume (ANOVA effect of assay environment on chlorophyll, $F_{3,744} = 17.83$, $p < 0.0001$). However, selection environment did not (ANOVA effect of selection environment on chlorophyll, $F_{4,168} = 0.90$, $p = 0.47$). Resistance type affected chlorophyll content (ANOVA effect of resistance type on chlorophyll, $F_{2,153} = 8.54$, $p < 0.0001$). Susceptible populations that had been evolving in the control environment contained high amounts of chlorophyll relative to their cell size when assayed under all three selection environments (low light, low salt and low phosphate) (Figure S5).

3.3. Selection and Assay Environments Affect Competitive Ability of O. tauri

In addition to measuring growth rate, size and chlorophyll content, we also tested if costs of resistance could be observed during pairwise competition between each population of S, R, and RP. We measured relative competitive ability, by competing each population against a common competitor harboring a GFP reporter, which allowed us to distinguish between the evolved population and the GFP line. Both selection environment and assay environment affected competitive ability against a roGFP-labeled strain (ANOVA effect of selection environment on competitiveness, $F_{4,622} = 16.41$, $p = < 0.0001$; ANOVA effect of assay environment on competitiveness, $F_{3,622} = 10.96$, $p < 0.0001$). Most populations were poor competitors relative to the roGFP line (Figure 5). Lines evolved in low light and low salt were the best competitors. Lines that were assayed in the same environment that they had evolved in were better competitors than control lines that were assayed in the selection environments. This shows that these lines adapted to their selection environment and that growth rate is not necessarily the most appropriate measure of adaptation in this study, which is consistent with other studies in *Ostreococcus* spp. [44]. Interestingly, populations in the control environment were the worst competitors, regardless of resistance type, with a 0.56 mean fold change, showing that all populations were out-competed by the roGFP line. This indicates that the control environment did in fact exert less selection on the populations than did the other environments.

Resistance type alone did not affect competitive ability (ANOVA effect of resistance type of competitiveness, $F_{2,622} = 1.22$, $p = 0.30$). Although competitive ability differed between resistance types, the response was not consistent across assay environments, with no one resistance type consistently being a better or poorer competitor.

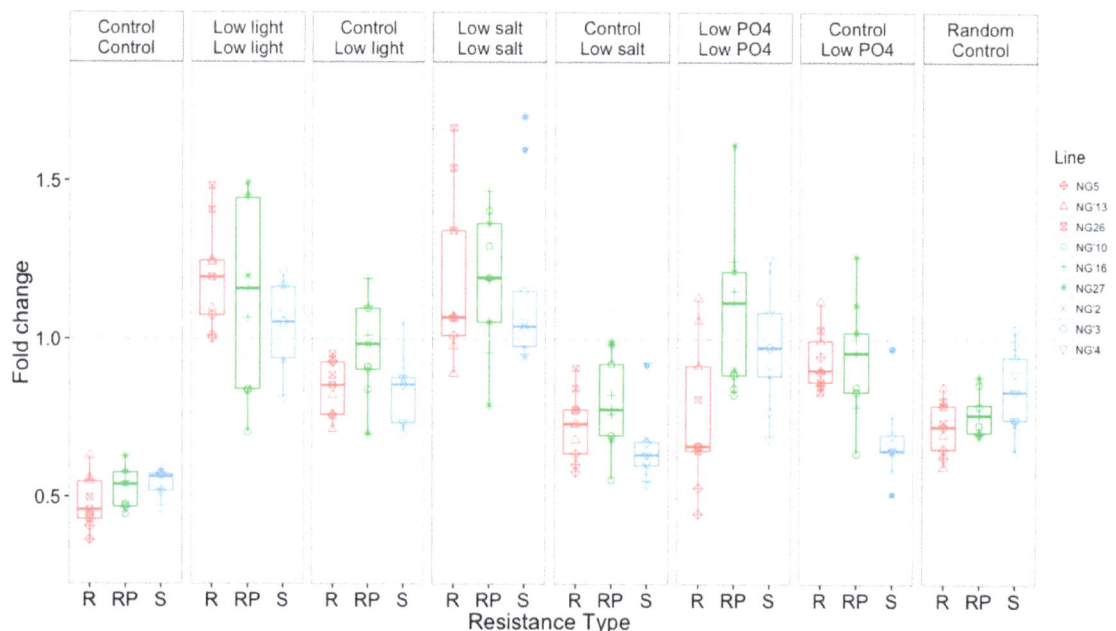

Figure 5. Competitive ability, as measured by fold difference in growth relative to a roGFP-modified *O. tauri* line, of evolved populations and control populations assayed in the selection environments. R = resistant, RP= resistant producer, S = susceptible. Each panel represents one assay, with populations evolved in the selection environment (top label) and competitiveness measured in the assay environment (bottom label). The dashed line represents no change (i.e., equal proportions of roGFP and competitor populations).

4. Discussion

We examined whether cost of resistance varied with the abiotic environment in which *O. tauri* populations evolved. A cost of resistance can manifest in different ways depending on the interaction between host and virus and on the way in which resistance is acquired (e.g., entry of the virus into the cell, and ability of the virus to replicate within the cell and cause lysis). This means that it is often difficult to detect a cost of resistance, so we measured three host responses: ability to maintain resistance, population growth rate and competitive ability.

4.1. Susceptibility to OtV5 Did Not Change after Evolution

After evolution in a new environment, OtV5 was still able to lyse susceptible (S) *O. tauri* populations under all environmental conditions tested, whereas R and RP populations remained resistant under all environments, despite the absence of selection pressure for viral resistance (Figure 1). Resistance to pathogens often comes at a fitness cost, such that a proportion of susceptible individuals remain in the population, thereby allowing viruses to persist [21]. If resistance does carry a fitness cost, populations should revert to susceptibility over time, in the prolonged absence of viruses, even if that cost is low, because susceptible cells have a fitness advantage in the absence of viruses [29]. Our study indicated that if there is a cost to simply maintaining resistance in *O. tauri*, it is small. Over the time scale of our experiment, the fitness advantage of susceptible types in the absence of viruses would have to be about 0.005 for a mutation conferring susceptibility in a resistant background to be fixed in the population following a spontaneous reversion of a resistant cell (where we calculate s from $\frac{1}{2}s/(1-e^{-2sN})$, and assume a starting frequency of $1/N$ [45]).

It is possible that there is a genetic constraint preventing the loss of resistance, making the transition from resistant to susceptible phenotypes rare even if resistance is costly. This is consistent with recent studies showing that the resistance mechanism in *O. tauri* is an intracellular response [35] and probably also involves rearrangements of chromosome 19 [36]. The presence of a genetic constraint on losing resistance would favor compensatory mutations that lead to alleles being selected that reduce the cost of resistance [46,47]. Studies evolving *E. coli* in the absence of bacteriophage observed that the cost of resistance to the T4 bacteriophage decreased after 400 generations due to compensatory adaptations [46]. A second possibility is that the cost of resistance to one strain of OtV means increased susceptibility to other virus strains. For example, cyanobacteria can rapidly evolve viral resistance when coevolving with viruses, however increased resistance to one virus can lead to a narrower resistance range thereby making cells more susceptible to other virus strains [27,28]. *O. tauri*-virus interactions can be complex with some OtVs being very specific to host *O. tauri* strains while others are generalists that can infect many strains [26,48]. Our experiment focused only on OtV5 and did not examine evolution of host resistance range.

At the end of the evolution experiment, OtV5 lysed susceptible (S) populations in all environments, but the extent of lysis differed between environments (Figure 1). This could be because one or more resistance mutations had appeared and risen to a detectable frequency in some populations. It is unclear whether incomplete lysis was due to some resistant cells evolving in the susceptible populations, or whether susceptible populations had evolved to make virus entry harder but still possible. Inoculations were performed from frozen stocks, thus OtV5 was not coevolving with the host, enabling us to measure evolution in the *O. tauri* populations relative to the ancestral virus population. We cannot rule out the possibility that there was a slow loss of infective virus titer in the cryopreserved stock, leading to fewer infectious viruses in the inoculum and therefore a lower multiplicity of infection. Physiological changes in susceptible populations arising as an adaptive response to abiotic environmental change did not prevent viral lysis, indicating that viral adsorption was not completely inhibited. This was even evident in the control populations, suggesting that although these populations did not experience a change in environment, they may have evolved changes in cell surface proteins, since were still evolving for the full length of the experiment. However, the biotic environment plays a larger role in resistance acquisition, since resistance to viruses is selected for by the virus [49]. Chemostat experiments to monitor population dynamics in *Chlorella* and *Paramecium bursaria Chlorella* Virus 1 (PBCV-1) showed that control populations maintained in the absence of viruses did not evolve resistance to the ancestor virus, suggesting that resistance arises from host-virus interactions [23]. In contrast, sensitive *E. coli* cells evolved complete resistance to λ infection and resistant cells increased susceptibility to T6* infection after 45,000 generations in the absence of phage [29]. In our experiment, low phosphate was the only environment in which the cell numbers of all lines fell below the starting cell density (Figure 1), suggesting that this environment either affected the infectivity of OtV5 directly or the cells' response to infection. Other studies report the opposite, with reduced virus infection of algae under low phosphate, possibly due to the requirement of phosphate for viral replication [9,10,13]. Though phosphate levels were low in our experiment, they were sufficient for population growth to be positive, and were higher than found in the Mediterranean Sea [50]. Conflicting results highlight the complexity of host-virus interactions in different study systems as well as different growth conditions.

There was a selection pressure against viral production on RP lines, but not on host resistance across all RP lines in all environments. Similarly, Yau et al. reported that over a two year period RP populations maintained under standard laboratory conditions stopped producing viruses [36]. If RP populations are indeed made up of a majority of resistant cells with a small proportion of susceptible cells arising that lyse upon OtV5 infection, thus maintaining the production of viruses in the media, then we would expect resistance to be selected for in the presence of viruses. Resistance in *O. tauri* is expected to be caused by over-expression of glycosyltransferase genes on chromosome 19 [36]. In this study, the selection environment did not affect the time it took for a selective sweep of resistance to occur in the RP lines, supporting the conclusion that there was little or no selection against resistance, that there is a genetic constraint on losing resistance, or that compensatory mutations enabled resistance to be maintained.

4.2. Resistance Type and Environment Affect Evolutionary Response of O. tauri to Environmental Change

We did not observe a growth cost of *O. tauri* being resistant to viral lysis, since R populations had the fastest growth overall whereas S populations had the slowest growth. Data on the growth effects of resistance in marine algae are rare. A 20% reduction in growth was reported in the ubiquitous cyanobacterium *Synechococcus* [22], however it is unknown whether viral resistance generally carries a growth cost in eukaryotic algae. Even with no or minimal costs of resistance, the chromosomal rearrangement associated with resistance in *O. tauri* means that the different resistance types have different genetic backgrounds. Therefore, evolution could take different trajectories in hosts with different resistance types due to epistatic interactions between resistance and adaptive changes. For example, trade-off shape varied in response to environmental change and physiological changes of bacteriophage resistant *E. coli*, leading to variation in sensitivity to environmental change across different strains [51]. In our study, when considering the direct response to evolution (which compares the growth rate of the evolved population in its selection environment with the plastic response of the control line in that selection environment), resistance type did not drive direct response. This indicates that the growth response of the three resistance types was similar within environments. If there is an effect of genetic background being introduced by resistance, it is not evident at the level of growth rate under these conditions.

Selection environment affected population growth, with populations evolved in the control environment having the highest growth rates in all assay environments (Figure 4). The decrease in growth in response to our selection environments is consistent with them being of lower quality than the control environment, by design, so that selection was stronger in the non-control environments. Variation in the direct response to evolution was explained by selection environment. Populations evolved in low phosphate had the lowest growth, which is expected when cells are nutrient limited. Interestingly, populations that had evolved in the control environment grew more rapidly in low phosphate than populations that had evolved in low phosphate. This may be because populations that had been evolving in the control environment had enough phosphate reserves within the cell to grow normally for a short period, since growth was only assayed for seven days. Overall, growth rates of populations evolved in the control environment were greater when assayed in the selection environments than the populations that had evolved in those environments, showing that increased growth could be initiated as a stress response, and that cells in the control environment (which was nutrient-replete, and at the optimal temperature and usual salinity for these lines of *O. tauri*) were in better condition overall. The extent of a cost of resistance can be highly dependent on environment. For example, cost of resistance differs when fitness of *E. coli* is measured under different nutrient resources and concentrations [20,52]. We show here that growth rate measurements may not be sensitive enough to detect very small differences between populations conferring a cost of resistance in *O. tauri*, as has also been observed in short term experiments using a single [35] and multiple environments [43]. Studies in bacteria also found that resistant strains grew at the same rate as susceptible strains [21,46]. Our results indicate that, regardless of resistance type, *O. tauri* is able to adapt to environmental change including low light, low salt and low phosphate. However, all populations in the high temperature environment went extinct, despite the modest (2 °C) increase, suggesting that although *O. tauri* can tolerate and grow at higher temperatures over the short-term, sustained temperature increases may exert stronger selection than predicted from short-term studies. It is not possible to infer as of yet whether resistance affects growth rate in natural habitats or whether a cost of resistance is instead associated with tradeoffs that are not related to the abiotic environment, such as resistance to other viral strains.

In contrast to cell division rates, resistance type affected cell size and chlorophyll content, but selection environment did not. Cells in RP populations were sometimes larger in size and S populations were slightly smaller. Often, small size is associated with a response to nutrient limitation, increased temperature and light limitation in phytoplankton [53–57], however all lines in this study showed slightly increased cell size in low phosphate. An increased cell volume has been observed in coccolithophores in response to phosphate limitation suggesting the adaptive strategy is to reduce phosphorous requirements rather than increasing surface area to volume ratio [58].

RP populations had less chlorophyll in most environments, however overall there was substantial variation in chlorophyll content, especially in S populations. When assayed in the control environment, populations that had evolved in low light, low salt, low phosphate and the random environment had lower chlorophyll than did control populations assayed in these same environments. The response of populations evolved in the control environment increasing their relative chlorophyll content when assayed in low light is consistent with responses to light limitation in other green algae [59–61]. Here, we show that response of chlorophyll content to environmental change is variable, both with environment and with resistance type. Previous studies in marine microalgae have reported lower reduced chlorophyll content under nutrient limitation [62] and higher chlorophyll content under some optimal salinities [63,64].

4.3. Resistance Type Did Not Affect Competitive Ability Regardless of Environment

Reduced competitive ability is often one of the main restrictions for resistance spreading through a population, however resistance type did not affect the competitive ability of evolved populations in our experiment. We found that environment did affect competitive ability, and similarly in bacteria, the environment that populations evolve in, such as the limiting sugar source or spatial heterogeneity, can affect competitive ability, both with and without coevolving phage [51,52,65]. Other studies have reported a trade-off between competitive ability and resistance, whereas here we found no evidence for reduced resistance with increased competitive ability. The nature of a cost of resistance will depend on the genetic or physiological changes to the cell. For example, *E. coli* mutants showed high variability in competitiveness which was associated with resistance strategy, with cross-resistance to phage T7 significantly decreasing competitive fitness by approximately 3-fold [21]. In contrast, competitions with cyanobacteria showed that total resistance (the total number of viruses to which a host strain was resistant) did not affect competitive ability [22]. These examples reveal that the magnitude of the reduced competitiveness trade-off can depend on the specific resistance strategy.

Evolved populations in the non-control environments were better competitors than control populations that had been exposed to the selection environments for the first time (plastic response), indicating that all lines had adapted to their selection environment. Thus, growth rate is not the most appropriate measure of adaptation in *O. tauri*, since the plastic response was to increase population growth rates, and the evolutionary response was to reverse this plastic increase in growth rates, and this strategy was associated with an increase in competitive fitness. Similar results have been reported previously in *Ostreococcus* spp. where populations with high growth rates in monoculture were poorer competitors than those with lower growth rates in monoculture [44].

5. Conclusions

Here, we show that there was no detectable cost of resistance to OtV5 as measured by growth rate or competitive ability for *O. tauri* evolved in several different environments, and that resistance to viruses did not affect adaptation to environmental change. Additionally, we found no reversion of R or RP populations to S as tested by exposure to OtV5, whereas lysis occurred in all S populations. Additionally, all RP lines stopped producing viruses within nine weeks of the experiment. This suggests that a shift from susceptibility to resistance is more common than a shift from resistance to susceptibility, regardless of selection environment, at least for the range of environments used here. Our experiment shows that the conditions under which a cost of resistance may occur or affect adaptation in *O. tauri* are not clear in the laboratory. More work is needed to understand the factors that affect host–virus interactions in the marine environment to better understand evolutionary and ecological responses of marine eukaryotic microalgae to environment change.

Acknowledgments: This work was supported by a BBSRC EASTBIO Doctoral Training Programme grant [BB/J01446X/1] awarded to SEH and BBSRC awards [BB/D019621 and BB/J009423] to Andrew Millar and others. SC is supported by a Royal Society University Research Fellowship and an ERC starting grant (260266). PFV is supported by a strategic award from the Wellcome Trust for the Centre for Immunity, Infection and Evolution (http://ciie.bio.ed.ac.uk; grant reference no. 095831), a Chancellor's Fellowship from the School of Biological Sciences—University of Edinburgh, and by a Society in Science—Branco Weiss fellowship ([http://www.society-in-science.org)] http://www.society-in-science.org), administered by ETH Zürich. We thank the lab group at Observatoire Océanologique, Banyuls-sur-Mer, for help with virus culturing techniques and for providing the *O. tauri* lines. We thank A. Millar for facilities and funding to create the roGFP line and K. Kis for technical assistance and maintaining the roGFP line. We are grateful to R. Lindberg and H. Kuehne for help with laboratory work and useful discussions.

Author Contributions: S.E.H., P.F.V. and S.C. conceived and designed the experiments; S.E.H. and K.K. performed the experiments; and S.E.H., P.F.V. and S.C. analyzed the data. All authors contributed to writing the paper.

References

1. Suttle, C.A. Marine viruses—Major players in the global ecosystem. *Nat. Rev. Microbiol.* **2007**, *5*, 801–812. [CrossRef] [PubMed]

2. Wilhelm, S.W.; Suttle, C.A. Viruses and nutrient cycles in the sea. *Bioscience* **1999**, *49*, 781–788. [CrossRef]

3. Koskella, B.; Brockhurst, M.A. Bacteria-phage coevolution as a driver of ecological and evolutionary processes in microbial communities. *FEMS Microbiol. Rev.* **2014**, *38*, 916–931. [CrossRef] [PubMed]

4. Dennehy, J.J. What Can Phages Tell Us about Host-Pathogen Coevolution? *Int. J. Evol. Biol.* **2012**, *2012*, 396165. [CrossRef] [PubMed]

5. Flint, S.J.; Enquist, L.W.; Krug, L.; Racaniella, V.; Skalka, A. *Principles of Virology: Molecular Biology, Pathogenesis, Virus Ecology*, 4th ed.; ASM Press: Washington, DC, USA, 2000.

6. Doney, S.C.; Ruckelshaus, M.; Emmett Duffy, J.; Barry, J.P.; Chan, F.; English, C.A.; Galindo, H.M.; Grebmeier, J.M.; Hollowed, A.B.; Knowlton, N.; et al. Climate Change Impacts on Marine Ecosystems. *Ann. Rev. Mar. Sci.* **2012**, *4*, 11–37. [CrossRef] [PubMed]

7. Nagasaki, K.; Yamaguchi, M. Effect of temperature on the algicidal activity and the stability of HaV (Heterosigma akashiwo virus). *Aquat. Microb. Ecol.* **1998**, *15*, 211–216. [CrossRef]

8. Wells, L.E.; Deming, J.W. Effects of temperature, salinity and clay particles on inactivation and decay of cold-active marine Bacteriophage 9A. *Aquat. Microb. Ecol.* **2006**, *45*, 31–39. [CrossRef]

9. Bellec, L.; Grimsley, N.; Derelle, E.; Moreau, H.; Desdevises, Y. Abundance, spatial distribution and genetic diversity of *Ostreococcus tauri* viruses in two different environments. *Environ. Microbiol. Rep.* **2010**, *2*, 313–321. [CrossRef] [PubMed]

10. Bratbak, G.; Egge, J.K.; Heldal, M. Viral mortality of the marine alga *Emiliania huxleyi* (Haptophyceae) and termination of algal blooms. *Mar. Ecol. Prog. Ser.* **1993**, *93*, 39–48. [CrossRef]

11. Bratbak, G.; Jacobsen, A.; Heldal, M.; Nagasaki, K.; Thingstad, F. Virus production in *Phaeocystis pouchetii* and its relation to host cell growth and nutrition. *Aquat. Microb. Ecol.* **1998**, *16*, 1–9. [CrossRef]

12. Wilson, W.H.; Carr, N.G.; Mann, N.H. The effect of phosphate status on the kinetics of cyanophage infection in the oceanic cyanobacterium *Synechococcus* sp. WH7803. *J. Phycol.* **1996**, *32*, 506–516. [CrossRef]

13. Maat, D.; Crawfurd, K.; Timmermans, K.; Brussard, C. Elevated CO_2 and phosphate limitation favor *Micromonas pusilla* through stimulated growth and reduced viral impact. *Appl. Environ. Microbiol.* **2014**, *80*, 3119–3127. [CrossRef] [PubMed]

14. Jacquet, S.; Bratbak, G. Effects of ultraviolet radiation on marine virus-phytoplankton interactions. *FEMS Microbiol. Ecol.* **2003**, *44*, 279–289. [CrossRef]

15. Jacquet, S.; Heldal, M.; Iglesias-Rodriguez, D.; Larsen, A.; Wilson, W.; Bratbak, G. Flow cytometric analysis of an *Emiliana huxleyi* bloom terminated by viral infection. *Aquat. Microb. Ecol.* **2002**, *27*, 111–124. [CrossRef]

16. Thyrhaug, R.; Larsen, A.; Brussaard, C.P.D.; Mcfadden, P. Cell Cycle Dependent Virus Production in Marine Phytoplankton 1. *Cell Cycle* **2002**, *343*, 338–343.

17. Larsen, J.B.; Larsen, A.; Thyrhaug, R.; Bratbak, G.; Sandaa, R.-A. Response of marine viral populations to a nutrient induced phytoplankton bloom at different pCO_2 levels. *Biogeosci. Discuss.* **2007**, *4*, 3961–3985. [CrossRef]

18. Chen, S.; Gao, K.; Beardall, J. Viral attack exacerbates the susceptibility of a bloom-forming alga to ocean acidification. *Glob. Chang. Biol.* **2014**, *21*, 629–636. [CrossRef] [PubMed]

19. Danovaro, R.; Corinaldesi, C.; Dell'Anno, A.; Fuhrman, J.A.; Middelburg, J.J.; Noble, R.T.; Suttle, C.A. Marine viruses and global climate change. *FEMS Microbiol. Rev.* **2011**, *35*, 993–1034. [CrossRef] [PubMed]

20. Bohannan, B.J.M.; Kerr, B.; Jessup, C.M.; Hughes, J.B.; Sandvik, G. Trade-offs and coexistence in microbial microcosms. *Antonie van Leeuwenhoek* **2002**, *81*, 107–115. [CrossRef] [PubMed]

21. Lenski, R.E. Experimental Studies of Pleiotropy and Epistasis in *Escherichia coli.* I. Variation in Competitive Fitness Among Mutants Resistant to Virus T4. *Evolution* **1988**, *42*, 425–432. [CrossRef]

22. Lennon, J.T.; Khatana, S.A.M.; Marston, M.F.; Martiny, J.B.H. Is there a cost of virus resistance in marine cyanobacteria? *ISME J.* **2007**, *1*, 300–312. [CrossRef] [PubMed]

23. Frickel, J.; Sieber, M.; Becks, L. Eco-evolutionary dynamics in a coevolving host-virus system. *Ecol. Lett.* **2016**, *19*, 450–459. [CrossRef] [PubMed]

24. Seed, K.D.; Faruque, S.M.; Mekalanos, J.J.; Calderwood, S.B.; Qadri, F.; Camilli, A. Phase Variable O Antigen Biosynthetic Genes Control Expression of the Major Protective Antigen and Bacteriophage Receptor in Vibrio cholerae O1. *PLoS Pathog.* **2012**, *8*, e1002917. [CrossRef] [PubMed]

25. León, M.; Bastías, R. Virulence reduction in bacteriophage resistant bacteria. *Front. Microbiol.* **2015**, *6*, 1–7. [CrossRef] [PubMed]

26. Clerissi, C.; Desdevises, Y.; Grimsley, N. Prasinoviruses of the marine green alga *Ostreococcus tauri* are mainly species specific. *J. Virol.* **2012**, *86*, 4611–4619. [CrossRef] [PubMed]

27. Marston, M.F.; Pierciey, F.J.; Shepard, A.; Gearin, G.; Qi, J.; Yandava, C.; Schuster, S.C.; Henn, M.R.; Martiny, J.B.H. Rapid diversification of coevolving marine Synechococcus and a virus. *Proc. Natl. Acad. Sci. USA* **2012**, *109*, 4544–4549. [CrossRef] [PubMed]

28. Avrani, S.; Wurtzel, O.; Sharon, I.; Sorek, R.; Lindell, D. Genomic island variability facilitates *Prochlorococcus*-virus coexistence. *Nature* **2011**, *474*, 604–608. [CrossRef] [PubMed]

29. Meyer, J.R.; Agrawal, A.A.; Quick, R.T.; Dobias, D.T.; Schneider, D.; Lenski, R.E. Parallel changes in host resistance to viral infection during 45,000 generations of relaxed selection. *Evolution* **2010**, *64*, 3024–3034. [CrossRef] [PubMed]

30. Avrani, S.; Lindell, D. Convergent evolution toward an improved growth rate and a reduced resistance range in *Prochlorococcus* strains resistant to phage. *Proc. Natl. Acad. Sci. USA* **2015**, *112*, E2191–E2200. [CrossRef] [PubMed]

31. Avrani, S.; Schwartz, D.A.; Lindell, D. Virus-host swinging party in the oceans: Incorporating biological complexity into paradigms of antagonistic coexistence. *Mob. Genet. Elem.* **2012**, *2*, 88–95. [CrossRef] [PubMed]

32. Field, C.B. Primary Production of the Biosphere: Integrating Terrestrial and Oceanic Components. *Science* **1998**, *281*, 237–240. [CrossRef] [PubMed]

33. Bellec, L.; Grimsley, N.; Moreau, H.; Desdevises, Y. Phylogenetic analysis of new Prasinoviruses (*Phycodnaviridae*) that infect the green unicellular algae *Ostreococcus*, *Bathycoccus* and *Micromonas*. *Environ. Microbiol. Rep.* **2009**, *1*, 114–123. [CrossRef] [PubMed]

34. Derelle, E.; Ferraz, C.; Escande, M.-L.; Eychenié, S.; Cooke, R.; Piganeau, G.; Desdevises, Y.; Bellec, L.; Moreau, H.; Grimsley, N. Life-cycle and genome of OtV5, a large DNA virus of the pelagic marine unicellular green alga *Ostreococcus tauri*. *PLoS ONE* **2008**, *3*, e2250. [CrossRef] [PubMed]

35. Thomas, R.; Grimsley, N.; Escande, M.-L.; Subirana, L.; Derelle, E.; Moreau, H. Acquisition and maintenance of resistance to viruses in eukaryotic phytoplankton populations. *Environ. Microbiol.* **2011**, *13*, 1412–1420. [CrossRef] [PubMed]

36. Yau, S.; Hemon, C.; Derelle, E.; Moreau, H.; Piganeau, G.; Grimsley, N. A Viral Immunity Chromosome in the Marine Picoeukaryote, *Ostreococcus tauri*. *PLoS Pathog.* **2016**, *12*, e1005965. [CrossRef] [PubMed]

37. Keller, M.D.; Selvin, R.C.; Claus, W.; Guillard, R.R.L. Media for the culture of oceanic ultraphytoplankton. *J. Phycol.* **1987**, *23*, 633–638.

38. Bates, D.; Mächler, M.; Bolker, B.; Walker, S. Fitting Linear Mixed-Effects Models using lme4. *J. Stat. Softw.* **2014**, *67*, 51.

39. Kuznetsova, A.; Brockhoff, P.B.; Christensen, R.H.B. lmerTest: Tests in Linear Mixed Effects Models. Cran, R Package. 2015. Available online: http://CRAN.R-project.org/package=lmerTest (accessed on 15 January 2017).

40. Schwarzländer, M.; Fricker, M.; Müller, C.; Marty, L.; Brach, T.; Novak, J.; Sweetlove, L.; Hell, R.; Meyer, A. Confocal imaging of glutathione redox potential in living plant cells. *J. Microsc.* **2008**, *231*, 299–316. [CrossRef] [PubMed]

41. Corellou, F.; Schwartz, C.; Motta, J.-P.; Djouani-Tahri, E.B.; Sanchez, F.; Bouget, F.-Y. Clocks in the green lineage: Comparative functional analysis of the circadian architecture of the picoeukaryote *Ostreococcus*. *Plant Cell* **2009**, *21*, 3436–3449. [CrossRef] [PubMed]

42. Van Ooijen, G.; Knox, K.; Kis, K.; Bouget, F.-Y.; Millar, A.J. Genomic Transformation of the Picoeukaryote *Ostreococcus tauri*. *J. Vis. Exp.* **2012**, *65*, e4074. [CrossRef] [PubMed]

43. Heath, S.E.; Collins, S. Mode of resistance to viral lysis affects host growth across multiple environments in the marine picoeukaryote *Ostreococcus tauri*. *Environ. Microbiol.* **2016**, *18*, 4628–4639. [CrossRef] [PubMed]

44. Schaum, E.; Collins, S. Plasticity predicts evolution in a marine alga. *Proc. R. Soc. B* **2014**, *281*. [CrossRef] [PubMed]

45. Bell, G. *Selection: The Mechanism of Evolution*, 2nd ed.; Oxford University Press: Oxford, UK, 2008.

46. Lenski, R.E. Experimental Studies of Pleiotropy and Epistasis in *Escherichia coli*. II. Compensation for Maldaptive Effects Associated with Resistance to Virus T4. *Evolution* **1988**, *42*, 433–440. [CrossRef]

47. Björkman, J.; Nagaev, I.; Berg, O.G.; Hughes, D.; Andersson, D.I. Effects of environment on compensatory mutations to ameliorate costs of antibiotic resistance. *Science* **2000**, *287*, 1479–1482. [PubMed]

48. Bellec, L.; Clerissi, C.; Edern, R.; Foulon, E.; Simon, N.; Grimsley, N.; Desdevises, Y. Cophylogenetic interactions between marine viruses and eukaryotic picophytoplankton. *BMC Evol. Biol.* **2014**, *14*, 59. [CrossRef] [PubMed]

49. Luria, S.; Delbrück, M. Mutations of Bacteria from Virus Sensitivity to Virus Resistance. *Genetics* **1943**, *28*, 491–511. [PubMed]

50. Karafistan, A.; Martin, J.M.; Rixen, M.; Beckers, J.M. Space and time distributions of phosphate in the Mediterranean Sea. *Deep. Sea Res. I* **2002**, *49*, 67–82. [CrossRef]

51. Jessup, C.M.; Bohannan, B.J.M. The shape of an ecological trade-off varies with environment. *Ecol. Lett.* **2008**, *11*, 947–959. [CrossRef] [PubMed]

52. Bohannan, B.J.M.; Travisano, M.; Lenski, R.E. Epistatic Interactions Can Lower the Cost of Resistance to Multiple Consumers. *Evolution* **1999**, *53*, 292–295. [CrossRef]

53. Finkel, Z.V.; Beardall, J.; Flynn, K.J.; Quigg, A.; Rees, T.A.V.; Raven, J.A. Phytoplankton in a changing world: Cell size and elemental stoichiometry. *J. Plankton Res.* **2010**, *32*, 119–137. [CrossRef]

54. Peter, K.H.; Sommer, U. Interactive effect of warming, nitrogen and phosphorus limitation on phytoplankton cell size. *Ecol. Evol.* **2015**, *5*, 1011–1024. [CrossRef] [PubMed]

55. Atkinson, D.; Ciotti, B.J.; Montagnes, D.J.S. Protists decrease in size linearly with temperature: Ca. 2.5% $°C^{-1}$. *Proc. Biol. Sci.* **2003**, *270*, 2605–2611. [CrossRef] [PubMed]

56. Morán, X.A.G.; López-Urrutia, Á.; Calvo-Díaz, A.; Li, W.K.W. Increasing importance of small phytoplankton in a warmer ocean. *Glob. Chang. Biol.* **2010**, *16*, 1137–1144. [CrossRef]

57. Geider, R.; Platt, T.; Raven, J. Size dependence of growth and photosynthesis in diatoms: A synthesis. *Mar. Ecol. Ser.* **1986**, *30*, 93–104. [CrossRef]

58. Šupraha, L.; Gerecht, A.C.; Probert, I.; Henderiks, J. Eco-physiological adaptation shapes the response of calcifying algae to nutrient limitation. *Sci. Rep.* **2015**, *5*, 16499. [CrossRef] [PubMed]

59. Ryther, J.; Menzel, D. Light adaptation by marine phytoplankton. *Limnol. Oceanogr.* **1959**, *4*, 492–497. [CrossRef]

60. Wozniak, B.; Hapter, R.; Dera, J. Light curves of marine plankton photosynthesis in the Baltic. *Oceanologia* **1989**, *29*, 61–78.

61. Renk, H.; Ochocki, S. Photosynthetic rate and light curves of phytoplankton in the southern Baltic. *Oceanologia* **1998**, *40*, 331–344.

62. Riemann, B.; Simonsen, P.; Stensgaard, L. The carbon and chlorophyll content of phytoplankton from various nutrient regimes. *J. Plankton Res.* **1989**, *11*, 1037–1045. [CrossRef]

63. McLachlan, J. The effect of salinity on growth and chlorophyll content in representative classes of unicellular marine algae. *Can. J. Microbiol.* **1961**, *7*, 399–406. [CrossRef]

64. Sigaud, T.C.S.; Aidar, E. Salinity and temperature effects on the growth and chlorophyll-a content of some planktonic aigae. *Bol. Inst. Oceanogr.* **1993**, *41*, 95–103. [CrossRef]

65. Brockhurst, M.A.; Rainey, P.B.; Buckling, A. The effect of spatial heterogeneity and parasites on the evolution of host diversity. *Proc. Biol. Sci.* **2004**, *271*, 107–111. [CrossRef] [PubMed]

Isolation and Characterization of a Double Stranded DNA Megavirus Infecting the Toxin-Producing Haptophyte *Prymnesium parvum*

Ben A. Wagstaff [1], Iulia C. Vladu [1], J. Elaine Barclay [1], Declan C. Schroeder [2], Gill Malin [3] and Robert A. Field [1,*]

[1] Department of Biological Chemistry, John Innes Centre, Norwich Research Park, Norwich NR4 7UH, UK; ben.wagstaff@live.co.uk (B.A.W.); vladu.iulia@yahoo.com (I.C.V.); elaine.barclay@jic.ac.uk (J.E.B.)

[2] Marine Biological Association of the UK, Plymouth PL1 2PB, UK; dsch@mba.ac.uk

[3] Centre for Ocean and Atmospheric Studies, School of Environmental Sciences, University of East Anglia, Norwich Research Park, Norwich NR4 7TJ, UK; g.malin@uea.ac.uk

* Correspondence: rob.field@jic.ac.uk

Academic Editors: Mathias Middelboe and Corina Brussaard

Abstract: *Prymnesium parvum* is a toxin-producing haptophyte that causes harmful algal blooms globally, leading to large-scale fish kills that have severe ecological and economic implications. For the model haptophyte, *Emiliania huxleyi*, it has been shown that large dsDNA viruses play an important role in regulating blooms and therefore biogeochemical cycling, but much less work has been done looking at viruses that infect *P. parvum*, or the role that these viruses may play in regulating harmful algal blooms. In this study, we report the isolation and characterization of a lytic nucleo-cytoplasmic large DNA virus (NCLDV) collected from the site of a harmful *P. parvum* bloom. In subsequent experiments, this virus was shown to infect cultures of *Prymnesium* sp. and showed phylogenetic similarity to the extended *Megaviridae* family of algal viruses.

Keywords: *Prymnesium parvum*; haptophyte; algal bloom; algal virus; *Megaviridae*; NCLDV

1. Introduction

The last two decades have seen a boom in the study of marine viruses and the role that they play in regulating both bacterial and unicellular eukaryote bloom dynamics [1,2]. Although phages and the bacteria that they infect have been studied for many years, the more recently discovered *Acanthamoeba polyphaga mimivirus* (APMV) and its *Megaviridae* relatives have brought about a new age in photosynthetic protist virology. It has recently been shown that dsDNA viruses infecting algae do not form monophyletic lineages [3], with divergence occurring even within the host division. A good example of this evolutionary divergence can be found in viruses that infect the coccolithophore *Emiliania huxleyi* (EhV) [4,5] and the prymnesiophyte *Phaeocystis globosa* (PgV) [6], which along with other algal viruses have been proposed to form an extended branch of the *Megaviridae* [7]. It is widely accepted that these viruses not only play a crucial role in ecosystem dynamics [8,9], but also contribute significantly to biogeochemical cycles [10,11]. A lesser studied impact, however, lies in the role that such viruses may play in the termination of toxic eukaryotic algal blooms. Lytic viruses that infect the toxic raphidophyte *Heterosigma akashiwo* have been extensively studied [12–19] but, because of the elusive nature of *H. akashwio* toxicity to fish, none of these studies sought to investigate the role of viral infection on levels of algal toxicity.

The toxin-producing haptophyte *Prymnesium parvum* forms dense blooms in marine, brackish and inland waters, devastating fish populations through the release of natural product toxins [20,21].

The haptophytes are a diverse division of microalgae that include the bloom-forming *Emiliania huxleyi* and *Phaeocystis globosa*, both of which play crucial roles in oceanic carbon and sulfur cycles [22,23]. Virus infection of these organisms has been studied in some detail, with the genome of the dsDNA *Phaeocystis globosa* virus (PgV-16T) being recently described [3]. From a metabolomics perspective, *Phaeocystis pouchetti* lysis by a strain-specific virus has been shown to cause substantial release of dimethyl sulphide and its major precursor dimethylsulphoniopropionate [24], an action that is believed to contribute significantly to the global sulfur cycle. Although much effort has gone into studying the relationship between *E. huxleyi* and its infecting viruses, viruses infecting toxin-producing algal species within the haptophyte family are much less well studied. These include the euryhaline species *Prymnesium* spp. and *Chrysochromulina* spp., whose blooms can often result in severe economic damage through loss of fish stocks [25,26]. Viruses that infect the non-toxic *P. kappa* have recently been described, but to date no viruses have been isolated and characterized that infect the toxin-producing *P. parvum* species, even though Schwierzke et al. have previously suggested a role for viruses in regulating natural *P. parvum* populations [27].

In this study, we isolated a novel lytic virus of *P. parvum* 946/6, *Prymnesium parvum* DNA virus BW1 (henceforth referred to as PpDNAV), from the site of a recent harmful bloom event of this species in Norfolk, England. We show that the virus has a typical narrow host range; using morphological characterisation and phylogenetics, we also show that the virus lies in the recently described clade of algal megaviruses.

2. Materials and Methods

2.1. Prymnesium parvum Culture Conditions

For choice of host cell, *P. parvum* 946/6 was obtained from the Culture Collection of Algae and Protozoa (CCAP—www.ccap.ac.uk). The additional 14 strains used for host range screening were obtained from the Marine Biological Association Culture Collection (https://www.mba.ac.uk/culture-collection/). Batch cultures were maintained at 22 °C on a 14:10 light cycle at 100 μmol·photons·m^{-2}·s^{-1}. Cultures were grown in f/2–Si medium at a salinity of 7–8 practical salinity unit (PSU). Under these conditions, cell densities of ~3 $\times$ 10^6 cells·mL^{-1} could be achieved after 12–16 days of growth.

2.2. Isolation of Lytic Virus Particles

PpDNAV was isolated from surface water samples taken at various locations on Hickling Broad, Norfolk, England on 9 February 2016. In brief, 4 $\times$ 100 mL water samples from various locations around the Broad were centrifuged at 3000$\times$ *g* and the supernatant subsequently filtered through 0.45 μm pore-size filters (Sartorius AG, Goettingen, Germany). The resulting solutions were then concentrated 100- to 200-fold using 100 kDa MW cut off spin filters (Amicon Ultra 15, Merck Millipore, Watford, UK) to give 0.5 to 1 mL of viral concentrate, which was stored at 4 °C in the dark until use. Small volumes (0.2 mL) of concentrate from each location were added to 1.8 mL of exponentially growing cultures of *P. parvum* 946/6. Blank culture medium was used as a control. Cultures were visually inspected for signs of cell lysis (culture clearing) after 7–10 days where the control cultures continued to grow. Culture clearing was then followed up by Transmission Electron Microscopy (TEM) analysis of the culture lysates. Clonal populations of PpDNAV were obtained by taking the supernatant of a lysed culture, and exhaustively diluting with media. These diluted samples (0.2 mL) were added to 1.8 mL of an exponentially growing culture of *P. parvum* 946/6. The highest dilution that still produced cell lysis after seven days was taken through to the next round. This was repeated at least three times and resulted in a population of PpDNAV free of morphologically different viruses, as judged by TEM.

2.3. Transmission Electron Microscopy

For TEM analysis of virus-like particles (VLPs) in culture supernatant, 2 mL of a virus-lysed culture was filtered through 0.45 μm filters and 10 μL of the filtrate was adsorbed onto a 400 mesh copper palladium grid with a carbon-coated pyroxylin support film before being negatively stained with 2% aqueous uranyl acetate [28]. The grids were viewed in a FEI Tecnai 20 transmission electron microscope (Eindhoven, The Netherlands) at 200 kV and digital TIFF images were taken with an AMT XR60B digital camera (Deben, Bury St Edmunds, UK).

For analysing intracellular VLPs, 1 mL of infected cultures of *P. parvum* 946/6 was taken at 24 and 48 h post-infection (p.i.). These were centrifuged at $3000 \times g$ to pellet algal cells and the supernatant was discarded. The pellet was washed twice with sterile medium. The pelleted cells were then resuspended in 2.5% (v/v) aqueous glutaraldehyde solution and left overnight. This suspension was then centrifuged at $3000 \times g$ to pellet the algal cells. Half the volume of the supernatant was then discarded and an equal volume of warm (60 °C) low gelling temperature agarose (Sigma Aldrich, Haverhill, UK) was added, before resuspension of the cells and placing on ice to solidify. The solidified samples were then put into 2.5% (v/v) glutaraldehyde with 0.05 M sodium cacodylate, pH 7.3 [29] and left overnight. Using a Leica EM TP machine (Leica Microsystems, Cambridge, UK), the samples were washed in 0.05 M sodium cacodylate and then post-fixed with 1% (w/v) OsO4 in 0.05 M sodium cacodylate for 60 min at room temperature. After washing and dehydration with ethanol, the samples were gradually infiltrated with LR White resin (London Resin Company, London, UK) according to the manufacturer's instructions. After polymerization, the resulting material was sectioned with a diamond knife using a Leica EM UC6 ultramicrotome (Leica Microsystems). Ultrathin sections of approximately 90 nm were picked up on 200 mesh gold grids that had been coated in pyroxylin and carbon. The grids were then contrast-stained with 2% (w/v) uranyl acetate for 1 h and 1% (w/v) lead citrate for 1 min, washed in distilled water and air-dried. The grids were then viewed with a FEI Tecnai 20 transmission electron microscope (Eindhoven, The Netherlands) at 200 kV and digital TIFF images were produced.

2.4. Host Specificity

Fifteen different strains of *Prymnesium* were tested in triplicate for signs of cell lysis by PpDNAV using the infection methodology described above. Cell lysis, as observed by culture clearing, was noted for five of the 15 strains tested (Table 1).

Table 1. Host range of PpDNAV. + lysed culture, − culture not lysed.

Genus/Species	Strain Code	Lysis with PpDNAV
Prymnesium parvum	946/6	+
Prymnesium parvum	94A	-
Prymnesium parvum	94C	+
Prymnesium parvum	579	-
Prymnesium patelliferum	527A	+
Prymnesium patelliferum	527C	+
Prymnesium patelliferum	527D	-
Prymnesium sp.	522	-
Prymnesium sp.	569	-
Prymnesium sp.	592	+
Prymnesium sp.	593	-
Prymnesium sp.	595	-
Prymnesium sp.	596	-
Prymnesium sp.	597	-
Prymnesium sp.	598	-

2.5. Infection Cycle

The virus–algae lytic cycle was investigated by accurately recording algal cell abundance during an infection cycle. A late-log phase culture of *P. parvum* 946/6 was infected with PpDNAV (0.1% v/v) and triplicate 2 mL aliquots were taken at various time points post infection (p.i.). These were diluted with 0.2 μm filtered seawater prior to counting using a Multisizer 3 Analyser (Beckman Coulter, High Wycombe, UK) fitted with a 100 μm aperture tube. The control culture continued to grow throughout the experiment, whilst the infected algal culture was lysed rapidly after 48 h.

2.6. Chloroform Sensitivity

To test the virus sensitivity to chloroform, an adaptation of the method of Martínez Martínez et al. was employed [30]. Briefly, 1 mL of 0.45 µm-filtered PpDNAV was added to an equivalent volume of chloroform and shaken vigorously for 5 min. The resulting mixture was then centrifuged at $4000 \times g$ in a benchtop centrifuge for 5 min to separate the organic and polar layers. The aqueous phase was transferred by pipetting to a clean microcentrifuge tube and incubated at 37 °C for 1 h to remove residual chloroform. As a control, 1 mL of chloroform was added to 1 mL of f/2 medium. Chloroform-treated PpDNAV, chloroform-treated medium and untreated PpDNAV were added to *P. parvum* 946/6 as described above in the infectivity experiment protocol; signs of lysis, as judged by culture clearing, were recorded after one week.

2.7. Viral DNA Extraction, Sequencing, and Phylogenetic Analyses

For DNA extraction, 1 L of late log phase *P. parvum* 946/6 was infected with axenic PpDNAV (0.1% *v/v*). Lysis was allowed to occur over a period of five days, by which point almost all cells had been lysed. The culture was centrifuged at $6500 \times g$ to pellet cell debris, before being filtered through 0.22 µm filters to remove remaining cell debris or contaminating bacteria. The filtrate was incubated for 72 h with 100 µg/mL carbenicillin before being concentrated to 30 mL using 100 kDa mw cut-off spin filters. Ultracentrifugation at $150,000 \times g$ was used to pellet viral particles, and these were re-suspended in 2 mL of $\varrho = 1.4$ CsCl and layered onto a CsCl gradient which was resolved at $150,000 \times g$ for 18 h. Fractions from $\varrho = 1.3$ to $\varrho = 1.4$ were pooled and DNA extracted using a PureLink Viral RNA/DNA Kit, according to the manufacturer's protocol.

An amount of 1 µg of purified viral DNA was then sent to The Earlham Institute, UK, for Illumina MiSeq sequencing (Illumina, Inc., San Diego, CA, USA) and assembly. The initial assembly was then analysed using GeneMarkS [31] which identified 332 protein-coding sequences. BLASTp analysis was then performed against the National Center for Biotechnology Information (NCBI) GenBank nonredundant (nr) protein sequence database [32] to identify major capsid protein and DNA Pol B candidates. Nucleic acid and amino acid sequences for the major capsid protein (MCP) and DNA Polymerase B (DNA polB) were submitted to Genbank with the accession codes KY509047 and KY509048, respectively.

Phylogenetic analysis was performed using the obtained sequences for MCP and DNA polB, as well as other related sequences from previously discovered algal viruses, identified using BLASTp. These sequences were aligned using the default settings of multiple sequence alignment software version 7 (MAFFT) [33], and trees were constructed from the neighbour-joining method [34] (midpoint-rooted) using Molecular Evolutionary Genetics Analysis version 7.0 (MEGA7) [35].

3. Results

3.1. Isolation of Lytic Virus Particles

PpDNAV isolation was conducted from water samples collected at Hickling Broad, Norfolk, England. Among four water samples from which viral lysates were prepared, lysis of *P. parvum* 946/6 occurred with three samples (Figure 1). Transmission electron micrographs of the viral lysates showed that icosahedral VLPs were present in all three samples, but samples 1 and 2 also contained significant levels of phage-like particles; we suspect that these were a result of infection with the low levels of bacteria that were present in the non-axenic *P. parvum* 946/6 cultures. To avoid further downstream separation of viruses, we chose to continue working with sample 4 only (sourced at—52°44'19.12" N, Long—1°34'39.49" E), which appeared by TEM to be free of phages. After a triplicate dilution series, the resulting monoclonal viral lysate still lysed the host cells and TEM of thin-sectioned cells confirmed the presence of VLPs (Figure 2A,B); thereby fulfilling Koch's postulates.

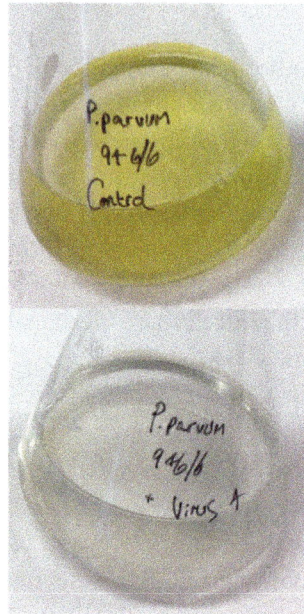

Figure 1. (**Top**) control culture; (**Bottom**) 'Cleared' culture 96 h post viral infection.

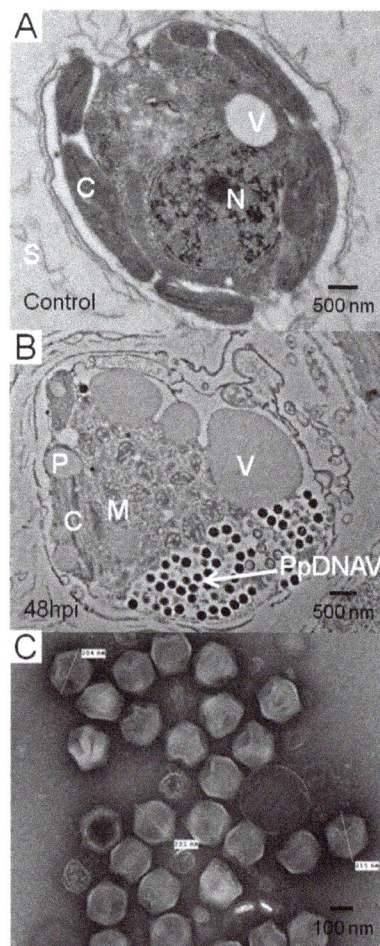

Figure 2. (**A**) Thin-sections of healthy *P. parvum* 946/6 cells; (**B**) Thin-sections of *P. parvum* 946/6 48 h post infection. (**C**) Free *Prymnesium parvum* DNA virus (PpDNAV) particles in culture supernatant 72 h post infection. C: chloroplast; V: contractile vacuole; N: nucleus; S: scales; M: mitochondria, P: pyrenoid.

3.2. Virus Morphology, Host Range, and Infectious Properties

Transmission electron microscopy of isolated and intracellular (thin sectioned) viruses revealed an icosahedral capsid with an average diameter of 221 nm ($n = 71$) (Figure 2). Although no external viral lipid membrane was evident, some viral particles showed an internal white 'halo' between the capsid and the DNA of PpDNAV, suggestive of the virus having an internal membrane. The presence of a viral factory or viroplasm [36] in the host cytoplasm, and in some cases an imperfect vertex or a tail-like structure were also observed (Figures S1 and S2). As seen in Figure 2B, establishment of a viral factory in the host cytoplasm also results in a loss of the nuclear envelope and therefore loss of the nucleus. This was observed in the majority of infected cells examined at 48 h p.i.

Fifteen different strains of *Prymnesium* were screened for sensitivity to PpDNAV (Table 1). PpDNAV was found to be sensitive to chloroform, whereby the chloroform-treated virus no longer caused lysis of *P. parvum* 946/6 (Figure S3). This supports the notion of a viral membrane in this system.

The lytic cycle of the virus was explored to determine both the incubation period and eclipse period (Figure 3). At 48 h p.i., the cells had clearly lost mobility and sedimented at the base of the culture flask. Re-suspension of the cells by shaking led to similar cell counts as seen at 24 h p.i., as determined by Coulter counting. The time before symptoms of viral infection, the incubation period, was therefore judged to be 24 h. The eclipse period reflects the time between infection and appearance of mature virus particles within the host; as new mature virions were first observed 48 h p.i., the eclipse period was judged to be 24–48 h. At 72 h p.i., the onset of cell lysis had occurred. PpDNAV appeared to lyse >95% of host cells by 120 h p.i., whilst uninfected control cultures continued to grow over the full course of the experiment.

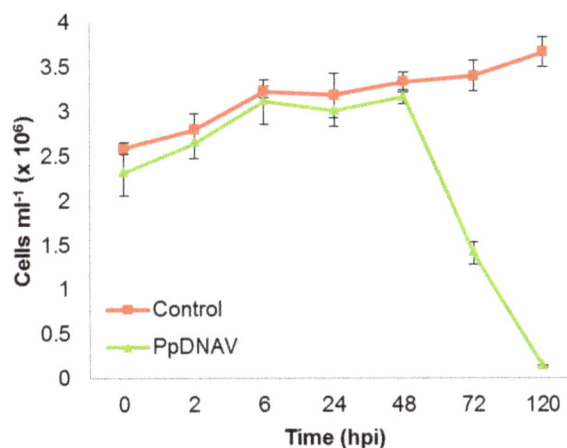

Figure 3. PpDNAV infection cycle propagated on *P. parvum* 946/6. Graph shows the average number of algal cells in control cultures (squares) and PpDNAV infected cultures (circles). Error bars represent the standard error for triplicate cultures.

3.3. Genome Sequencing and Phylogenetic Analysis

Predicted proteins from the initial genome assembly included the MCP1 protein (KY509047) and DNA polB (KY509048) which were used for phylogenetic analysis. The 525 aa sequence for MCP1 was found to have 91% sequence similarity to the major capsid protein 1 of *Phaeocystis globosa* virus (YP_008052475.1) and 84% similarity to MCP1 of Organic Lake Phycodnavirus 2 (ADX06358.1) with E-values of 0.0 in each case. This alignment allowed construction of a phylogenetic tree (Figure S4) that shows clustering with other megaviruses, including PgV-16T.

For DNA polB (KY509048), the 1281 aa sequence displayed 77% sequence similarity to DNA polB of PgV-16T (YP_008052566.1) and 64% similarity to DNA polB of Organic Lake phycodnavirus 2 (ADX06483.1). The resulting phylogenetic tree (Figure 4) shows a similar clustering of PpDNAV to the algal *Megaviridae* family, but also illustrates an obvious divergence between algal viruses that fall within the *Megaviridae* family and those that do not; with EhV-86 and *Heterosigma akashiwo* virus (HaV)-1 rightfully placed outside of the *Megaviridae* clade.

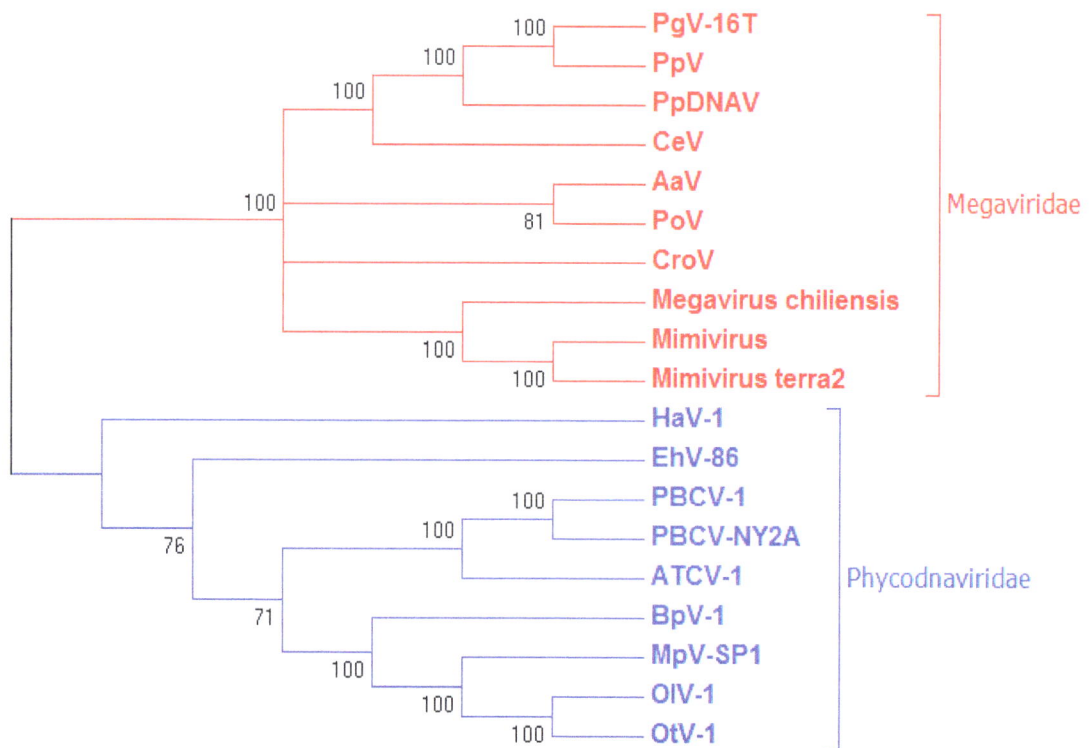

Figure 4. Phylogenetic clustering of PpDNAV with other large algal *Megaviridae*. Alignment was performed using the default settings of multiple sequence alignment software version 7 (MAFFT) [33], and the neighbour-joining method (midpoint-rooted) [34] was used to construct a tree from 19 viral DNA Polymerase Beta (polB) sequences using Molecular Evolutionary Genetics Analysis version 7.0 (MEGA7) [35]. The final tree was based on 630 ungapped positions, 500 resampling permutations, and was collapsed for bootstrap values <50. The tree shows that PpDNAV clusters with the well-defined clade of *Megaviridae* and the algal-infecting *Megaviridae* (red), and not with the *Phycodnaviridae* (blue).

4. Discussion

Haptophytes are abundant in marine waters but can also thrive in brackish inland waters. Whilst a significant amount of work has been done on the marine dwelling coccolithophore *Emiliania huxleyi* and its associated viruses, little work has looked at the toxin-producing members of the haptophytes. In the present study, we isolated and characterized a novel megavirus, PpDNAV, from brackish inland waters where harmful blooms of *Prymnesium parvum* frequently occur [37]. We showed that this lytic virus was able to infect *P. parvum* 946/6, later expanded to five out of 15 *Prymnesium* strains tested. Morphological and phylogenetic analysis of two core dsDNA virus conserved genes suggests that this virus belongs to the extended *Megaviridae* family of algal-infecting viruses.

Transmission electron microscopy of negatively stained virus particles from a lysed culture supernatant revealed icosahedral capsids with an average diameter of 221 nm. Many particles appeared to have one imperfect vertex, with some showing material protruding from what appeared to be a stargate [38] (Figure S1). These likely represent particles in an advanced stage of packing or unpacking genetic material [38], and suggested early on that PpDNAV lies in the extended *Megaviridae* branch of algal viruses. Thin sections of infected *P. parvum* 946/6 cells showed evidence for a viroplasm as the site of replication, where empty capsids could be seen closer to the centre of the viroplasm (Figure S2). This further supported the inclusion of PpDNAV in the extended *Megaviridae* family [39]. The infectivity of PpDNAV is chloroform sensitive (Figure S3), and the lack of an obvious external lipid membrane observed by TEM may suggest that internal membrane/s are present; although chloroform sensitivity cannot always be used to confirm lipid membrane presence [40].

New mature virions were first observed by electron microscopy at 48 h p.i., so the eclipse period of the virus in infected algal cells was estimated to be 24–48 h. At 48 h, the algal cell count as recorded by Coulter counting was still the same as at 24 h, but a complete sedimentation of cells had occurred, suggesting a loss in motility and a likely shutdown of important cellular processes. By 72 h, a rapid decline in cell abundance could be observed, showing that the loss of motility precedes the host lysis event, as is seen for some other flagellated algae [18]. In its natural environment, this may lead to accumulation of viral particles at the sediment surface rather than dispersed in the water column.

The algal host species specificity of PpDNAV was assessed against *P. parvum* 946/6, which had been kept in 7–8 PSU f/2 medium for two years, and 14 other *Prymnesium* strains which had been maintained in a full strength seawater medium. Initially, PpDNAV only infected *P. parvum* 946/6, but after ~6 months of sub-culturing of the other 14 strains in 7–8 PSU f/2 medium, the host range broadened to five out of the 15 strains. We speculate that the change in salinity contributed to the change in sensitivity to PpDNAV; recent work by Nedbalová et al. [41] suggests that a change in membrane lipid composition in different salinities may account for this situation. This somewhat less restricted host range is similar to that found for *Haptolina ericina* virus (HeV RF02), *Prymnesium kappa* virus (PkV RF01) and *Prymnesium kappa* virus (PkV RF02) [42]. Taken together, this suggested that PpDNAV was a member of the algal *Megavirus* family [43].

Phylogenetic analysis using sequences for MCP1 and DNA polB of PpDNAV confirmed morphological findings, showing that PpDNAV clusters amongst the algal viruses belonging to the *Megaviridae* family, such as PgV-16T [3,6], *Chrysochromulina ericina* virus (CeV) [44], *Pyramimonas orientalis* virus (PoV) [45] and the recently reassigned *Aureococcus anophagefferens* virus (AaV) [46]. With the exception of *Emliania huxleyi* virus (EhV-86), which appears to branch independently, the close clustering of viruses infecting haptophytes, as well as the chlorella viruses clustering together, supports the notion that viruses co-evolve with their hosts [6,46,47].

Of the algal viruses compared in this study, only HaV-1 is known to infect a toxin producing host [12–14]. However, the toxic metabolites responsible for bloom toxicity are not established in *Heterosigma akashiwo*, making studies of viral impact on toxicity difficult. On the other hand, reports of toxic *P. parvum* metabolites are numerous and include fatty acids [48], glycerolipids [49] and very large ladder-frame polyether toxins, known collectively as the prymnesins [50–52]. Reports of cases of toxic and non-toxic blooms of *Prymnesium* and other harmful algal species [37] has led to speculation that an ecological trigger exists for toxicity. While efforts have been made to associate nutrients, pH and other conditions to bloom toxicity [21], the identity of the full spectrum of toxicity-causing agents remains to be establshed; there may conceivably be a role for viral infection in *Prymnesium* cell lysis and hence toxin release. We now have the opportunity to use this algae–virus system in clearing up some of these unanswered questions. Further studies into the effect of viral infection and host algal cell lysis on toxic bloom events need to be explored in order to fully understand the underlying mechanisms behind production and release of toxins from Prymnesium. In addition, as further sequences of algal viruses become available, new opportunities will open up for accurate monitoring of viral population fluctuations with respect their host. Furthermore, the increase in characterized viruses will provide more information when analysing metagenomic data sets such as those generated by the *Tara* Oceans expedition [53,54]. Hence the discovery and characterization of PpDNAV in this study will aid this burgeoning field of scientific endeavour.

Acknowledgments: These studies were supported by the BBSRC Institute Strategic Programme on Understanding and Exploiting Metabolism (MET) [BB/j004561/1] and the John Innes Foundation. B.A.W. was supported by a BBSRC industrial CASE PhD studentship supported by Environment Agency. We thank Willie Wilson for helpful advice of viral morphology. We also thank Steve Lane, Andy Hindes, John Currie, Jenny Pratscher and Colin Murrell for their ongoing support.

Author Contributions: B.A.W. and R.A.F. conceived and designed the experiments; B.A.W. performed the experiments, with help from I.C.V. for chloroform sensitivity and culture maintenance, and J.E.B. for thin-section preparation and EM imaging; D.C.S. provided 14 *Prymnesium* strains for host range screening and reviewed the manuscript; G.M. also reviewed the manuscript; B.A.W., G.M. and R.A.F. analysed the data; B.A.W. and R.A.F. wrote the manuscript.

References

1.   Thingstad, T.F. Elements of a theory for the mechanisms controlling abundance, diversity, and biogeochemical role of lytic bacterial viruses in aquatic systems. *Limnol. Oceanogr.* **2000**, *45*, 1320–1328. [CrossRef]

2.   Fuhrman, J.A. Marine viruses and their biogeochemical and ecological effects. *Nature* **1999**, *399*, 541–548. [CrossRef] [PubMed]

3.   Santini, S.; Jeudy, S.; Bartoli, J.; Poirot, O.; Lescot, M.; Abergel, C.; Barbe, V.; Wommack, K.E.; Noordeloos, A.A.; Brussaard, C.P.; et al. Genome of *Phaeocystis globosa* virus PgV-16T highlights the common ancestry of the largest known DNA viruses infecting eukaryotes. *Proc. Natl. Acad. Sci. USA* **2013**, *110*, 10800–10805. [CrossRef] [PubMed]

4.   Wilson, W.H.; Tarran, G.A.; Schroeder, D.; Cox, M.; Oke, J.; Malin, G. Isolation of viruses responsible for the demise of an *Emiliania huxleyi* bloom in the English Channel. *J. Mar. Biol. Assoc. UK* **2002**, *82*, 369–377. [CrossRef]

5.   Schroeder, D.C.; Oke, J.; Malin, G.; Wilson, W.H. Coccolithovirus (Phycodnaviridae): Characterisation of a new large dsDNA algal virus that infects *Emiliana huxleyi*. *Arch. Virol.* **2002**, *147*, 1685–1698. [CrossRef] [PubMed]

6.   Brussaard, C.P.D.; Short, S.M.; Frederickson, C.M.; Suttle, C.A. Isolation and Phylogenetic Analysis of Novel Viruses Infecting the Phytoplankton *Phaeocystis globosa* (*Prymnesiophyceae*). *Appl. Environ. Microbiol.* **2004**, *70*, 3700–3705. [CrossRef] [PubMed]

7.   Moniruzzaman, M.; Gann, E.R.; LeCleir, G.R.; Kang, Y.; Gobler, C.J.; Wilhelm, S.W. Diversity and dynamics of algal Megaviridae members during a harmful brown tide caused by the pelagophyte, *Aureococcus anophagefferens*. *FEMS Microbiol. Ecol.* **2016**, *92*, fiw058. [CrossRef] [PubMed]

8.   Short, S.M. The ecology of viruses that infect eukaryotic algae. *Environ. Microbiol.* **2012**, *14*, 2253–2271. [CrossRef] [PubMed]

9.   Brussaard, C.P.D. Viral Control of Phytoplankton Populations—A Review. *J. Eukaryot. Microbiol.* **2004**, *51*, 125–138. [CrossRef] [PubMed]

10. Suttle, C.A. Marine viruses—Major players in the global ecosystem. *Nat. Rev. Microl.* **2007**, *5*, 801–812. [CrossRef] [PubMed]

11. Wilhelm, S.W.; Suttle, C.A. Viruses and Nutrient Cycles in the Sea: Viruses play critical roles in the structure and function of aquatic food webs. *Bioscience* **1999**, *49*, 781–788. [CrossRef]

12. Nagasaki, K.; Yamaguchi, M. Isolation of a virus infectious to the harmful bloom causing microalga *Heterosigma akashiwo* (*Raphidophyceae*). *Aquat. Microb. Ecol.* **1997**, *13*, 135–140. [CrossRef]

13. Keizo, N.; Mineo, Y. Intra-species host specificity of HaV (*Heterosigma akashiwo* virus) clones. *Aquat. Microb. Ecol.* **1998**, *14*, 109–112. [CrossRef]

14. Keizo, N.; Mineo, Y. Effect of temperature on the algicidal activity and the stability of HaV (*Heterosigma akashiwo* virus). *Aquat. Microb. Ecol.* **1998**, *15*, 211–216. [CrossRef]

15. Lawrence, J.E.; Chan, A.M.; Suttle, C.A. A novel virus (HaNIV) causes lysis of the toxic bloom-forming alga *Heterosigma akashiwo* (*Raphidophyceae*). *J. Phycol.* **2001**, *37*, 216–222. [CrossRef]

16. Lawrence, J.E.; Chan, A.M.; Suttle, C.A. Viruses causing lysis of the toxic bloom-forming alga *Heterosigma akashiwo* (*Raphidophyceae*) are widespread in coastal sediments of British Columbia, Canada. *Limnol. Oceanogr.* **2002**, *47*, 545–550. [CrossRef]

17. Tai, V.; Lawrence, J.E.; Lang, A.S.; Chan, A.M.; Culley, A.I.; Suttle, C.A. Characterization of HaRNAV, a single-stranded RNA virus causing lysis of *Heterosigma akashiwo* (*Raphidophyceae*). *J. Phycol.* **2003**, *39*, 343–352. [CrossRef]

18. Janice, E.L.; Curtis, A.S. Effect of viral infection on sinking rates of *Heterosigma akashiwo* and its implications for bloom termination. *Aquat. Microb. Ecol.* **2004**, *37*, 1–7. [CrossRef]

19. Lawrence, J.E.; Brussaard, C.P.D.; Suttle, C.A. Virus-Specific Responses of *Heterosigma akashiwo* to Infection. *Appl. Environ. Microbiol.* **2006**, *72*, 7829–7834. [CrossRef] [PubMed]

20. Granéli, E.; Edvardsen, B.; Roelke, D.L.; Hagström, J.A. The ecophysiology and bloom dynamics of *Prymnesium* spp. *Harmful Algae* **2012**, *14*, 260–270. [CrossRef]

21. Manning, S.R.; La Claire, J.W. Prymnesins: Toxic Metabolites of the Golden Alga, *Prymnesium parvum* Carter (*Haptophyta*). *Mar. Drugs* **2010**, *8*, 678–704. [CrossRef] [PubMed]

22. Schoemann, V.; Becquevort, S.; Stefels, J.; Rousseau, V.; Lancelot, C. *Phaeocystis* blooms in the global ocean and their controlling mechanisms: A review. *J. Sea Res.* **2005**, *53*, 43–66. [CrossRef]

23. Leblanc, K.; Hare, C.E.; Feng, Y.; Berg, G.M.; DiTullio, G.R.; Neeley, A.; Benner, I.; Sprengel, C.; Beck, A.; Sanudo-Wilhelmy, S.A.; et al. Distribution of calcifying and silicifying phytoplankton in relation to environmental and biogeochemical parameters during the late stages of the 2005 North East Atlantic Spring Bloom. *Biogeosciences* **2009**, *6*, 2155–2179. [CrossRef]

24. Malin, G.; Wilson, W.H.; Bratbak, G.; Liss, P.S.; Mann, N.H. Elevated production of dimethylsulfide resulting from viral infection of cultures of *Phaeocystis pouchetii*. *Limnol. Oceanogr.* **1998**, *43*, 1389–1393. [CrossRef]

25. Edvardsen, B.; Paasche, E. *Bloom Dynamics and Physiology of Prymnesium and Chrysochromulina in Physiological Ecology of Harmful Algal Blooms*; Springer-Verlag: Berlin/Heidelberg, Germany, 1998.

26. Roelke, D.L.; Barkoh, A.; Brooks, B.W.; Grover, J.P.; Hambright, K.D.; LaClaire, J.W.; Moeller, P.D.R.; Patino, R. A chronicle of a killer alga in the west: Ecology, assessment, and management of *Prymnesium parvum* blooms. *Hydrobiologia* **2016**, *764*, 29–50. [CrossRef]

27. Schwierzke, L.; Roelke, D.L.; Brooks, B.W.; Grover, J.P.; Valenti, T.W.; Lahousse, M.; Miller, C.J.; Pinckney, J.L. *Prymnesium parvum* Population Dynamics During Bloom Development: A Role Assessment of Grazers and Virus. *J. Am. Water. Resour. Assoc.* **2010**, *46*, 63–75. [CrossRef]

28. Brenner, S.; Horne, R.W. A negative staining method for high resolution electron microscopy of viruses. *Biochim. Biophys. Acta* **1959**, *34*, 103–110. [CrossRef]

29. Gordon, G.B.; Miller, L.R.; Bensch, K.G. Fixation of tissue culture cells for ultrastructural cytochemistry. *Exp. Cell Res.* **1963**, *31*, 440–443. [CrossRef]

30. Martínez Martínez, J.; Boere, A.; Gilg, I.; van Lent, J.W.M.; Witte, H.J.; van Bleijswijk, J.D.L.; Brussaard, C.P.D. New lipid envelope-containing dsDNA virus isolates infecting *Micromonas pusilla* reveal a separate phylogenetic group. *Aquat. Microb. Ecol.* **2015**, *74*, 17–28. [CrossRef]

31. Besemer, J.; Lomsadze, A.; Borodovsky, M. GeneMarkS: A self-training method for prediction of gene starts in microbial genomes. Implications for finding sequence motifs in regulatory regions. *Nucleic Acids Res.* **2001**, *29*, 2607–2618. [CrossRef] [PubMed]

32. Altschul, S.F.; Gish, W.; Miller, W.; Myers, E.W.; Lipman, D.J. Basic local alignment search tool. *J. Mol. Biol.* **1990**, *215*, 403–410. [CrossRef]

33. Katoh, K.; Toh, H. Recent developments in the MAFFT multiple sequence alignment program. *Brief. Bioinform.* **2008**, *9*, 286–298. [CrossRef] [PubMed]

34. Saitou, N.; Nei, M. The neighbor-joining method: A new method for reconstructing phylogenetic trees. *Mol. Biol. Evol.* **1987**, *4*, 406–425. [CrossRef] [PubMed]

35. Kumar, S.; Stecher, G.; Tamura, K. MEGA7: Molecular Evolutionary Genetics Analysis version 7.0 for bigger datasets. *Mol. Biol. Evol.* **2016**, *33*, 1870–1874. [CrossRef] [PubMed]

36. Den Boon, J.A.; Diaz, A.; Ahlquist, P. Cytoplasmic Viral Replication Complexes. *Cell Host Microbe* **2010**, *8*, 77–85. [CrossRef] [PubMed]

37. Holdway, P.A.; Watson, R.A.; Moss, B. Aspects of the ecology of *Prymnesium parvum* (Haptophyta) and water chemistry in the Norfolk Broads, England. *Freshwat. Biol.* **1978**, *8*, 295–311. [CrossRef]

38. Zauberman, N.; Mutsafi, Y.; Halevy, D.B.; Shimoni, E.; Klein, E.; Xiao, C.; Sun, S.; Minsky, A. Distinct DNA Exit and Packaging Portals in the Virus *Acanthamoeba polyphaga mimivirus*. *PLoS Biol.* **2008**, *6*, e114. [CrossRef] [PubMed]

39. Mutsafi, Y.; Fridmann-Sirkis, Y.; Milrot, E.; Hevroni, L.; Minsky, A. Infection cycles of large DNA viruses: Emerging themes and underlying questions. *Virology* **2014**, *466–467*, 3–14. [CrossRef] [PubMed]

40. Feldman, H.A.; Wang, S.S. Sensitivity of various viruses to chloroform. *Proc. Soc. Exp. Biol. Med.* **1961**, *106*, 736–738. [CrossRef] [PubMed]

41. Nedbalová, L.; Střížek, A.; Sigler, K.; Řezanka, T. Effect of salinity on the fatty acid and triacylglycerol composition of five haptophyte algae from the genera Coccolithophora, Isochrysis and Prymnesium determined by LC-MS/APCI. *Phytochemistry* **2016**, *130*, 64–76. [CrossRef] [PubMed]

42. Johannessen, T.V.; Bratbak, G.; Larsen, A.; Ogata, H.; Egge, E.S.; Edvardsen, B.; Eikrem, W.; Sandaa, R.-A. Characterisation of three novel giant viruses reveals huge diversity among viruses infecting *Prymnesiales* (*Haptophyta*). *Virology* **2015**, *476*, 180–188. [CrossRef] [PubMed]

43. Wilhelm, S.W.; Coy, S.R.; Gann, E.R.; Moniruzzaman, M.; Stough, J.M.A. Standing on the Shoulders of Giant Viruses: Five Lessons Learned about Large Viruses Infecting Small Eukaryotes and the Opportunities They Create. *PLoS Pathog.* **2016**, *12*, e1005752. [CrossRef] [PubMed]

44. Gallot-Lavallée, L.; Pagarete, A.; Legendre, M.; Santini, S.; Sandaa, R.-A.; Himmelbauer, H.; Ogata, H.; Bratbak, G.; Claverie, J.-M. The 474-Kilobase-Pair Complete Genome Sequence of CeV-01B, a Virus Infecting *Haptolina* (*Chrysochromulina*) *ericina* (*Prymnesiophyceae*). *Genome Announc.* **2015**, *3*, e01413–e01415. [CrossRef]

45. Sandaa, R.-A.; Heldal, M.; Castberg, T.; Thyrhaug, R.; Bratbak, G. Isolation and Characterization of Two Viruses with Large Genome Size Infecting *Chrysochromulina ericina* (*Prymnesiophyceae*) and *Pyramimonas orientalis* (*Prasinophyceae*). *Virology* **2001**, *290*, 272–280. [CrossRef] [PubMed]

46. Moniruzzaman, M.; LeCleir, G.R.; Brown, C.M.; Gobler, C.J.; Bidle, K.D.; Wilson, W.H.; Wilhelm, S.W. Genome of brown tide virus (AaV), the little giant of the Megaviridae, elucidates NCLDV genome expansion and host-virus coevolution. *Virology* **2014**, *466–467*, 60–70. [CrossRef] [PubMed]

47. Mirza, S.F.; Staniewski, M.A.; Short, C.M.; Long, A.M.; Chaban, Y.V.; Short, S.M. Isolation and characterization of a virus infecting the freshwater algae *Chrysochromulina parva*. *Virology* **2015**, *486*, 105–115. [CrossRef] [PubMed]

48. Henrikson, J.C.; Gharfeh, M.S.; Easton, A.C.; Easton, J.D.; Glenn, K.L.; Shadfan, M.; Mooberry, S.L.; Hambright, K.D.; Cichewicz, R.H. Reassessing the ichthyotoxin profile of cultured *Prymnesium parvum* (golden algae) and comparing it to samples collected from recent freshwater bloom and fish kill events in North America. *Toxicon* **2010**, *55*, 1396–1404. [CrossRef] [PubMed]

49. Kozakai, H.; Oshima, Y.; Yasumoto, T. Isolation and Structural Elucidation of Hemolysin from the Phytoflagellate *Prymnesium parvum*. *Agric. Biol. Chem.* **1982**, *46*, 233–236. [CrossRef]

50. Igarashi, T.; Satake, M.; Yasumoto, T. Prymnesin-2: A Potent Ichthyotoxic and Hemolytic Glycoside Isolated from the Red Tide Alga *Prymnesium parvum*. *J. Am. Chem. Soc.* **1996**, *118*, 479–480. [CrossRef]

51. Igarashi, T.; Satake, M.; Yasumoto, T. Structures and Partial Stereochemical Assignments for Prymnesin-1 and Prymnesin-2: Potent Hemolytic and Ichthyotoxic Glycosides Isolated from the Red Tide Alga *Prymnesium parvum*. *J. Am. Chem. Soc.* **1999**, *121*, 8499–8511. [CrossRef]

52. Rasmussen, S.A.; Meier, S.; Andersen, N.G.; Blossom, H.E.; Duus, J.Ø.; Nielsen, K.F.; Hansen, P.J.; Larsen, T.O. Chemodiversity of Ladder-Frame Prymnesin Polyethers in *Prymnesium parvum*. *J. Nat. Prod.* **2016**, *79*, 2250–2256. [CrossRef] [PubMed]

53. Roux, S.; Brum, J.R.; Dutilh, B.E.; Sunagawa, S.; Duhaime, M.B.; Loy, A.; Poulos, B.T.; Solonenko, N.; Lara, E.; Poulain, J.; et al. Ecogenomics and potential biogeochemical impacts of globally abundant ocean viruses. *Nature* **2016**, *537*, 689–693. [CrossRef] [PubMed]

54. Karsenti, E.; Acinas, S.G.; Bork, P.; Bowler, C.; De Vargas, C.; Raes, J.; Sullivan, M.; Arendt, D.; Benzoni, F.; Claverie, J.-M.; et al. A Holistic Approach to Marine Eco-Systems Biology. *PLoS Biol.* **2011**, *9*, e1001177. [CrossRef] [PubMed]

6

Change in *Emiliania huxleyi* Virus Assemblage Diversity but Not in Host Genetic Composition during an Ocean Acidification Mesocosm Experiment

Andrea Highfield [1,†], Ian Joint [1,†], Jack A. Gilbert [2,3], Katharine J. Crawfurd [4] and Declan C. Schroeder [1,*]

[1]   The Marine Biological Association, The Laboratory, Citadel Hill, Plymouth PL1 2PB, UK; ancba@mba.ac.uk (A.H.); ianjoi@mba.ac.uk (I.J.)

[2]   The Microbiome Centre, Department of Surgery, University of Chicago, Chicago, IL 60637, USA; gilbertjack@uchicago.edu

[3]   Division of Bioscience, Argonne National Laboratory, 9700 South Cass Avenue, Argonne, IL 60439, USA

[4]   Department of Biological Oceanography, NIOZ–Royal Netherlands Institute for Sea Research, P.O. Box 59, 1790 AB Den Burg, Texel, The Netherlands; kate.crawfurd@gmail.com

*   Correspondence: dsch@mba.ac.uk

†   These authors contributed equally to this work.

Academic Editors: Corina P.D. Brussaard and Mathias Middelboe

Abstract: Effects of elevated pCO_2 on *Emiliania huxleyi* genetic diversity and the viruses that infect *E. huxleyi* (EhVs) have been investigated in large volume enclosures in a Norwegian fjord. Triplicate enclosures were bubbled with air enriched with CO_2 to 760 ppmv whilst the other three enclosures were bubbled with air at ambient pCO_2; phytoplankton growth was initiated by the addition of nitrate and phosphate. *E. huxleyi* was the dominant coccolithophore in all enclosures, but no difference in genetic diversity, based on DGGE analysis using primers specific to the calcium binding protein gene (*gpa*) were detected in any of the treatments. Chlorophyll concentrations and primary production were lower in the three elevated pCO_2 treatments than in the ambient treatments. However, although coccolithophores numbers were reduced in two of the high-pCO_2 treatments; in the third, there was no suppression of coccolithophores numbers, which were very similar to the three ambient treatments. In contrast, there was considerable variation in genetic diversity in the EhVs, as determined by analysis of the major capsid protein (*mcp*) gene. EhV diversity was much lower in the high-pCO_2 treatment enclosure that did not show inhibition of *E. huxleyi* growth. Since virus infection is generally implicated as a major factor in terminating phytoplankton blooms, it is suggested that no study of the effect of ocean acidification in phytoplankton can be complete if it does not include an assessment of viruses.

Keywords: *Emiliania huxleyi*; CO_2; ocean acidification; climate change; *Coccolithovirus*; EhV

1. Introduction

The rise in anthropogenic CO_2 in the atmosphere and subsequent dissolution in the oceans has changed the carbonate: bicarbonate: dissolved CO_2 equilibrium, lowering seawater pH—a trend that is predicted to continue [1]. Change of pH is of particular significance for marine organisms that have calcium carbonate structures, such as corals and coccolithophores, because less alkaline conditions and pH-dependent shifts in equilibrium of the carbonate system will lead to higher dissolution rates of carbonate. Coccolithophores are ubiquitous and have global significance in regulating the carbon cycle in the oceans [2]. They form massive blooms, whose wide distribution and abundance is readily detected in satellite imagery. Given this wide distribution, it is important to determine if the lower pH of a future ocean will affect the success of coccolithophores and if there will be an impact on marine food webs and biogeochemical cycles.

The effect of changing pH on the important coccolithophore, *Emiliania huxleyi*, has been the focus of much research in recent years. However, results have been variable and consensus has been difficult to reach. In laboratory experiments, both negative and positive effects of increasing pCO_2 have been described (see, for example, [3–5]). Another important approach has been to use large volume enclosures—mesocosms—to investigate a range of conditions that might apply to the future ocean. Unlike laboratory-based experiments, which usually focus on a single organism in the experimental design, mesocosms include all components of the pelagic system from viruses to zooplankton. By maintaining the possibility of complex interactions between different components of the food web, it has been assumed that mesocosms should offer advantages over single-organism culture experiments. However, results have also been rather variable. Early experiments suggested negative effects of higher pCO_2 on production and calcification in *E. huxleyi* [6], but other studies have indicated that the effect of increased pCO_2 is minimal for other coccolithophore species [7]. Time series analysis of natural populations has been another approach and a recent analysis of coccolithophore abundance in the North Sea concluded that increasing pCO_2 on decadal scales has resulted in larger coccolithophore populations [8]. The contradictory results make it difficult to robustly predict how natural populations will respond to pH change in a future ocean.

We suggest that real understanding of the effect of pH change/higher pCO_2 requires more detailed information than has been obtained to date, particularly in relation to phytoplankton genetic variability and virus infection. In this study, the response of a population of *E. huxleyi* to increased pCO_2 at the early stages of a phytoplankton bloom in a mesocosm experiment has been investigated. In addition, changes in diversity of the viruses that infect *E. huxleyi* (EhVs) were followed during the experiment with diversity distinguished on the basis of a major capsid protein (*mcp*) gene as a molecular marker. Virus diversity is known to be high [9,10], and viruses are important components of the pelagic system that require attention in both laboratory and mesocosm experiments. All *E. huxleyi*-infecting viruses that have been characterised to date are dsDNA viruses, classified in the family *Phycodnaviridae*. They are significant mortality agents of *E. huxleyi*, implicated in the termination of large-scale blooms. Viruses have a proven role in structuring and maintaining host population diversity [11–14] and virus infection can have significance for the cycling of carbon and trace elements. The 'viral shunt' releases nutrients as well as dissolved and particulate organic matter from lysed organisms into the organic carbon pool [15,16]. This material, and the rate of supply by viral lysis, of substrates for heterotrophic microbial communities, has implications for species succession, biogeochemical cycles and feedback mechanisms.

Given that diversity of both *E. huxleyi* and EhV assemblages can be variable, and that different *E. huxleyi* and EhV assemblages may come to eventually dominate natural communities, it is important to know the impact of elevated pCO_2. *E. huxleyi* blooms are typically dominated by certain alleles/genotypes, and by asexual reproduction, with rarer alleles/genotypes tending to fluctuate [17]. As such, the impact of elevated CO_2 on the composition of *E. huxleyi* populations can easily be monitored by studying these entities. Virus infection may be an explanation for some of the variability reported from different mesocosm experiments that were designed to investigate potential effects of higher pCO_2. In this study, the aim was to understand how pCO_2 change might influence *E. huxleyi* and EhV population structure and the diversity of host and virus. We suggest that viral infection can result in variability between replicate enclosures.

2. Materials and Methods

2.1. Experimental Set-Up and Sampling

The mesocosm experiment was done in the Raunefjorden at the University of Bergen Espegrand field station, Norway (latitude: 60°16' N; longitude: 5°13' E) during May 2006. The experiment had two phases. The first phase, until 15 May, followed the development of a phytoplankton bloom and the second phase studied the decline of the bloom; only the first phase of the experiment is considered here. Six polyethylene enclosures of 2 m diameter and 3.5 m depth containing 11 m^3 water were moored ca. 200 m from the shore and filled simultaneously with fjord water, salinity 31.4, and temperature 10.4 °C. Over a 40 h period from 4–6 May, 3 enclosures were bubbled with air enriched with CO_2 to 760 ppmv whilst the other 3 enclosures were bubbled with air at ambient pCO_2. The pCO_2 in the air mixture was measured inline with a LI-COR 6262 CO_2/H_2O analyser (LI-COR, Inc., Lincoln, NE, USA). After equilibration, the pH of each of the high pCO_2 treatments was 7.8 and the ambient treatment mesocosms were all pH 8.15. High precision alkalinity and pCO_2 measurements were made throughout the experiment and pH was calculated [18]. All mesocosms were covered with UV-transparent polyethylene to maintain the appropriate CO_2 concentration in the headspace above the enclosures, whilst allowing transmission of the complete spectrum of light and the exclusion of rainwater. Phytoplankton blooms were initiated on 6 May by the addition of 15 μmol$\cdot$L^{-1} $NaNO_3$ bringing the initial nitrate concentration to 16.1 μmol$\cdot$N$\cdot$L^{-1} and 1 μmol$\cdot$L^{-1} NaH_2PO_4 to give an initial phosphate concentration 1.19 μmol$\cdot$P$\cdot$L^{-1}. Silicate was not added because the aim was to test the effects of pH change on coccolithophores, but rather to stimulate diatom growth; the initial silicate concentration was 0.25 μmol$\cdot$Si$\cdot$L^{-1}.

2.2. Water Sampling

The majority of measurements were made on water samples taken at the same time each day, between 10 a.m. and 11 a.m. Water samples were collected in 5 L carboys and transported to the shore laboratory where they were processed in a temperature controlled room at ambient seawater temperature.

2.3. Nutrient and Phytoplankton Analysis

Nutrient concentrations were determined on duplicate water samples by colorimetric analysis using the methods of Brewer and Riley [19] for nitrate, Grasshoff [20] for nitrite, and Kirkwood [21] for phosphate. Chlorophyll concentration was determined fluorometrically each day during the experiment, using the method of Holm-Hansen et al. [22] on water samples filtered through GFF glass fibre filters to monitor phytoplankton development. Samples were also taken for HPLC analysis of phytoplankton pigments, with GFF filters being stored at −80 °C between the period of sampling and laboratory analysis. Coccolithophore numbers were enumerated by analytical flow cytometry.

The rate of carbon fixation was estimated from the incorporation of ^{14}C-bicarbonate following the method of Joint and Pomroy [23]. Surface water samples were collected from each mesocosm at dawn and transferred into five 60 mL clear polycarbonate bottles and a single black polycarbonate bottle; all bottles were cleaned following JGOFS protocols [24] to reduce trace metal contamination. Each bottle was inoculated with 37 kBq (1 μCi) $NaH^{14}CO_3$; bottles were incubated at the surface and depths of 0.5, 1, 2 and 3 m in the fjord adjacent to the mesocosm facility for 24 h. Samples were filtered through 0.2 μm pore-size polycarbonate filters, dried, and treated with fuming HCl to remove unfixed ^{14}C and the assimilated ^{14}C fraction was measured in a liquid scintillation counter. The efficiency of the LSC was determined with an external standard, channels ratio method. The quantity of ^{14}C added to the experimental bottles was determined by adding aliquots of the stock ^{14}C solution to a CO_2-absorbing scintillation cocktail, which was counted immediately in the LSC.

2.4. Extraction of DNA

Collected water was stored at 4 °C until it was filtered, which occurred within several hours. Five litres of water from each mesocosm were filtered through a Sterivex-GP Sterile Vented Filter Unit, 0.22 μm (Millipore, Merck KGaA, Darmstadt, Germany). Filters were snap frozen in liquid nitrogen and maintained at −80 °C until they were processed. In addition, 2 mL 1× PBS was applied to the filters to wash off biomass and this was pelleted by centrifugation. DNA was extracted using the Qiagen DNeasy blood and tissue kit (Qiagen, Valencia, CA, USA) according to the manufacturer's instructions.

2.5. Polymerase Chain Reaction (PCR) and Denaturing Gradient Gel Electrophoresis (DGGE) of E. huxleyi and EhV Populations

PCR/DGGE analyses of extracted DNA from the 6 mesocosms were carried out according to the protocol for *E. huxleyi* and *E. huxleyi* viruses (EhV), as detailed in Schroeder et al. [25] and Schroeder et al. [13], respectively, using primers specific to the calcium binding protein gene (*gpa*) for *E. huxleyi* and the major capsid protein (*mcp*) gene for EhV. PCR products for *gpa* and *mcp* were run on a 30%–50% denaturing gel according to Schroeder et al. in order to visualise the respective community structures [13]. DGGE profiles for EhV were analysed using Genetools (Syngene, Cambridge, UK) using rolling disk baseline correction and minimum peak detection; width 7, height 3, volume 2% and Savitsky–Golay filter 3 to discriminate and quantify different bands/peaks.

2.6. Statistical Analysis

Ambient and high CO_2 multi-dimensional analysis (MDA) ordinations were calculated using Primer (v6) [26] using Bray–Curtis resemblance matrices produced from the DGGE profiles where bands were detected according to their migration distance down the tracks using Genetools (Syngene, Cambridge, UK). Principal component analysis (PCA) were calculated in Primer using all data obtained in the experiment to investigate which components might define differences/similarities between samples.

3. Results

3.1. Bloom Evolution—pH, Nutrients and Primary Production

Following bubbling to achieve the target pHs in all mesocosms, the experimental phase was initiated on 6 May, by the addition of nitrate and phosphate. Initial pH of the non-modified treatment mesocosms was 8.14. Figure 1a shows the values of pH during the first nine days of the experiment that were calculated from high precision pCO_2 data (Figure 1b) [18]. For four days, pH and pCO_2 remained constant with little variation between replicate treatments. After 10 May, pH began to increase in all mesocosms, with declining pCO_2 values as the phytoplankton bloom developed. Figure 1c,d record the changes in nitrate and phosphate concentration, including the initial nutrient addition. Both nutrients declined in concentration after 10 May as phytoplankton biomass increased (Figure 1e). Chlorophyll a concentration increased rapidly in all enclosures (Figure 1f), reaching a maximum on 13 May. However, there were differences in the maximum concentrations attained; the three high pCO_2-treatment mesocosms had maximum chlorophyll concentrations of 6.23, 4.51 and 6.08 μg·L^{-1}, but chlorophyll concentrations were higher (10.71 and 11.22 μg·L^{-1}) in two of the ambient high pCO_2-treatment mesocosms (4 and 6). A slightly lower phytoplankton biomass developed in enclosure M5—one of the ambient pCO_2-treatment mesocosm—with a chlorophyll a concentration of 9.60 μg·L^{-1}. The chlorophyll concentration in this mesocosm also declined more rapidly after 13 May than in the other treatments.

Primary production rates were very consistent in the three high pCO_2-treatment mesocosms (Figure 1f), reaching maximum values on 12 May, with little variation between enclosures. In all the ambient pCO_2 mesocosms, primary production was >900 mg C m^{-2}·d^{-1} on 12 May and remained at this value for two days in M4 and M6. However, production in M5 was less than in the other two ambient pCO_2-treatments, which is consistent with the lower chlorophyll concentration in this enclosure.

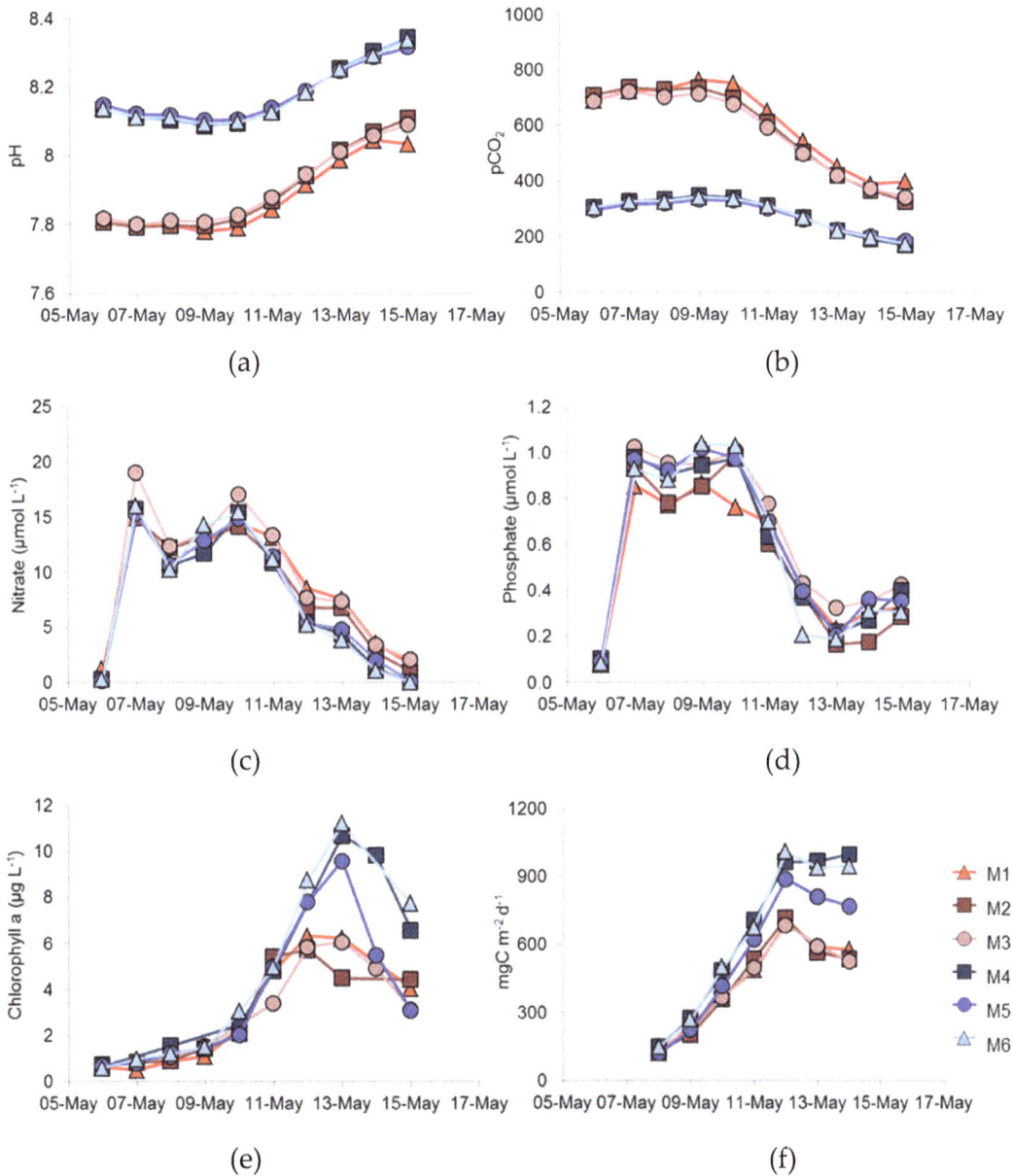

Figure 1. Temporal changes over the period of the experiment in (**a**) pH, which was calculated from measurements of (**b**) pCO_2 in μatmospheres; (**c**) nitrate concentration, $\mu mol \cdot N \cdot L^{-1}$; (**d**) phosphate concentration $\mu mol \cdot P \cdot L^{-1}$; (**e**) chlorophyll concentration $\mu g \cdot L^{-1}$; and (**f**) depth-integrated primary production as $mg \cdot C \cdot m^{-2} \cdot d^{-1}$. Enclosures M1 (▲), M2 (■), M3 (●), M4 (■), M5 (●), M6 (▲).

3.2. E. huxleyi Genetic Composition during the Mesocosm Experiment

E. huxleyi was a significant component of the phytoplankton assemblage that developed in each enclosure. Diatom numbers were insignificant because silicate was not added to the initial nutrient addition, being three orders of magnitude less abundant in light microscope analysis than the total flagellate fraction, which includes coccolithophores. Hopkins et al. [18] reported the dominance of large picoeukaryotes in each mesocosm assemblage but with the flagellates contributing greatest to phytoplankton biomass.

All enclosures showed steady increases in coccolithophore numbers (as assessed by flow cytometry) immediately after nutrient addition. Numbers reached 600–1000 cells mL^{-1} on 9 May, which is typical of numbers seen during the early-stages of *E. huxleyi* blooms (Figure 2) [12]. In the ambient-pCO_2 treatments (M4, M5, M6) coccolithophore numbers increased until 12 May, but with a slight pause in growth, numbers increased further to a maximum of 2500–3000 cells mL^{-1} on 14 May. Cell numbers then declined to 1000–2000 cells mL^{-1} on 15 May. Coccolithophore biomass was different in the three replicate high pCO_2-treatment mesocosms. In two enclosures, numbers plateaued at about 1000 cells mL^{-1}, but, in the third mesocosm, numbers were higher and, indeed, very similar (3100 cells mL^{-1}) to the peak biomass in the ambient pCO_2-treatment mesocosms (Figure 2). Cell numbers declined in all six mesocosms after 14 May, even in those enclosures with lower cell numbers.

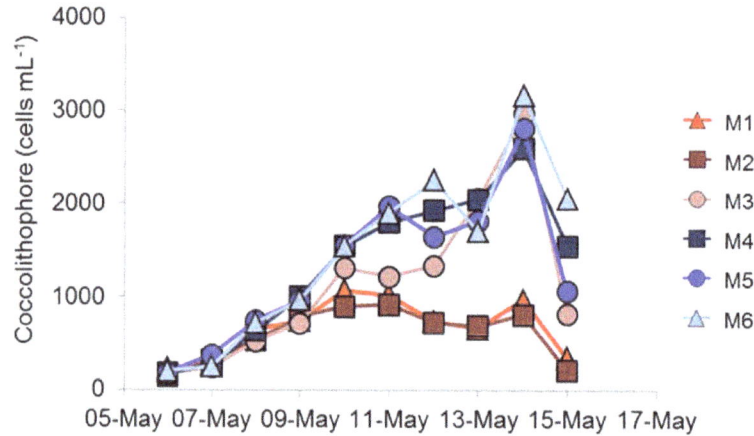

Figure 2. Total coccolithophore numbers assessed by flow cytometry. Enclosures M1 (▲), M2 (■), M3 (●), M4 (■), M5 (●), M6 (▲).

Traditional microscopy, neither light nor electron, is capable of distinguishing between *E. huxleyi* genotypes with Figure 3 showing that identical morphology (typical type A) was present in both pCO_2 treatments throughout the experiment.

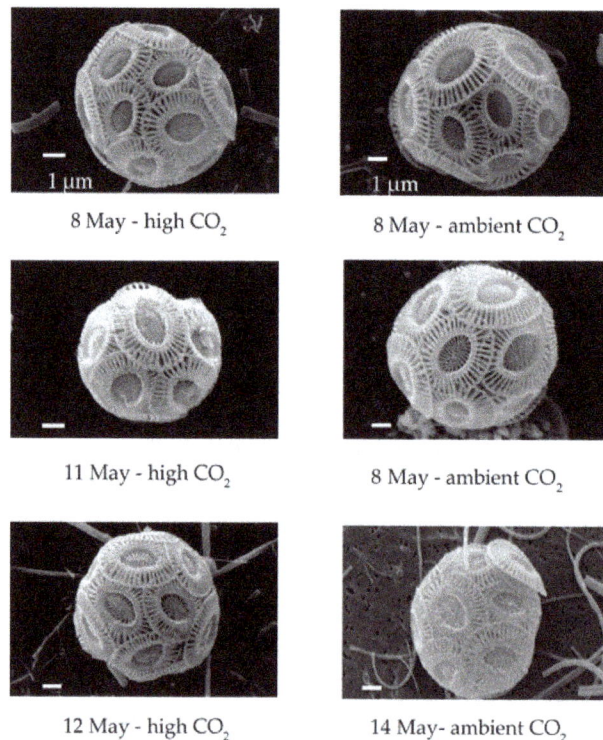

8 May - high CO_2

8 May - ambient CO_2

11 May - high CO_2

8 May - ambient CO_2

12 May - high CO_2

14 May - ambient CO_2

Figure 3. TEM images of identical *Emiliania huxleyi* morphologies (typical type A) present in both pCO_2 treatments throughout the experiment.

However, molecular analysis showed a large genetic diversity that was not revealed by microscopy. DGGE analysis of the *E. huxleyi* population using the *gpa* marker detected two to three dominant bands throughout the experiment as indicated by the arrows in Figure S1. This gene has been verified for *E. huxleyi* diversity analysis, with a limited number of genotypes known to exist [17] that can largely be separated by DGGE [25]. There was some small-scale variability in the *E. huxleyi* population between samples, as indicated by migration profiles. Overall, the *E. huxleyi* populations had similar genetic composition in all six mesocosms and no major differences were identified between treatments or replicates.

3.3. EhV Population Analysis

Flow cytometry revealed the presence of large DNA viruses in all enclosures (data not shown), indicating background levels less than 10^5 viruses per ml as described previously in other Bergen based mesocosm experiments [12,13]. Although there was little variation in *E. huxleyi* genotypes throughout the experiment, the virus (EhV) that infects this alga did show considerable variation. DGGE analysis (Figure 4) indicated that the EhV population was more diverse and had a more variable genetic structure than the host. Whilst DGGE has its limitations due to co-migration events meaning that a single band can be comprised of >1 OTU, it is still accepted as a useful tool for looking at changes in microbial communities. In particular, DGGE has proven to be a robust and reliable technique for the study of EhVs. Limitations often described in the literature, centre on PCR-DGGE designed to target a large taxonomic group where the scale of diversity is massive, e.g., 16S rRNA. Focussing on a smaller taxonomic unit improves resolution [27] as does careful design and optimisation of primers. A two-stage PCR has been well optimised for EhV with the use of a GC-clamp to improve resolution. Although not all EhVs can be discriminated from one another, such as EhV-201 and EhV-205 that only differ by 1 bp in the target *mcp* region, virus isolates, including EhVs 203, 201, 202, 163, 84 and 86, can clearly be separated on a DGGE gel, with EhV-84 and EhV-86 differing from each other by only 3 bp [13]. Furthermore, DGGE gels for EhV have been found to be highly reproducible with the samples being run on >1 separate gels, generating the same migration profile. This is also corroborated by previous work of Sorensen et al. [14] and Martinez-Martinez et al. [12], where replicate gels routinely produce the same profile.

In the early stages of the mesocosm bloom development before 9 May, when coccolithophore numbers were <800 cells mL^{-1}, there was high variability in the EhV population, both between replicates and on different days sampled, for example within mesocosm 4 on 7 May there were four bands, on 8 May seven bands and on 9 May eight bands, with four of these being unique; the percentage similarity was less than 40%. Whilst we cannot ascertain that each band represents a single out, we can still infer the changes observed and the overall temporal patterns of diversity. A Bray–Curtis similarity analysis (Figure 5) showed that, as coccolithophore numbers increased, exceeding 1.5×10^3 cells mL^{-1} in ambient enclosures, there was less variability in the EhV population in the ambient pCO$_2$-treatment mesocosms, which shared more similarity (>43%) between replicates. This was compared to as little as 1% in the early stages of the experiment when coccolithophore numbers were less than 1×10^3 cells mL^{-1} (Figure 5).

Within the 2 high-pCO$_2$ treatment enclosures that had the lowest coccolithophore cell number (M1 and M2), the DGGE profiles showed low similarity in the EhV population between dates and replicates (Figure 4). Stabilisation of the EhV population was not evident as the experiment proceeded in the high pCO$_2$ samples, mesocosms 1 and 2, and similarity between samples remained low (9%). In contrast, in mesocosm M3, all samples shared at least 55% similarity, indicating a stable EhV population across the time series. One of the dominant bands on the DGGE profile in M3 (marked with a triangle in Figure 4a) was an EhV genotype that also dominated in the ambient-pCO$_2$ treatments in enclosures M4, M5 and M6.

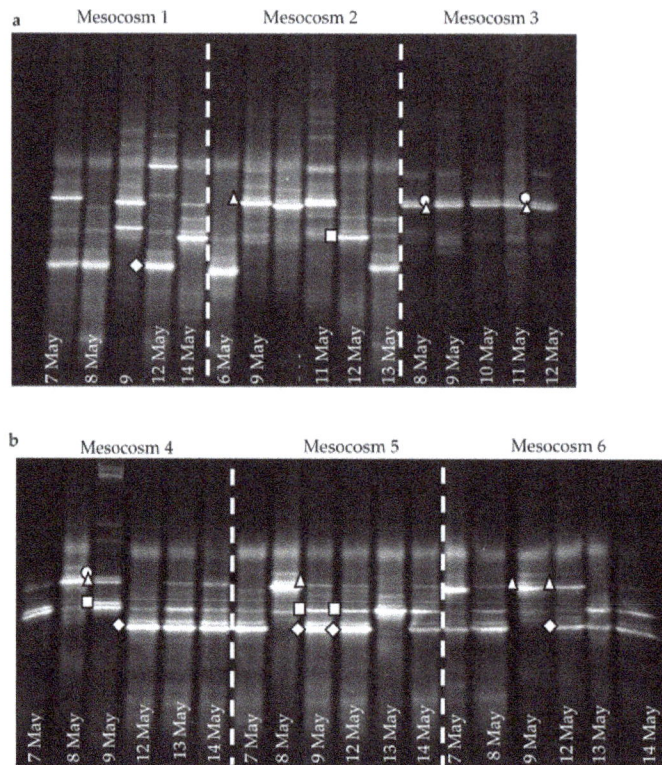

Figure 4. DGGE gels of EhV *mcp*-PCR products during the experiment from (**a**) high pCO$_2$-treatment mesocosms, 1, 2, and 3 and (**b**) ambient pCO$_2$-treatment mesocosms 4, 5, and 6. Bands that migrated at the same position when run on the same gel are indicated with the same symbol.

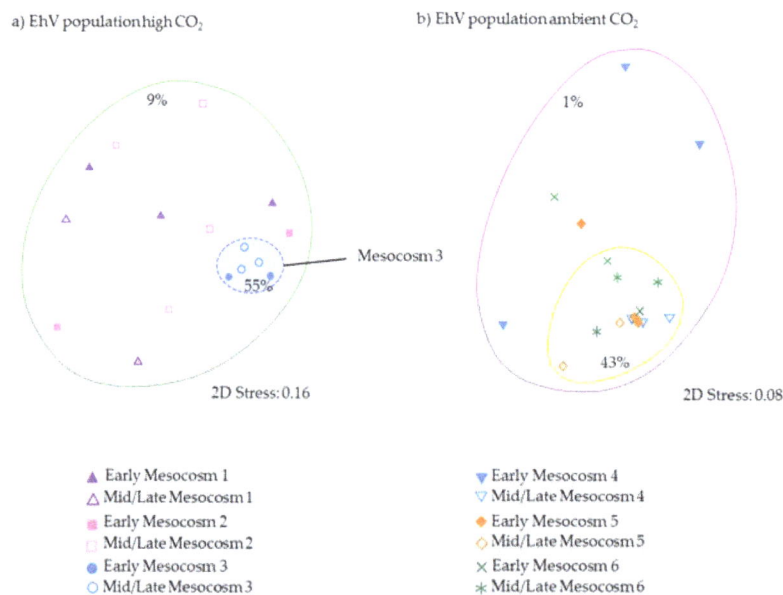

Figure 5. Bray–Curtis multidimensional plots based on the DGGE profiles (Figure 4) for EhV from (**a**) the high pCO$_2$-treatment mesocosms 1, 2, and 3 and (**b**) ambient pCO$_2$-treatment mesocosms 4, 5, and 6. "Early stage" corresponds to 7–9 May when coccolithophore numbers were <1000 cells mL^{-1} in ambient enclosures and "mid/late stage" corresponds to 12–14 May when coccolithophore numbers exceeded 1500 cells mL^{-1} in ambient enclosures. Contours indicate the percentage similarity, as indicated.

Samples from the 9 and 12 May from all six mesocosms were additionally run on the same gel. The shapes on the gels (Figure 4) indicate bands that migrated to the same position and hence can be inferred to be the same EhV sequence.

4. Discussion

A number of mesocosm experiments have been performed to investigate the potential effects of increased pCO_2 and reduced pH on complex pelagic assemblages—from bacteria to zooplankton. Mesocosm enclosures have the advantage of capturing more of the intrinsic complexity of a pelagic assemblage than is possible in a laboratory experiment because of the very large volumes of water (several thousand litres) that are involved. They also, by the nature of enclosure, eliminate problems of dispersal that make the study of variability in natural environments so complex. They are fundamentally attractive to experimentalists because they offer a means to manipulate large water volumes, with their associated planktonic assemblages, in order to test the effects of environmental problems such as eutrophication or ocean acidification.

In this study, we aimed to investigate how phytoplankton might respond in a future high-pCO_2 ocean by comparing natural phytoplankton assemblage development in enclosures at ambient pCO_2, (initial condition ca. 300 µatm.) and enriched pCO_2 conditions (initially ca. 700 µatm.)—Figure 1a,b; see also Hopkins et al. [18]. During the course of the experiment, utilisation of CO_2 by phytoplankton reduced pCO_2 and the pCO_2/pH values were continually changing. An obvious difference between the treatments was that less phytoplankton biomass, as indicted by chlorophyll a concentration, developed in the high- compared to the ambient-pCO_2 conditions (Figure 1e). Not only did less biomass develop, but primary production (Figure 1f) was also lower under high-, rather than under ambient-pCO_2, suggesting that increased pCO_2 might have a deleterious effect on the total phytoplankton biomass. In contrast, other experiments have suggested that dissolved inorganic carbon uptake would be enhanced under elevated pCO_2 conditions [28]. Different phytoplankton types are likely to respond differently to pCO_2 and Riebesell et al. suggested that diatoms showed enhanced uptake, whereas coccolithophores did not [28]. There was no suggestion in that study, though, that coccolithophore growth might be reduced under elevated pCO_2 conditions.

The dominant species of coccolithophore within the mesocosms was *E. huxleyi*, with coccolithophores being reported as contributing 6% and 12% to the total flagellate biomass for M1 and M6, respectively [18]. Manipulating the development of *E. huxleyi* blooms within mesocosm enclosures is well established at the mesocosm facility at Raunefjorden, and is well documented. Addition of nitrate and phosphate, with the omission of silicate, usually results in a bloom of *E. huxleyi* at this site, particularly during May/June [11,12,14] and an increase in *E. huxleyi* abundance happened in the present study. Previous mesocosm experiments of this nature have described a dominance of coccolithophores; however, differences in methodology most likely resulted in the dominance of large picoeukaryotes [18]. In the present study, a single nutrient enrichment was undertaken at the beginning of the study, whereas daily enrichments are often used e.g., Jacquet et al. [29]. Nevertheless, the maximum number of *E. huxleyi* cells that developed was significantly lower in the high- compared to ambient-pCO_2 conditions (Figure 2), suggesting that *E. huxleyi* might be particularly susceptible to variations in pCO_2.

Replication between enclosures was rather variable. The peak chlorophyll concentrations were very similar in the three ambient pCO_2 replicates (Figure 1e), but the bloom decayed more rapidly in M5 than in M4 and M6; there were similar differences in primary production (Figure 1f). However, the largest difference between replicates was in coccolithophore numbers (Figure 2), which showed significant differences between the three high-pCO_2 treatment mesocosms with such differences between replicates in other key phytoplankton groups not reported by Hopkins et al. [18]. Enclosure M3 had coccolithophore cell numbers at the peak of the bloom that were very similar to the three ambient pCO_2 mesocosms, unlike the lower numbers in M1 and M2. Although cell numbers were different in M3, no major differences in *E. huxleyi* genotype or phenotype were detected between treatments or over the course of the experiment, suggesting a stable community within all enclosures over the duration of the experiment. Stable *E. huxleyi* populations have been found in previous mesocosm experiments [12,14] and, indeed, in naturally occurring *E. huxleyi* blooms [9,17]. In this study, we have no evidence to support the hypothesis that higher pCO_2 conditions might benefit certain *E. huxleyi* genotypes; we could detect no restructuring of the *E. huxleyi* population in the different pCO_2 treatments.

Contradictory results are constant features of experiments to investigate the effect of pCO_2/pH change on coccolithophores. In laboratory culture experiments, Riebesell et al. [30], Zondervan et al. [31] and Richier et al. [32] all reported increased production by *E. huxleyi* under elevated pCO_2 conditions, but Sciandra et al. [33] and Langer et al. [4] found decreased production. In two different CO_2-manipulated mesocosm experiments in the Raunefjorden, Engel et al. [6] and Paulino et al. [34] found little difference in *E. huxleyi* cell concentrations over the course of their experiments. However, Engel et al. [6] calculated that the net specific growth rate of *E. huxleyi* was reduced at 710 µatm compared with 410 µatm. In a long-term batch culture experiment conducted over one year, Lohbeck et al. [5] found that *E. huxleyi* cultures maintained at ambient pCO_2 (400 µatm) went through 530 generations over the one year experimental period, but the same strain cultured at 1100 µatm achieved only 500 generations, and, at 2200 µatm, growth was even lower, with only 430 generations. *E. huxleyi* would appear to be more sensitive to pCO_2 change than other phytoplankton.

In the context of the present study, the reduced primary production in the high-pCO_2 treatment enclosures is consistent with the finding of Lohbeck et al. [5] that higher pCO_2/lower pH reduces primary production of an *E. huxleyi* dominated phytoplankton assemblage, although, in this study, other phytoplankton groups, e.g., picoeukaryotes, cryptophytes and cyanobacteria, will have also contributed to this. However, it is not consistent with the suggestion of Rivero-Calle et al. [8] that increasing pCO_2 is one of the factors most responsible for the decadal increase in abundance of coccolithophores in the North Atlantic. The relationship between pCO_2/pH change and success or, otherwise of cocccolithophores, remains confusing and requires clearer examination of mechanisms that might lead to phytoplankton changes in the future ocean.

Given that the coccolithophores numbers in enclosure M3 were very different from the other two high pCO$_2$ treatments, and yet the *E. huxleyi* genetic diversity was very similar in all three enclosures, could viral infection by EhV be a contributing factor to explain the observed variations within and between treatments? In the three ambient-pCO_2 enclosures, the EhV population followed a pattern that has been seen in other mesocosm experiments—high variability during the early stages of phytoplankton bloom development, with a smaller number of genotypes coming to dominate as the *E. huxleyi* numbers increase [12,14]. Although phytoplankton bloom development in the current study was short (<10 days), there was sufficient time for the virus populations to change because EhV populations are inherently dynamic [9] and known to change on very short time scales [14]. Daily changes in EhV composition can be expected since Sorensen et al. [14] showed that EhVs can appear/disappear from the water column in a matter of hours.

In the high-pCO_2 treatment enclosures M1 and M2, where *E. huxleyi* population densities did not exceed 1.1×10^3 mL^{-1}, the EhV population did not stabilise (Figures 4 and 5) and EhV diversity was typical of early or non-bloom conditions [10]; that is, it was a highly dynamic and diverse EhV population. EhVs are known to have different host ranges [35] as well as different characteristics of infection, such as burst size and latent period. Therefore, any changes in environmental conditions that directly affect these traits could ultimately select for different genotypes, hence restructuring the EhV population.

In contrast, the third high-pCO_2 treatment enclosure M3, where coccolithophore cell densities reached similar values to ambient enclosures, and the EhV population structure was very different. Two EhVs dominated over the course of the experiment (Figure 4) and the population was much less changeable compared to the other high-pCO_2 enclosures and, indeed, to the three ambient enclosures. The EhV assemblage in M3, right from the early stages of the experiment, reflected what would be expected in the later stages of a bloom. Even in the early stages of the bloom, the EhV assemblage was stable and clustered closely in the MDS plot (Figure 5) with samples taken later in the bloom.

Given that pH and pCO_2 were so similar throughout the experiment in the three high-pCO_2 enclosures, why did enclosure M3 have lower coccolithophore numbers and a lower and stable EhV diversity? Other studies have shown that, under non-bloom conditions, many different EhV genotypes are present and abundance fluctuates on short time scales [14]. In addition, during the initial phase of an *E. huxleyi* bloom, EhV populations remain diverse and are often highly dynamic. Sorensen et al. [14] showed that, as a bloom developed in a mesocosm experiment and viruses numbers proliferated, one viral genotype dominated, and suggested that this dominant virus caused the termination of the bloom. However, the dominant EhV is not always the same, even when host genotypes do not vary. Martínez-Martínez et al. [12] and Sorensen et al. [14] found that, although *E. huxleyi* populations were dominated by the same genotypes in different years at Raunefjorden (2000, 2003 and 2008), viruses changed and the dominant EhV in 2008 was different to the EhVs that dominated during the 2000 and 2003 mesocosm experiments. The reason is not known, but these authors suggest that a slight change in environmental conditions might have favoured dominance by a different virus genotype. Whilst the number of mesocosms sampled might be perceived as limited, the fact that Martinéz-Martinéz [35] described how the same DGGE profile was generated from four replicate mesocosms; in his studies, we can assume that the changes that we are observing are genuine.

Other mesocosm experiments have studied how altered pCO_2 influences natural virus communities. Larsen et al. [36], using flow cytometry analysis and pulsed field gel electrophoresis (PFGE), found slightly more (but statistically insignificant) EhVs under present-day pCO_2 mesocosms than in high-pCO_2 treatments; this was not a consequence of lower *E. huxleyi* cell densities in the high-pCO_2 treatments. The authors speculated that elevated-pCO_2 may affect host–virus interactions or influence viral replication, since a 26 kb genome virus was only detected in ambient conditions, and was absent from high-pCO_2 treatments, and a 105 kb genome virus was only detected in the highest pCO_2 treatment of 1050 ppm. Unfortunately, the taxonomic affiliation of the viruses was not verified, which limits comparison with the present study.

Some laboratory experiments have investigated the effect of higher pCO_2 on marine phytoplankton viruses. Carreira et al. [37] studied interaction between *E. huxleyi* and the virus EhV-99B1. *E. huxleyi* growth rate was not affected by the different pCO_2 treatments, but the burst size of EhV-99B1 was lower in present-day, compared with higher and lower (pre-industrial) pCO_2. In addition, release of EhVs was delayed in high-pCO_2 treatments. Other virus groups that have also been tested for sensitivity to elevated pCO_2. Chen et al. [38] found lower burst size of the *Phaeocystis globosa* virus (PgV) at high-pCO_2, and Traving et al. [39] found that the cyanophage S-PM2, which infects *Synechococcus* sp, had reduced burst size at lower pH. However, the extracellular phase, quantified as infectivity loss rates/decay, did not change. These experiments illustrate that pCO_2 can influence virus–host interactions, albeit to a relatively minor extent. However, extrapolation from these laboratory-scale experiments to natural virus communities would involve considerable uncertainty.

Another study considered a much longer time scale. Coolen [40] investigated *E. huxleyi* and EhV diversity in Black Sea sediments over a 7000 year period, showing that EhV diversity was highest during periods of change in hydrological and nutrient regimes. Shifts in EhV genotypic diversity typically coincided with Holocene environmental change with some viruses having limited persistence, yet others were found to persist for over a century. This study alluded to the impact that a change in CO_2/pH could have on future EhV populations.

Might the different EhV populations that dominated in each enclosure be an explanation for the differences observed? It is generally accepted that virus infection is a major reason why *E. huxleyi* cells stop growing and blooms are terminated [14]. Certainly, nutrients were still available, albeit at low concentrations when biomass peaked in each enclosure (Figure 1c,d). If viral mortality was the major limit on bloom development, then enclosures M1 & M2 must have been infected with more aggressive EhVs than in the other treatments. The effect of pCO_2 treatment, per se, on *E. huxleyi* cells cannot be responsible for the reduced growth in the two high-pCO$_2$ treatments because an identical pH/pCO$_2$ treatment in M3 did not reduce the size of the bloom. Therefore, pCO$_2$ change must have resulted in different viral diversity, if infection is indeed the main cause of lower cell numbers and smaller *E. huxleyi* bloom.

How might reduced EhV diversity in M3 have resulted in higher coccolithophore numbers developing than in the two other high pCO$_2$ treatments where growth was curtailed? One explanation would be that the EhVs that lead to rapid termination of *E. huxleyi* growth [14] were not present in sufficient numbers in M3 to suppress growth of the *E. huxleyi* population in this enclosure. Genetic diversity in natural populations of EhVs, especially in pre-bloom conditions [10], coupled with the rapid rate at which individual EhVs can come to dominate [9,14] means that the matrix of EhVs that could be selected for is large. It is not clear why only two EhVs were dominant in enclosure M3, but it is probably significant that viral infection in this enclosure did not suppress growth of *E. huxleyi* compared to M1 and M2.

In this study, it is difficult to distinguish whether pCO$_2$ change is affecting the viruses specifically, or their hosts independently, and/or the interactions between them. Both the external virus population (the virus particles present in the water used to fill the mesocosms) and internal virus population (present in infected *E. huxleyi* cells) are important for the ultimate progression of the *E. huxleyi* bloom and EhV population. By comparing data from Schroeder et al. [13] and Martínez Martínez [12] it can be seen that the diversity of EhVs amplified from water samples within a mesocosm bloom can be very different from that amplified from *E. huxleyi* cells, particularly at the early stage of the bloom—that is, external and internal EhV assemblages can have very different composition. Thus, studies that aim to explain the effect of elevated pCO$_2$ must incorporate into their experimental design ways to distinguish between a direct effect on external virus particles, or on *E. huxleyi* cells, or on EhVs that have already infected *E. huxleyi* cells. Alternatively, other unknown factors that are not related to pCO$_2$ cannot be dismissed and might be the cause of the different coccolithophore response in enclosure M3.

Our study demonstrates the need for further investigations on the effects of elevated pCO$_2$ on the *E. huxleyi* EhV system since there are ecological impacts of virus competition on biogeochemical cycles. Nissimov et al. [41] investigated competition between the two EhVs, finding that EhV-207 had a competitive advantage over EhV-86. It would thus be of value to determine how external factors, such as elevated CO$_2$, would affect relative competitive ability of EhVs; would EhV-207 still outcompete EhV-86?

In this study, we have shown that elevated pCO$_2$ can affect the structure of EhV assemblages. The data do not allow us to distinguish if this is a direct impact of pCO$_2$ on the viruses themselves, but it is clear that caution is required in interpreting results from manipulation experiments and that deep analysis is required to truly understand how complex assemblages respond. For example, analysis of only the chlorophyll concentration or primary production data would not have revealed that there were differences in the triplicate high-pCO$_2$ treatments: cell counts and microscopy would not have revealed the difference in *E. huxleyi* diversity in the replicate enclosures, and only analysis of virus genotypes could have revealed how different the viruses were in apparently identical replicate enclosures. We suggest that viruses cannot be ignored in any study of the potential effects of ocean acidification on phytoplankton productivity in the future high-CO$_2$ ocean.

Acknowledgments: Twenty-seven people participated in the mesocosm experiment and we thank all of them for their contributions. In particular, we thank Dorothee Bakker for high precision measurements of the carbonate system, Isabel Mary and Andrew Whiteley for flow cytometer measurements of coccolithophore number and Cecilia Balestreri for laboratory support. Special thanks are due to the staff at the Espegrand field station for their assistance. The mesocosm experiment was supported by NERC, through Grant No. NE/C507902, as part of the Post-Genomics and Proteomics Programme.

Author Contributions: I.J. and D.C.S. conceived and designed the experiments; A.H. and K.J.C. performed the experiments; A.H., I.J. and D.C.S. analyzed the data; J.G. contributed reagents/materials/analysis tools; and A.H. and I.J. wrote the paper.

References

1.   Orr, J.C.; Fabry, V.J.; Aumont, O.; Bopp, L.; Doney, S.C.; Feely, R.A.; Gnanadesikan, A.; Gruber, N.; Ishida, A.; Joos, F.; et al. Anthropogenic ocean acidification over the twenty-first century and its impact on calcifying organisms. *Nature* **2005**, *437*, 681–686. [CrossRef] [PubMed]

2.   Balch, W.M. Re-evaluation of the physiological ecology of the coccolithophores. In *Coccolithophores: From Molecular Processes to Global Impact*; Thierstein, H.R., Young, J.R., Eds.; Springer: New York, NY, USA, 2004; pp. 165–190.

3.   Iglesias-Rodriguez, M.D.; Halloran, P.R.; Rickaby, R.E.M.; Hall, I.R.; Colmenero-Hidalgo, E.; Gittins, J.R.; Green, D.R.H.; Tyrrell, T.; Gibbs, S.J.; von Dassow, P.; et al. Phytoplankton calcification in a high-CO_2 world. *Science* **2008**, *320*, 336–340. [CrossRef] [PubMed]

4.   Langer, G.; Nehrke, G.; Probert, I.; Ly, J.; Ziveri, P. Strain-specific responses of *Emiliania huxleyi* to changing seawater carbonate chemistry. *Biogeosciences* **2009**, *6*, 2637–2646. [CrossRef]

5.   Lohbeck, K.T.; Riebesell, U.; Reusch, T.B.H. Gene expression changes in the coccolithophore *Emiliania huxleyi* after 500 generations of selection to ocean acidification. *Proc. R. Soc. B* **2014**, *281*. [CrossRef] [PubMed]

6.   Engel, A.; Zondervan, I.; Aerts, K.; Beaufort, L.; Benthien, A.; Chou, L.; Delille, B.; Gattuso, J.-P.; Harlay, J.; Heemann, C.; et al. Testing the direct effect of CO_2 concentration on a bloom of the coccolithophorid *Emiliania huxleyi* in mesocosm experiments. *Limnol. Oceanogr.* **2005**, *50*, 493–507. [CrossRef]

7.   Meyer, J.; Riebesell, U. Reviews and Syntheses: Responses of coccolithophores to ocean acidification: A meta-analysis. *Biogeosciences* **2015**, *12*, 1671–1682. [CrossRef]

8.   Rivero-Calle, S.; Gnanadesikan, A.; Del Castillo, C.E.; Balch, W.M.; Guikema, S.D. Multidecadal increase in North Atlantic coccolithophores and the potential role of rising CO_2. *Science* **2015**, *350*, 1533–1537. [CrossRef] [PubMed]

9.   Highfield, A.; Evans, C.; Walne, A.; Miller, P.I.; Schroeder, D.C. How many *Coccolithovirus* genotypes does it take to terminate an *Emiliania huxleyi* bloom? *Virology* **2014**, *466*, 138–145. [CrossRef] [PubMed]

10. Rowe, J.M.; Fabre, M.F.; Gobena, D.; Wilson, W.H.; Wilhelm, S.W. Application of the major capsid protein as a marker of the phylogenetic diversity of *Emiliania huxleyi* viruses. *FEMS Microbiol. Ecol.* **2011**, *76*, 373–380. [CrossRef] [PubMed]

11. Bratbak, G.; Wilson, W.; Heldal, M. Viral control of *Emiliania huxleyi* blooms? *J. Mar. Syst.* **1996**, *9*, 75–81. [CrossRef]

12. Martínez Martínez, J.M.; Schroeder, D.C.; Larsen, A.; Bratbak, G.; Wilson, W.H. Molecular dynamics of *Emiliania huxleyi* and co-occurring viruses during two separate mesocosm studies. *Appl. Environ. Microbiol.* **2007**, *73*, 554–562. [CrossRef] [PubMed]

13. Schroeder, D.C.; Oke, J.; Hall, M.; Malin, G.; Wilson, W.H. Virus succession observed during an *Emiliania huxleyi* bloom. *Appl. Environ. Microbiol.* **2003**, *69*, 2484–2490. [CrossRef] [PubMed]

14. Sorensen, G.; Baker, A.C.; Hall, M.J.; Munn, C.B.; Schroeder, D.C. Novel virus dynamics in an *Emiliania huxleyi* bloom. *J. Plankton Res.* **2009**, *31*, 787–791. [CrossRef] [PubMed]

15. Gobler, C.J.; Hutchins, D.A.; Fisher, N.S.; Cosper, E.M.; Sanudo-Wilhelmy, S.A. Release and bioavailability of C, N, P, Se, and Fe following viral lysis of a marine chrysophyte. *Limnol. Oceanogr.* **1997**, *42*, 1492–1504. [CrossRef]

16. Wilhelm, S.W.; Suttle, C.A. Viruses and nutrient cycles in the sea—Viruses play critical roles in the structure and function of aquatic food webs. *Bioscience* **1999**, *49*, 781–788. [CrossRef]

17. Krueger-Hadfield, S.A.; Balestreri, C.; Schroeder, J.; Highfield, A.; Helaouet, P.; Allum, J.; Moate, R.; Lohbeck, K.T.; Miller, P.I.; Riebesell, U.; et al. Genotyping an *Emiliania huxleyi* (Prymnesiophyceae) bloom event in the North Sea reveals evidence of asexual reproduction. *Biosciences* **2014**, *11*, 5215–5234.

18. Hopkins, F.E.; Turner, S.M.; Nightingale, P.D.; Steinke, M.; Bakker, D.; Liss, P.S. Ocean acidification and marine trace gas emissions. *Proc. Natl. Acad. Sci. USA* **2010**, *107*, 760–765. [CrossRef] [PubMed]

19. Brewer, P.G.; Riley, J.P. The automatic determination of nitrate in seawater. *Deep Sea Res.* **1965**, *12*, 765–772.

20. Grasshoff, K. *Methods of Seawater Analysis*; Verlag Chemie: Weinheim, Germany, 1976.

21. Kirkwood, D.S. Simultaneous determination of selected nutrients in seawater. In *ICES*; National Marine Biological Library: Plymouth, UK, 1989.

22. Holm-Hansen, O.; Lorenzen, C.J.; Holmes, R.W.; Strickland, J.D.H. Fluorometric determination of chlorophyll. *ICES J. Mar. Sci.* **1965**, *30*, 3–15. [CrossRef]

23. Joint, I.; Pomroy, A. Phytoplankton biomass and production in the southern North Sea. *Mar. Ecol. Prog. Ser.* **1993**, *99*, 169–182. [CrossRef]

24. IOC; Paris, France. Protocols for the Joint Global Ocean Flux Study (JGOFS) core measurements. In *IOC Manuals and Guides No. 29*; JGOFS International Project Office: Bergen, Norway, 1994; p. 126.

25. Schroeder, D.C.; Biggi, G.F.; Hall, M.; Davy, J.; Matinez Martinez, J.; Richardson, A.J.; Malin, G.; Wilson, W.H. A genetic marker to separate *Emiliania huxleyi* (Prynesiophyceae) morphotypes. *J. Phycol.* **2005**, *41*, 874–879. [CrossRef]

26. Clarke, K.R.; Gorley, R.N. *PRIMER V6: User Manual/Tutorial*; PRIMER-E Ltd.: Plymouth, UK, 2006.

27. Marzorati, M.; Wittebolle, L.; Boon, N.; Daffonchio, D.; Verstraete, W. How to get more out of molecular fingerprints: Practical tools for microbial ecology. *Environ. Microbiol.* **2008**, *10*, 1571–1581. [CrossRef] [PubMed]

28. Riebesell, U.; Schulz, K.G.; Bellerby, R.G.J.; Botros, M.; Fritsche, P.; Meyerhöfer, M.; Neill, C.; Nondal, G.; Oschlies, A.; Wohlers, J.; et al. Enhanced biological carbon consumption in a high CO_2 ocean. *Nature* **2007**, *450*, 545–549. [CrossRef] [PubMed]

29. Jacquet, S.; Heldal, M.; Iglesias-Rodriguez, D.; Larsen, L.; Wilson, W.H.; Bratbak, G. Flow cytometric analysis of an *Emiliana huxleyi* bloom terminated by viral infection. *Aquat. Microb. Ecol.* **2002**, *27*, 111–124. [CrossRef]

30. Riebesell, U.; Zondervan, I.; Rost, B.; Tortell, P.D.; Zeebe, R.E.; Morel, F.M. Reduced calcification of marine plankton in response to increased atmospheric CO_2. *Nature* **2000**, *407*, 364–367. [CrossRef] [PubMed]

31. Zondervan, I.; Rost, B.; Riebesell, U. Effect of CO_2 concentration on the PIC/POC ratio in the coccolithophore *Emiliania huxleyi* grown under light-limiting conditions and different day lengths. *J. Exp. Mar. Biol. Ecol.* **2002**, *272*, 55–70. [CrossRef]

32. Richier, S.; Fiorini, S.; Kerros, M.E.; von Dassow, P.; Gattuso, J.P. Response of the calcifying coccolithophore *Emiliania huxleyi* to low pH/high pCO_2: From physiology to molecular level. *Mar. Biol.* **2011**, *158*, 551–560. [CrossRef] [PubMed]

33. Sciandra, A.; Harlay, J.; Lefèvre, D.; Lemée, R.; Rimmelin, P.; Denis, M.; Gattuso, J.P. Response of coccolithophorid *Emiliania huxleyi* to elevated partial pressure of CO_2 under nitrogen limitation. *Mar. Ecol. Prog. Ser.* **2003**, *261*, 111–122. [CrossRef]

34. Paulino, A.I.; Egge, J.K.; Larsen, A. Effects of increased atmospheric CO_2 on small and intermediate sized osmotrophs during a nutrient induced phytoplankton bloom. *Biogeosciences* **2008**, *5*, 739–748. [CrossRef]

35. Martínez, J.M. Molecular ecology of marine algal viruses. Ph.D. Thesis, University of Plymouth, Plymouth, UK, 2006.

36. Larsen, J.B.; Larsen, A.; Thyrhaug, G.; Bratbak, G.; Sandaa, R.A. Response of marine viral populations to a nutrient induced phytoplankton bloom at different pCO_2 levels. *Biogeosciences* **2008**, *5*, 523–533. [CrossRef]

37. Carreira, C.; Heldal, M.; Bratbak, G. Effect of increased pCO_2 on phytoplankton-virus interactions. *Biogeochemistry* **2013**, *114*, 391–397. [CrossRef]

38. Chen, S.; Gao, K. Viral attack exacerbates the susceptibility of a bloom-forming alga to ocean acidification. *Glob. Chang. Biol.* **2015**, *21*, 629–636. [CrossRef] [PubMed]

39. Traving, S.J.; Clokie, M.R.J.; Middelboe, M. Increased acidification has a profound effect on the interactions between the cyanobacterium *Synechococcus* sp. WH7803 and its viruses. *FEMS Microb. Ecol.* **2013**, *87*, 133–141. [CrossRef] [PubMed]

40. Coolen, M.J.L. 7000 years of *Emiliania huxleyi* viruses in the Black Sea. *Science* **2011**, *333*, 451–452. [CrossRef] [PubMed]

41. Nissimov, J.I.; Napier, J.A.; Allen, M.J.; Kimmance, S.A. Intragenus competition between *coccolithoviruses*: And insight on how a select few can come to dominate many. *Environ. Microbiol.* **2015**, *18*, 133–145. [CrossRef] [PubMed]

A Student's Guide to Giant Viruses Infecting Small Eukaryotes: From *Acanthamoeba* to *Zooxanthellae*

Steven W. Wilhelm *, Jordan T. Bird, Kyle S. Bonifer, Benjamin C. Calfee, Tian Chen, Samantha R. Coy, P. Jackson Gainer, Eric R. Gann, Huston T. Heatherly, Jasper Lee, Xiaolong Liang, Jiang Liu, April C. Armes, Mohammad Moniruzzaman, J. Hunter Rice, Joshua M. A. Stough, Robert N. Tams, Evan P. Williams and Gary R. LeCleir

The Department of Microbiology, The University of Tennessee, Knoxville, TN 37996, USA; jbird9@tennessee.edu (J.T.B.); kbonifer@tennessee.edu (K.S.B.); bcalfee@tennessee.edu (B.C.C.); Tchen18@tennessee.edu (T.C.); srose16@utk.edu (S.R.C.); pgainer@utk.edu (P.J.G.); egann@tennessee.edu (E.R.G.); hheather@tennessee.edu (H.T.H.); jlee175@tennessee.edu (J.L.); xliang5@tennessee.edu (X.L.); jliu36@tennessee.edu (J.L.); amitch51@tennessee.edu (A.C.A.); mmoniruz@tennessee.edu (M.M.); jrice18@utk.edu (J.H.R.); jstough@tennessee.edu (J.M.A.S.); rtams@tennessee.edu (R.N.T.); ewilli99@tennessee.edu (E.P.W.); glecleir@tennessee.edu (G.R.L.)
* Correspondence: wilhelm@utk.edu

Academic Editors: Mathias Middelboe and Corina Brussard

Abstract: The discovery of infectious particles that challenge conventional thoughts concerning "what is a virus" has led to the evolution a new field of study in the past decade. Here, we review knowledge and information concerning "giant viruses", with a focus not only on some of the best studied systems, but also provide an effort to illuminate systems yet to be better resolved. We conclude by demonstrating that there is an abundance of new host–virus systems that fall into this "giant" category, demonstrating that this field of inquiry presents great opportunities for future research.

Keywords: giant viruses; nucleocytoplasmic large DNA viruses (NCLDVs); *Mimiviridae*

1. Introduction: Defining Giant Viruses

In their editorial introduction to the "Giant Viruses" special issue of Virology, Fischer and Condit [1] stated "It is commonly agreed upon that these are double-stranded DNA (dsDNA) viruses with genome sizes beyond 200 kb pairs, and particles that do not pass through a 0.2-μm pore-size filter". This definition illustrates the two striking features of giant viruses: their genome and particle size are both larger than has been historically considered for viruses. Beyond their breaking of previous paradigms, how giant viruses are defined remains contentious. Our goal in assembling this synthesis is to provide a "primer" for students of microbiology whom are interested in knowing more about these atypical viruses, and to establish a set of boundaries for their discussion. While not exhaustive, this overview addresses many of the main ideas that, for now, are current within a rapidly expanding field.

Some definitions of giant viruses focus only on genome size with lower limits ranging from undefined [2] to stringent (280 kb or 300 kb) cutoffs [3,4]. Other efforts have focused on the virus particle, suggesting they should be larger than 100 nm [2] or need be easily visible by light microscopy (>300 nm) [5]. One problem with establishing a particular definition for either genome or particle size is that, as additional large viruses are isolated, the rationale may no longer be justified (e.g., Aureococcus anophagefferens virus (AaV), a close phylogenetic relative of *Mimivirus*, is only ~140 nm in diameter) [6]. Indeed, a previous definition proposed a genome minimum of 280 kb due to a notable inflection point in a rank order plot of virus genome size [3]. However, in re-examining the largest 100 complete virus genomes in the National Center for Biotechnology Information's (NCBI) genome database, this gap is no longer present and a change in slope now occurs at ~400 kb (Figure 1A). This undersampling of giant viruses has resulted in a lack of sufficient information to describe their general characteristics [7,8]. While the vagaries of this definition will fade over time, herein we consider viruses 'giant' if their genome is larger than 200 kb. Moreover, this review will focus primarily on giants that infect single-celled eukaryotes.

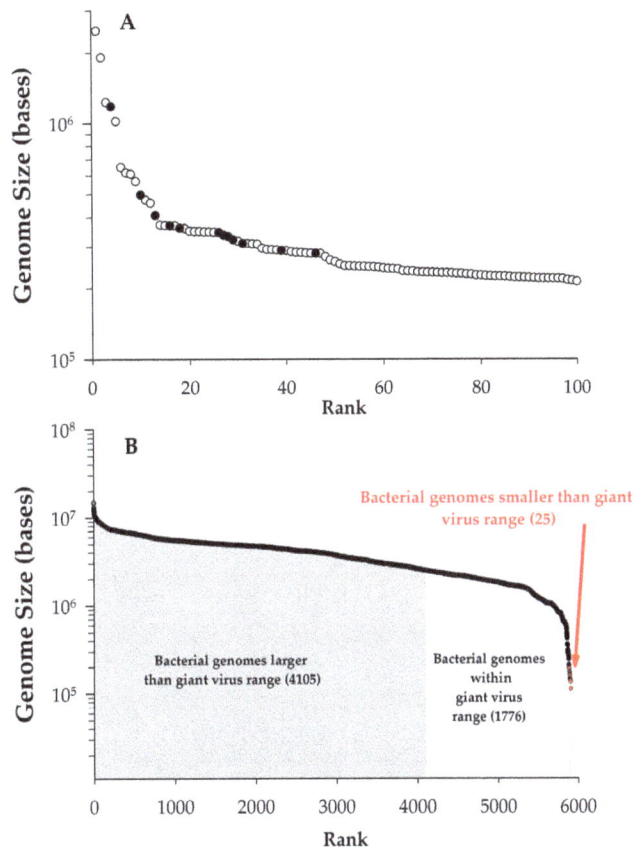

Figure 1. The scale of giant virus genomes. (**A**). Genome size vs. rank plot for the largest 100 complete viral genomes as of January 2016 from National Center for Biotechnology Information (NCBI). Data points noted (●) were previously used in discussion by Claverie et al. [3] to define giants viruses as having genomes > 280 kb, open circles (○) represent additional data; (**B**). Genome size vs. rank order of completed bacterial genomes in NCBI as of January 2016. Sizes are color-coded to match the ranges of giant virus genomes.

Using a cutoff of genomic content >200 kb pairs (kbp), ~2.2% (115/5356) of all of the completed virus genomes in NCBI fall within the realm of giants (Figure 1A). To date, all of these giants have genomes consisting of double-stranded DNA: the largest complete genome for other nucleic acid-type viruses is that of the double-stranded RNA (dsRNA) *Dendrolimus punctatus* cypovirus 22 (32.75 kbp) [9]. Perhaps more surprising is that this genome size range for giant viruses overlaps with more than ~one third of the complete prokaryotic genomes in NCBI (Figure 1B), as well as the genome sizes of several small eukaryotes [10]. This includes the smallest free-living archaeon (*Methanothermus fervidus*, 1.2 Mb) and the smallest free-living bacterium (*Candidatus* Actinomarina minuta, estimated ~700 kbp) [11]. While we will not consider them beyond the occasional passing mention in this article, it should be noted that several bacteriophages have genomes exceeding the 200 kbp genome size (see Table 1), and therefore qualify as giants. These phages infect both Gram-positive and -negative bacteria, including cyanobacteria [12,13].

Table 1. Comparison of host and viral genome size and GC content. All data was collected from the NCBI repository.

Giant Virus	Size Virus (Mb)	Virus GC (%)	ORFs*	Accession	Host	Size Host (Mb)	Host GC (%)	Host-Virus Genome Size	Host-Virus GC	Accession
Pandoravirus salinus	2.5	61.7	2541	NC_022098.1	Acanthamoeba castellanii	46.7	58.3	18.9	-3.4	AHJI00000000.1
Pandoravirus dulcis	1.9	63.7	1487	NC_021858.1	A. castellanii	46.7	58.3	24.5	-5.4	AHJI00000000.1
Acanthamoeba polyphaga mimivirus	1.2	28.0	1018	NC_014649.1	A. polyphaga	120.4	59.3	102.0	31.3	CDFK00000000.1
Acanthamoeba polyphaga moumouvirus	1.0	24.6	915	NC_020104.1	A. polyphaga	120.4	59.3	118.1	34.7	CDFK00000000.1
Mollivirus sibericum	0.7	60.1	523	NC_027867.1	A. castellanii	42.0	58.4	64.6	-1.7	AHJI00000000.1
Pithovirus sibericum	0.6	35.8	467	NC_023423.1	A. castellanii	42.0	58.4	68.9	22.6	AHJI00000000.1
Emiliania huxleyi virus 86	0.4	40.2	478	NC_007346.1	Emiliania huxleyi	167.7	65.7	409.0	25.5	AHAL00000000.1
Marseillevirus marseillevirus	0.4	44.7	457	NC_013756.1	A. polyphaga	120.4	59.3	325.5	14.6	CDFK00000000.1
Aureococcus anophagefferens virus	0.4	28.7	384	NC_024697.1	A. anophagefferens	56.7	69.5	153.1	40.8	NZ_ACJI00000000.1
Melbournevirus	0.4	44.7	403	NC_025412.1	A. castellanii	42.0	58.4	113.6	13.7	AHJI00000000.1
Paramecium bursaria Chlorella virus NY2A	0.4	40.7	411	NC_009898.1	Chlorella variabilis NC64A	46.2	67.1	124.8	26.4	ADIC00000000.1
Brazilian marseillevirus	0.4	43.3	491	NC_029692.1	A. castellanii	42.0	58.4	116.7	15.1	AHJI00000000.1
Lausannevirus	0.4	42.9	444	NC_015326.1	A. castellanii	42.0	58.4	120.1	15.5	AHJI00000000.1
Ectocarpus siliculosus virus 1	0.3	51.7	240	NC_002687.1	Ectocarpus siliculosus	195.8	53.5	575.9	1.8	CABU00000000.1
Paramecium bursaria Chlorella virus AR158	0.3	40.8	366	NC_009899.1	C. variabilis NC64A	46.2	67.1	135.8	26.3	ADIC00000000.1
Paramecium bursaria Chlorella virus 1	0.3	40.0	376	NC_000852.5	C. variabilis NC64A	46.2	67.1	139.9	27.1	ADIC00000000.1
Micromonas pusilla virus 12T	0.2	39.8	265	NC_020864.1	Micromonas pusilla	22.0	65.9	104.6	26.1	NZ_ACCP00000000.1
Sample Bacteriophage										
Bacillus phage G	0.5	29.9	694	NC_023719.1	Bacillus megaterium	5.3	38.1	10.7	8.2	NZ_CP009920.1
Prochlorococcus phage P-SSM2	0.3	35.5	335	NC_006883.2	Prochlorococcus marinus	1.8	36.4	7.0	0.9	NC_005042.1
Ralstonia phage RSL1	0.2	58.0	345	NC_010811.2	Ralstonia solanacearum	5.6	66.5	24.3	8.5	NC_003295.1
Sinorhizobium phage phiN3	0.2	49.1	408	NC_028945.1	Sinorhizobium meliloti	3.7	62.7	17.4	13.6	NC_003047.1
Pseudomonas phage EL	0.2	49.3	201	NC_007623.1	Pseudomonas aeruginosa	6.3	66.6	29.8	17.3	NC_002516.2

* ORF = Open reading frame

As with observed ranges in genomic size, there is also a wide range of GC content of these viruses relative to the small eukaryotes they infect (Table 1). On average, mobile elements such as phage and plasmids are more AT-rich than their host, but usually by only ~5% [14]. In contrast, Emiliania huxleyi virus (EhV) and AaV, which infect eukaryotic algae, have GC contents that are 24.3% and 38.7% lower than their hosts nuclear genomes, respectively [15,16], while the chloroviruses (freshwater viruses infecting *Chlorella*) have GC contents that are ~21% lower than their host's nuclear genome. While not a defining feature of all large viruses, this GC difference raises interesting questions concerning the scavenging of nucleotides during the infection cycle. Construction of new viruses is in some cases thought to depend on materials "scavenged" from the host cell, yet in the case of these viruses there would seem to be a discrepancy in terms of what would be available for scavenging. An interest side note to this is that mitochondrial and chloroplast genomes are often observed to have such relative low GC content genomes, similar to these viruses [14,15], implying a potential for scavenged materials from organelles to be important in the construction of new virus particles.

The current size range for giant virus particles varies from our operationally defined ~200 nm to >1500 nm in diameter [5], although as noted, phylogenetic relatives to these giants exist that are only ~140 nm. Indeed, the upper limit of this range is larger than for several bacteria and archaea (Figure 1B), redefining how we think about the relative size of prokaryotes and viruses. These large particle diameters may be needed to house their large genomes (see below), but it has been argued that there are other evolutionary pressures for these virus particles to retain large physical sizes [5]. For example, viruses infecting *Acanthamoeba* are internalized via phagocytosis, and it has been shown that this process works less efficiently on smaller (<600 nm) particles [16]. Additionally, based on standard contact kinetics, a larger particle size may increase the probability of contact between the virus and its host in the environment [17].

In addition to a tremendous variation in genome and particle size, giant viruses also have highly diverse morphologies that can be broadly categorized into two groups: ovoid and icosahedral (Figure 2). These morphological differences correspond to the structural proteins that make the virion capsids; icosahedrons are built by homologous β-barrel jelly-roll Major Capsid Proteins (MCPs) with minor capsid proteins acting as scaffolds connecting trisymmetrons and the outer capsid to the inner membrane surrounding the viral genome [18]. In contrast, ovoid viruses encode phylogenetically distant (*Mollivirus*) to unconvincing (*Pandoravirus* and *Pithovirus*) homologs to MCP [19–21]. It is unclear how the virion shape provides a selective advantage, since both types have been isolated in similar habitats.

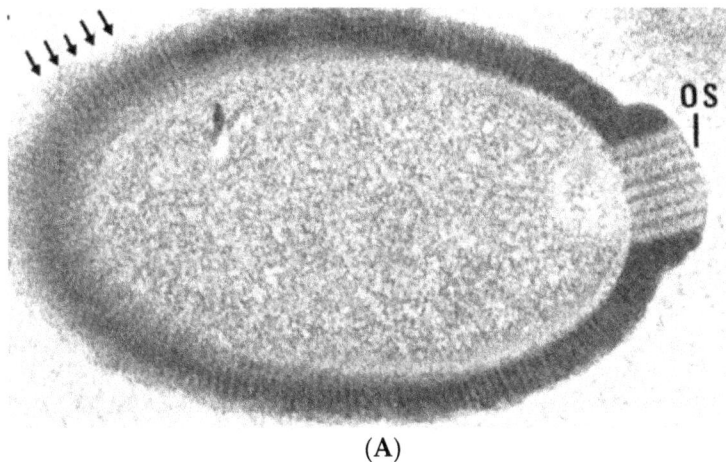

(A)

Figure 2. *Cont.*

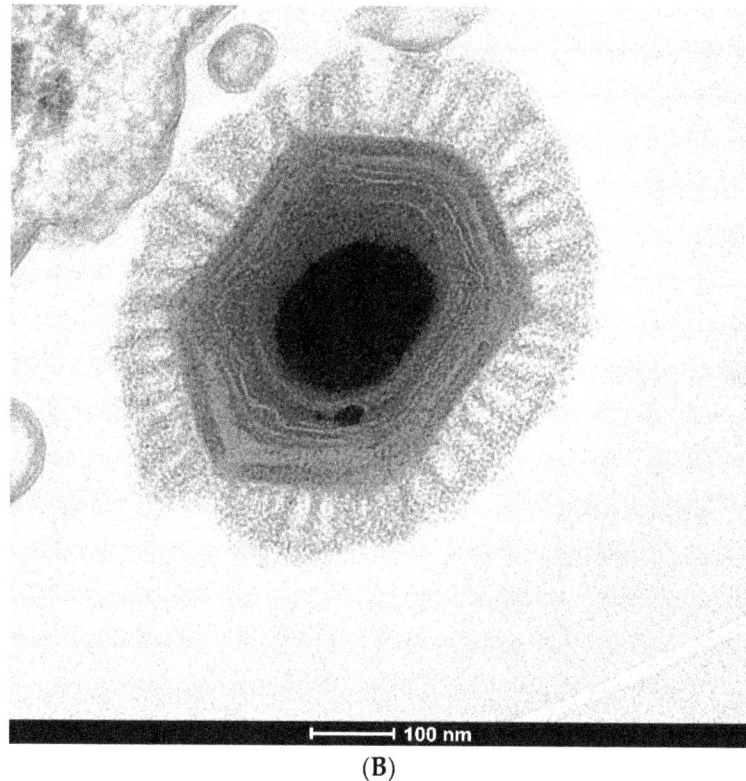

(B)

Figure 2. Transmission electron micrographs of giant virus particles. (**A**) *Pithovirus*, as seen in Michel et al. [22]. Originally identified as a KC5/2 parasite, the image shows the electron dense viral wall consisting of perpendicularly oriented fibers or microtubules (arrows), and a marked ostiole (os) located at the apical end of the cell. Reprinted with permission—original magnification at 85,000×; (**B**) *Megavirus chilensis*. Image courtesy of Professors Chantal Abergel and Jean-Michel Claverie.

Another mysterious aspect of these giant virus particles are the unique biochemical and morphological features. Virus–host interactions are thought to be facilitated in one of two ways: adsorption to the host cell wall, as is typical of algal host–virus systems [23], or phagocytosis by a protist host. These interactions often involve unique structures. For example, *Mimivirus* and its close relatives (*Megavirus, Marseillevirus, Lausannevirus*, and *Moumouviru*s) are characterized by proteinaceous fibers anchored to the icosahedron capsid [24,25] that are covered in glycolinkages [26–28]. It has been hypothesized that these fibers work in tandem with the large size of the viruses to facilitate phagocytosis, as they appear to have a similar composition to peptidoglycan and thus help mimic a bacterium (indeed, the name *Mimivirus* comes from "Mimicking Microbe" [29]). Additionally, the fibrous glycoproteins enable viral adsorption to diverse organisms ranging from bacteria and fungi to arthropods [30], implying a potential for both environmental dispersion and an incidental infection strategy in amoeba. *Phycodnaviridae* may also use unique structures to gain access to their host, though their mode of entry is typically by adsorption/injection, as opposed to phagocytosis. For example, the Chlorovirus capsid contains one spike located at a unique vertex of the icosahedral capsid that must be oriented towards the host cell surface to initiate infection [31]. Similarly, *Mimivirus* and its relatives utilize a five-pointed vertex called the 'stargate' structure that permits the first step in activating infection. Infection is initiated by fusion of the internal viral lipid membrane to the phagosomal membrane [24,32], which differs from algal viruses that fuse with the host cell membrane. This fusion event is observed in all giant viruses despite differences in structural features or infection strategies [20]. Whether these features are the result of homologous or convergent evolution remains to be determined, though given the breadth of physiological variation in the taxa, conservation of this mechanism is a compelling argument for monophyly.

2. Non-Structural Components of the Virion

An anomaly among the giant viruses are several viruses that include EhV, which have a lipid envelope outside of the capsid. These viruses include a *Phaeocystis globosa* virus (PgV-07T) [33] and several viruses infecting *Micromonas pussila* [34] which allows for a unique mode of infection and provides protection from environmental stressors [35]. This may be vital to the survival and transmission of these viruses, as they are ingested and transported across blooms by copepods [36]. An additional role of this lipid envelope and its associated proteins is an assumed association with recognition of the host and initiation of infection.

The nucleocytoplasmic large DNA viruses (NCLDVs) (described below) package a variety of proteins inside their capsids encoded by either viral or host genomes that are deployed immediately upon infection. For example, the seven proteomes of giant virus particles currently available (below) contain proteins predicted to combat oxidative stress, presumably because viral infections have been shown to generate Reactive Oxygen Species (ROS) that can inhibit viral replication [37]. Interestingly, *Pandoravirus salinus* carries one viral-encoded oxidoreductase, as well as three host-derived proteins predicted to combat ROS [19]. *P. bursaria chlorella* virus-1 (PBCV-1) and the more recently described *Megavirus chilensis* package homologous Cu-Zn superoxide dismutases [38,39]. In *M. chilensis*, this protein is remarkable for having the unique ability to fold and incorporate key metallic cofactors without the aid of chaperone proteins [39]. Additionally, *C. roenbergensis* virus (CroV) and *Mimivirus* both package novel sulfhydryl oxidases that may function in the formation of disulfide bonds [40,41]. These sulfhydryl oxidases as well as other protein disulfide isomerases present in CroV could aid in protein folding or viral entry similar to those found in retroviruses [42–44].

3. Gauging the Host Range of Giant Viruses in Nature

One concern regarding giant virus isolation using *Acanthamoeba* spp. is that while these are permissive, they may not be the natural hosts. Genomic analyses have been used in an attempt to determine natural hosts. In *Mimivirus*, most of the genes horizontally transferred from eukaryotes originated from amoeba, indicating amoebae are most likely the natural host of *Mimivirus*, but alternative hosts are still possible [45]. Indeed, their unique size and independence from host machinery may allow giant viruses to infect a wide range of hosts, which makes the search for the natural host more challenging. In addition to amoeba, NCLDVs have been reported to infect mice [46] and the symbiotic zooxanthelle of corals [47]. Giant viruses have also been isolated from human blood [48] and have been found in the human virome [49], indicating a potential role in human health (or at least a route of exposure). Indeed, the recent finding of *Acanthocystis turfacea* chlorella virus 1 (ATCV-1) from human oropharyngeal samples is intriguing: subsequent analyses have shown consistency between the presence of these viruses and reduced cognitive function in humans and mice [49].

4. Creating (an) Order from the Chaos: The Nucleocytoplasmic Large DNA Viruses

The NCLDV classification was created to define a monophyletic group of families that, when initially conceived, included *Asfarviridae*, *Phycodnaviridae*, *Poxviridae*, and *Iridoviridae* [50]. The rationale for this grouping was based on a conserved core of (1) nine genes hypothesized as representative of a common NCLDV ancestor and (2) a total of 22 more genes found in at least three of the four constituent viral families. The name is a reference to the replicative strategies of the included families as they replicate in both the nucleus and cytoplasm (phycodnaviruses, asfarvavirus and iridovirus) [51–53], or totally within the cytoplasm (poxviruses) [54]. A NCLDV ancestor has been hypothesized to have originated early in evolutionary history, possibly contemporaneously with early eukaryotic evolution, as suggested by the broad host range of NCLDV members [55]. However, the nature of this ancestral NCLDV remains unclear. Due to non-orthologous displacement of core genes [55] and potential reductive evolution [5] it is especially difficult to estimate the approximate genome size of any common ancestor and whether it would qualify as a giant virus when compared with modern giants. Indeed given theories on genome size variability, such as the genomic accordion [56,57], it is likely that predecessors of a variety of genome sizes existed. Moreover, it has been argued that mobile genetic elements encoded by virophages and transpovirions may have contributed significantly to the size of the NCLDV genome [58,59]. Therefore, the ancestral NCLDV may have been much smaller in genome size than modern representatives, and the mechanism by which it expanded its genome may have resulted in the wide range of genome sizes seen in current NCLDV members [60].

The NCLDV classification is not without its shortcomings. As new members are added to the group, the "nucleocytoplasmic" distinction of replicative strategies becomes less useful due to the increasing diversity of virion production. Many NCLDV families utilize a nucleocytoplasmic route for replication, including *Asfarviridae* [52], *Iridoviridae* and *Ascoviridae* [53], *Phycodnaviridae* [51], and *Pandoraviridae*. Other families, like *Poxviridae* [54], *Mimiviridae* [61], *Marseilleviridae* [62] and *Pithoviridae* [5], begin and complete their replication cycles exclusively in the cytoplasm, encoding the replication and transcription machinery necessary to produce virions without nuclear involvement. From a taxonomic perspective, the NCLDV group does not follow the naming conventions of (and is not recognized by) the International Committee on Taxonomy of Viruses (ICTV), as the classification lacks context within a larger hierarchy. To rectify this, Colson et al. proposed to reclassify NCLDVs within the new viral order *Megavirales* [63,64] based on the presence of conserved ancestral genes and a large icosahedral capsid composed of a homologous β-barrel jelly roll protein. This classification scheme, however, excludes the *Poxviridae* and *Ascoviridae*, [65] as well as *Pandoravirus*, *Pithovirus* and *Mollivirus*. In addition, the *Megavirales* classification required the capacity to assemble viral factories within the cytoplasm of host cells [62,66–69], a feature found in RNA viruses [70] but not seen in DNA viruses outside of the NCLDV group [64]. Currently (as of December 2016), the *Megavirales* is not considered a classification by the ICTV.

Most recently, the NCLDV genome size range has expanded to include genomes from 100 kb to 2.77 Mb encoding from 110 to 2556 genes [19,60]. The ten groups of NCLDV (*Phycodnaviridae*, *Poxviridae*, *Asfarviridae*, *Ascoviridae*, *Iridoviridae*, *Mimiviridae*, *Marseilleviridae*, *Pandoraviridae*, *Pithovirus*, and *Mollivirus*) infect a broad spectrum of hosts. In keeping with the NCLDV group's high degree of variability regarding particle size and host range, these viruses also display varying degrees of reliance on host metabolism and machinery, resulting in a limited number of highly conserved or "core" genes (e.g., see [71]). Yet despite these variances in NCLDV traits, common ground does exist. There are genes conserved amongst all available NCLDV genomes that are crucial for viral production or virion structure, such as the D5R packaging ATPase, D13L major capsid protein, and the B family DNA polymerase.

Comparative analyses of the genes conserved amongst different giant virus families has historically supported the monophyletic nature of the NCLDV group, and recent efforts to determine the clusters of orthologous groups (COGs) for giant viruses support their monophyly [72]. The conserved genes further provide potential markers that might be used in the discovery of novel NCLDVs and the determination of phylogenetic relationships between more closely related taxa [60]. For example, Moniruzzaman and colleagues [73] demonstrated an expanded level of diversity of the algal-specific members of the *Mimiviridae* by targeting the conserved MCP gene in this clade. However, this approach has its limits; the three recently discovered representatives of *Pandoravirus* lack the major capsid protein and the D5R helicase, as well as a number of other core NCLDV genes [19]. Indeed only 17 of the 49 inferred ancestral NCLDV genes were found in at least one of the *Pandoravirus* genomes, calling into question their inclusion in the giant virus clade despite their particle and genome size [74].

5. Viruses as a Possible Fourth Domain of Life

Initially viruses were defined by their intrinsic filterability away from cellular life forms [75–77], a definition subsequently refined to include their lack of ribosomes, a host-dependent metabolic strategy, and replication by means other than binary fission [78]. That the unique capabilities of NCLDVs still fit well within the latter definition, after fifty years of discovery and scientific scrutiny, highlights a fundamental difference between cellular organisms and viral particles. However, giant viruses do challenge these distinctions. Independent of their size, which invalidates the informal 0.2-µm filter cutoff, NCLDVs are remarkably cell-like in virion structure and gene content. In addition to their protein coat, membrane, and genome, *Mimivirus* and *Marseillevirus* particles contain messenger RNA molecules, making them the only viruses, to date, that contain both types of nucleic acid [79]. Moreover, several viruses encode genes involved in translational processes, such as varying numbers of aminoacyl-transfer RNA (tRNA) synthetases [69,80]. Indeed, we hypothesize these proteins may be useful in overcoming differences in GC content seen between some viruses and hosts (Table 1), but this has yet to be empirically demonstrated.

The discovery of translational machinery (including that mentioned above) encoded in select virus genomes allows for comparisons to traditionally "cellular" functions normally associated with the three domains of life. Sequence alignments comparing multiple genes involved in DNA replication and repair, transcription, and translation shared between cellular organisms and NCLDVs appear to show deeply branching relationships as ancient as the domain Eukarya. It was subsequently hypothesized that giant viruses evolved from a cellular common ancestor belonging to a currently extinct fourth domain of life, unique from Bacteria, Archaea, and Eukarya [63,81]. Seemingly in support of this hypothesis is the abundance of coding sequences (ORFans) in giant viral genomes with no known homologues in the other domains.

These ideas have proven somewhat controversial, as direct sequence comparison of genes conserved among cellular organisms with virus-encoded homologs is problematic. As selective pressures on similar genes within viruses and their hosts are likely different, accelerated sequence divergence in viruses may exaggerate their perceived distance from the derived gene [82]. Subsequent alignments accounting for compositional heterogeneity and homoplasy place giant virus genes with eukaryotes [83]. While it has been countered that giant virus genes do not evolve more quickly than their cellular counterparts, this has yet to be demonstrated outside of a single example within *Marseilleviridae* [5,84]. Indeed, an overabundance of viral open reading frames (ORFs) without known homologues is not a problem unique to giant viruses [85]. These observations and others have led to the alternative hypothesis: gene content within the different NCLDV families suggests that their genomes have been built up from smaller viruses over time, rather than by loss of unnecessary genes by an ancient cellular ancestor [72]. As some of the current NCLDVs replicate in phagotrophs like *Acanthamoeba* and *Cafeteria roenbergensis*, it was hypothesized that smaller viruses may incorporate genetic material from other organisms phagocytosed by the host.

6. Giant Viruses in the Environment

While surveys are not yet exhaustive, giant viruses appear to be found in all environments. Since the discovery of *Mimivirus* from a water cooling tower [86], giant viruses have been found in locations where amoebae normally thrive, including seawater, soil, aerosols, and man-made aquatic environments such as sewage, fountains and air conditioners [87], in addition to harsh, unexpected ecosystems such as permafrost [20]. Lastly, giant viruses or their DNA sequences have been observed in animals such as dinoflagellate-associated coral [47], arthropods, and humans [49,88].

A powerful tool in the identification of putative new viruses are environmental metagenomic studies (Table 2), though most have not focused specifically on giant viruses until recently [21]. Current research suggests giant viruses only comprise a small percentage of viruses (<1%) in most samples. However, virus densities can fluctuate based on contact with their host: for example, *Chlorella* viruses are much more abundant when their hosts, normally sequestered as endosymbionts of *Paramecium bursaria*, are made available as a consequence of predatory activity on the *Paramecium* [89]. Regardless, it is clear some families tend to be more common than others: in marine metagenomics samples *Phycodnaviridae*-related sequences were found to be highest in abundance, followed by *Mimiviridae* [90,91]. It is also clear that these viruses are persistent: the discovery of 30,000-year-old *Mollivirus* particles in permafrost suggests that giant viruses can survive, under the correct conditions, for long periods of time [21]. When combined with other tools such as flow cytometry sorting of either individual particles [92] or infected hosts [93], these new approaches will begin to shed significant light on the natural diversity of these populations.

Table 2. Comparison of giant virus reads to total viral reads in shotgun metagenomic studies from different environments.

Environment	Location	Abundance	Total Reads	Most Common Virus Families Present	Source
Marine	Indian Ocean	0.3%–1.4%	N/A	*Mimiviridae, Phycodnaviridae*	[91]
Antarctic soil	Antarctica	2.82%–7.71%	123/1595-177/6264	*Mimiviridae, Phycodnaviridae*	[94]
Coral	USA	1.2%	744/60485	*Mimiviridae, Phycodnaviridae*	[95]
Human (respiratory system)	Sweden	0.00002%	2/111931	*Mimiviridae*	[96]

To date, much of the focus on giant viruses has been on their genomics rather than their influence on the environments in which they persist. Several large, dsDNA viruses including EhV [97], PgV [98], AaV [6,97] and *Heterosigma akashiwo* virus (HaV) [99] are associated with algal blooms, although only a few have been directly shown to infect and lyse the phytoplankton involved with the bloom in situ [97,100]. Algal blooms occur on large geographical scales and result in significant influxes of atmospheric carbon into the world's oceans. Viruses, particularly bacteriophages, are known drivers of dissolved organic matter (DOM) release back into the environment via a process known as the "viral shunt" [101,102]. With the large biomass of algae associated with these blooms, virus-mediated collapse by giant viruses may also be an important driver of dissolved and particulate organic matter release. Giant viruses that infect algae may be likened to bacteriophages in terms of participating in the viral shunt, and the release of nutrients back into the environment may be an important part of the ecological cycle in aquatic systems [102].

A recent estimate suggested that giant viruses available in culture were infectious to at least 22 different algal species [103]. Globally, it has been proposed that there are more than 350,000 algal species [104]. Given the possibility that all algae may be infected with one or more viruses [105,106], the possibility of a collection of unknown giants remains very real, and indeed molecular data point to at least a broad diversity within the known groups [107,108]. Building on the above, it is clear from a survey of the literature that researchers identified candidate protist-giant virus systems well before *Mimivirus* was documented (Table 3). In the late 1960s and early 1970s, the expanded availability of transmission electron microscopes to researchers resulted in a series of observations concerning the presence of large virus-like particles inside algal cells [107,109]. In many cases, these virus–host systems have been largely ignored by the scientific community, creating a broad spectrum of opportunities for researchers to begin to cultivate these plankton in an effort to isolate and characterize new giant viruses. Given the expansive putative host-range that has been observed, it is likely that many of these viruses could fill in knowledge gaps concerning the diversity and potential function of these particles. Indeed, one example of how new hosts can be used to discover new viruses are the *Faustovirus*, recently discovered using *Vermamoeba* (a protist found in both humans and natural systems) as a screen [110]: unique to these viruses is a collection of genes three times larger than the other members of the *Asfarviridae* family.

And while it is obvious that there is a dearth of knowledge concerning giant viruses that infect algae in the environment, there is an even larger knowledge gap regarding giant viruses infecting heterotrophic eukaryotes. The most studied of these viruses is CroV, which infects the heterotrophic grazer *Cafeteria roenbergensis* [111]. Given this organism is a grazer of primary producers it is possible that infection of this organism by CroV could have effects on lower trophic level organisms. It has been shown that grazing can be an important driver of algal bloom decline [112], so it stands to reason that the effects giant viruses have on mixo/heterotrophic-plankton are critical to understanding bloom dynamics. Almost no information, at this time, is available to discuss the impacts of these infections, but they will most likely result in interesting discoveries and further our understanding of how giant viruses alter the microbial food web.

Table 3. A chronological list of organisms shown in the literature to contain viruses consistent with the giant virus size class.

Year	Organism	Particle Size	References
1970	*Aphelidium* sp. (fungal parasite of algae)	190–210 nm	[113]
1972	*Oedogonium* spp. "L" (Chlorophyceae)	240 nm	[114]
	Chorda tomentosa (Phaeophyceae)	170 nm	[115]
1973	*Ectocarpus* sp.; *Ectocarpus fasciculatus* (Phaeophyceae)	150 nm, 170 nm	[116,117]
	Aulacomonas submarina (Chlorophyceae)	200–230 nm	[118]
1974	*Pylaiella littoralis* (Phaeophyceae)	130–170 nm	[119]
	Pyramimonas orientalis (Prasinophyceae)	200 nm	[120]
1975 [†]	*Chara corallina* (Charophyceae)	18 nm × 532 nm	[121]
1978	*Sorocarpus uvaeformis* (Phaeophyceae)	170 nm	[122]
1979	*Gymnodinium uberrimum* (Dinophyceae)	385 nm	[123]
	Mallomonas sp. (Synurophyceae)	175 nm	[123]
1980	*Uronema gigas* (Chlorophyceae)	390 nm	[124]
1984	*Paraphysomonas corynephora* (Chrysophyceae)	150–180 nm, 270–300 nm	[125]
1993	Various Phaeodarian food vacuoles	300–750 nm	[126]

[†] Although in length this virus qualifies as a giant, its rod shaped morphology is more consistent with Tobacco mosaic virus than any member of the *Mimiviridae*.

7. Intimate Interactions with the Host: Eco-Evolutionary Consequences

Only recently have we come to appreciate the possibility of gene transfer between giant viruses and their hosts. A large proportion of giant virus genes comes from diverse sources, including from their eukaryotic hosts [127]. In EhV, seven genes involved in sphingolipid biosynthesis pathway were putatively transferred from the host algae [128]. Upon infection, the host sphingolipid biosynthesis pathway is downregulated concomitant with the upregulation of the corresponding viral genes, leading to increased production of viral glycosphingolipids (vGSLs) [129]. EhV particles are covered by vGSLs, and this unique lipid molecule ultimately induces programmed cell death (PCD) in infected hosts [130].

A genome wide phylogenetic study of AaV identified a number of genes having their highest phylogenetic affinity to host (*Aureococcus*) homologs, [71]. This agrees with observations made by earlier studies on several other giant viruses [127,131]. While gene acquisition may be one of the evolutionary strategies of giant viruses, how these genes confer ecological advantages remains largely unknown. As the vast majority of viruses harbor streamlined genomes with few genes, the enormous genetic resource of giant viruses poses a paradox in terms of energetic cost of replication. Closer inspection of a number of sequenced eukaryotic genomes revealed a large number of genes originated from giant viruses [132,133]. In a recent study, large genomic islands, putatively derived from both giant viruses and a virophage, were found in *Bigelowiella natans*, a Cryptomonad algae [134]. In another study, "core" genes from giant viruses were detected in eight protists and a metazoan (*Hydra magnipapillata*) genome [132]. Remarkably, a 400-kb region in the *H. magnipapillata* was putatively identified to be of viral origin [132]. Major capsid gene phylogeny indicated the genes were likely from a *Mimiviridae* family member. Giant virus particles and marker genes have also recently also been observed associated with zooxanthellae from the genus *Symbiodinium*, a dinoflagellate typically found closely associated with corals [47]. Giant virus-like genes were also found in several other protists [133,135] and some plant genomes, namely *Physcomitrella patens* and *Selaginella moellendorffii* [136]. The role of giant virus-derived genes in host remain an open question.

Host–virus interactions result in an evolutionary arms race—leading to the emergence of new diversity in the host and virus population [137]. Hosts of giant viruses have evolved a variety of defense mechanisms against giant viruses. An elegant example is the 'Cheshire cat' strategy adopted by *Emiliania huxleyi* [138]. The diploid calcified cells of *E. huxleyi* are susceptible to EhV infection, while the haploid stage is 'invisible' to infection. It has been suggested that during the decline of the Brown tide blooms, a virus-resistant population of the *Aureococcus* persists, maintaining a relatively high abundance of *Aureococcus* even after the demise of the bloom [73,97].

8. Virophage

Another interesting characteristic of some giant viruses is their susceptibility to infection by other bioactive particles, termed "virophage". The first virophage to be isolated was named Sputnik [59], which replicates within the viral factory used by *Mamavirus* within *Acanthamoeba castellanii*. Because of this, Sputnik only replicates within *A. castellanii* co-infected with Mamavirus. Infection by the virophage causes abnormal capsid structure of *Mamavirus*, increasing capsid size and causing abnormal fiber localization on its surface, suggesting a parasitic relationship between the two [59]. Co-incubation of Sputnik and *Mamavirus* decreased infective *Mamavirus* particle titers by approximately 70% and increased the survival rate of the *A. castellanii* [59]. Similar virophages have been found infecting other giant viruses as well [139–142]. The discovery of "viruses that infect viruses" has strengthened the argument that viruses are living entities [143]. Some classes of *Mimivirus* appear to have developed a CRISPR-CAS-like system suggested to combat these virophages, called the *Mimivirus* virophage resistant element (MIMIVIRE) [144]. Interestingly, a number of genes homologous to those in the MIMIVIRE system are present in other giant viruses, suggesting that the MIMIVIRE-like defense systems might not be exclusive to *Mimivirus* [134,145]. Other interpretations, however, are questioning these conclusions [146]. Much has yet to be learned in these systems, but virophages may act like both 'provirophage' and 'provirus', depending on the genomic context at multiple levels.

9. Conclusions

The discovery of *Mimivirus* has driven both the nascence and evolution of a new area of scientific inquiry. Giant viruses are now the topics of evolutionary, ecological and biotechnological inquiries. Moreover, broad-scale efforts to identify new virus–host systems, ranging from classic culture-based approaches to newer bioinformatics efforts to link viruses and their hosts [147] will soon provide a larger data base of information concerning the key features of these novel virus particles. Indeed, a survey of older literature (Table 3) clearly demonstrates that there are many virus–host systems that have been observed but are yet to be isolated and characterized. Moving forward, there is little doubt that the study of giant viruses will shed new light not only on virus–host relationships, but also on key evolutionary processes including the natural occurrence rates of transduction and horizontal gene transfer.

Acknowledgments: The authors wish to thank the Department of Microbiology at the University of Tennessee: this review was assembled as part of a graduate course (Virology 604) instructed by S.W.W. and G.R.L. Funds to support publication were received from the Kenneth & Blaire Mossman Endowment to the University of Tennessee (S.W.W.).

Author Contributions: S.W.W. and G.R.L. conceived this exercise for the Department of Microbiology graduate journal club in virology. All authors participated in research and crafting of the document.

References

1.   Fischer, M.G.; Condit, R.C. Editorial introduction to "giant viruses" special issue of virology. *Virology* **2014**, *466–467*, 1–2. [CrossRef] [PubMed]

2.   Durzyriska, J. Giant viruses: Enfants terribles in the microbal world. *Future Virol.* **2015**, *10*, 795–806. [CrossRef]

3.   Claverie, J.M.; Ogata, H.; Audic, S.; Abergel, C.; Suhre, K.; Fournier, P.E. Mimivirus and the emerging concept of "giant" virus. *Virus Res.* **2006**, *117*, 133–144. [CrossRef] [PubMed]

4.   Yamada, T. Giant viruses in the environment: Their origins and evolution. *Curr. Opin. Virol.* **2011**, *1*, 58–62. [CrossRef] [PubMed]

5.   Abergel, C.; Legendre, M.; Claverie, J.M. The rapidly expanding universe of giant viruses: Mimivirus, Pandoravirus, Pithovirus and Mollivirus. *FEMS Microbiol. Rev.* **2015**, *39*, 779–796. [CrossRef] [PubMed]

6.   Gastrich, M.D.; Leigh-Bell, J.A.; Gobler, C.; Anderson, O.R.; Wilhelm, S.W. Viruses as potential regulators of regional brown tide blooms caused by the alga, *Aureococcus anophagefferens*: A comparison of bloom years 1999–2000 and 2002. *Estuaries* **2004**, *27*, 112–119. [CrossRef]

7.   Serwer, P.; Hayes, S.J.; Thomas, J.A.; Hardies, S.C. Propagating the missing bacteriophages: A large bacteriophage in a new class. *Virol. J.* **2007**, *4*, 21. [CrossRef] [PubMed]

8.   Martínez, J.M.; Swan, B.K.; Wilson, W.H. Marine viruses, a genetic reservoir revealed by targeted viromics. *ISME J.* **2014**, *8*, 1079–1088. [CrossRef] [PubMed]

9.   Zhou, Y.; Qin, T.; Xiao, Y.; Qin, F.; Lei, C.; Sun, X. Genomic and biological characterization of a new cypovirus isolated from *Dendrolimus punctatus*. *PLoS ONE* **2014**, *9*, e113201. [CrossRef] [PubMed]

10. Corradi, N.; Pombert, J.-F.; Farinelli, L.; Didier, E.S.; Keeling, P.J. The complete sequence of the smallest known nuclear genome from the microsporidian *Encephalitozoon intestinalis*. *Nat. Commun.* **2010**, *1*, 77. [CrossRef] [PubMed]

11. Martínez-Cano, D.J.; Reyes-Prieto, M.; Martínez-Romero, E.; Partida-Martínez, L.P.; Latorre, A.; Moya, A.; Delaye, L. Evolution of small prokaryotic genomes. *Front. Microbiol.* **2015**, *5*, 742. [CrossRef] [PubMed]

12. Mesyanzhinov, V.V.; Robben, J.; Grymonprez, B.; Kostyuchenko, V.A.; Bourkaltseva, M.V.; Sykilinda, N.N.; Krylov, V.N.; Volckaert, G. The genome of bacteriophage φKZ of *Pseudomonas aeruginosa*. *J. Mol. Biol.* **2002**, *317*, 1–19. [CrossRef] [PubMed]

13. Sullivan, M.B.; Coleman, M.L.; Weigele, P.; Rohwer, F.; Chisholm, S.W. Three *Prochlorococcus* cyanophage genomes: Signature features and ecological interpretations. *PLoS Biol.* **2005**, *3*, e144. [CrossRef] [PubMed]

14. Ong, H.C.; Wilhelm, S.W.; Gobler, C.J.; Bullerjahn, G.; Jacobs, M.A.; McKay, J.; Sims, E.H.; Gillett, W.G.; Zhou, Y.; Haugen, E.; et al. Analyses of the complete chloroplast genome of two members of the pelagophyceae: *Aureococcus anophagefferens* CCMP1984 and *Aureoumbra lagunesis* CCMP1507. *J. Phycol.* **2010**, *46*, 602–615. [CrossRef]

15. Orsini, M.; Costelli, C.; Malavasi, V.; Cusano, R.; Alessandro, C.; Angius, A.; Cao, G. Complete sequence and characterization of mitochondrial and chloroplast genome of *Chlorella variabilis* NC64A. *Mitochondrial DNA A* **2015**, *27*, 3128–3130.

16. Korn, E.D.; Weisman, R.A. Phagocytosis of latex beads by *Acanthomoeba*. *J. Cell Biol.* **1967**, *34*, 219–227. [CrossRef]

17. Murray, A.G.; Jackson, G.A. Viral dynamics: A model of the effects of size, shape, motion and abundance of single-celled planktonic organisms and other particles. *Mar. Ecol. Prog. Ser.* **1992**, *89*, 103–116. [CrossRef]

18. Klose, T.; Rossmann, M.G. Structure of large dsDNA viruses. *Biol. Chem.* **2014**, *395*, 711–719. [CrossRef] [PubMed]

19. Philippe, N.; Legendre, M.; Doutre, G.; Coute, Y.; Poirot, O.; Lescot, M.; Arslan, D.; Seltzer, V.; Bertaux, L.; Bruley, C.; et al. Pandoraviruses: Amoeba viruses with genomes up to 2.5 mb reaching that of parasitic eukaryotes. *Science* **2013**, *341*, 281–286. [CrossRef] [PubMed]

20. Legendre, M.; Bartoli, J.; Shmakova, L.; Jeudy, S.; Labadie, K.; Adrait, A.; Lescot, M.; Poirot, O.; Bertaux, L.; Bruley, C.; et al. Thirty-thousand-year-old distant relative of giant icosahedral DNA viruses with a Pandoravirus morphology. *Proc. Natl. Acad. Sci. USA* **2014**, *111*, 4274–4279. [CrossRef] [PubMed]

21. Legendre, M.; Lartigue, A.; Bertaux, L.; Jeudy, S.; Bartoli, J.; Lescot, M.; Alempic, J.M.; Ramus, C.; Bruley, C.; Labadie, K.; et al. In-depth study of *Mollivirus sibericum*, a new 30,000-y-old giant virus infecting *Acanthamoeba*. *Proc. Natl. Acad. Sci. USA* **2015**, *112*, E5327–E5335. [CrossRef] [PubMed]

22. Pearson, H. 'Virophage' suggests viruses are alive. *Nature* **2008**, *454*, 677. [CrossRef] [PubMed]

23. Wilson, W.H.; Van Etten, J.L.; Allen, M.J. The phycodnaviridae: The story of how tiny giants rule the world. In *Lesser Known Large dsDNA Viruses*; VanEtten, J.L., Ed.; Springer Science & Business Media: Berlin, Germany, 2009; Volume 328, pp. 1–42.

24. Xiao, C.; Rossmann, M.G. Structures of giant icosahedral eukaryotic dsDNA viruses. *Curr. Opin. Virol.* **2011**, *1*, 101–109. [CrossRef] [PubMed]

25. Xiao, C.A.; Chipman, P.R.; Battisti, A.J.; Bowman, V.D.; Renesto, P.; Raoult, D.; Rossmann, M.G. Cryo-electron microscopy of the giant mimivirus. *J. Mol. Biol.* **2005**, *353*, 493–496. [CrossRef] [PubMed]

26. Chothi, M.P.; Duncan, G.A.; Armirotti, A.; Abergel, C.; Gurnon, J.R.; Van Etten, J.L.; Bernardi, C.; Damonte, G.; Tonetti, M. Identification of an l-rhamnose synthetic pathway in two nucleocytoplasmic large DNA viruses. *J. Virol.* **2010**, *84*, 8829–8838. [CrossRef] [PubMed]

27. Tonetti, M.; Chothi, M.P.; Abergel, C.; Seltzer, V.; Gurnon, J.; Van Etten, J.L. Glycosylation in nucleo-cytoplasmic large DNA viruses (NCLDV). *FEBS J.* **2011**, *278*, 420.

28. Piacente, F.; Gaglianone, M.; Laugieri, M.E.; Tonetti, M.G. The autonomous glycosylation of large DNA viruses. *Int. J. Mol. Sci.* **2015**, *16*, 29315–29328. [CrossRef] [PubMed]

29. Raoult, D. Viruses reconsidered. *Scientist* **2014**, *28*, 41.

30. Rodrigues, R.A.L.; Silva, L.K.D.; Dornas, F.P.; de Oliveira, D.B.; Magalhaes, T.F.F.; Santos, D.A.; Costa, A.O.; Farias, L.D.; Magalhaes, P.P.; Bonjardim, C.A.; et al. Mimivirus fibrils are important for viral attachment to the microbial world by a diverse glycoside interaction repertoire. *J. Virol.* **2015**, *89*, 11812–11819. [CrossRef] [PubMed]

31. Zhang, X.Z.; Xiang, Y.; Dunigan, D.D.; Klose, T.; Chipman, P.R.; Van Etten, J.L.; Rossmann, M.G. Three-dimensional structure and function of the *Paramecium bursaria* chlorella virus capsid. *Proc. Natl. Acad. Sci. USA* **2011**, *108*, 14837–14842. [CrossRef] [PubMed]

32. Suzan-Monti, M.; La Scola, B.; Raoult, D. Genomic and evolutionary aspects of mimivirus. *Virus Res.* **2006**, *117*, 145–155. [CrossRef] [PubMed]

33. Maat, D.S.; Bale, N.J.; Hopmans, E.C.; Baudoux, A.C.; Damste, J.S.S.; Schouten, S.; Brussaard, C.P.D. Acquisition of intact polar lipids from the prymnesiophyte *Phaeocystis globosa* by its lytic virus PgV-07t. *Biogeosciences* **2014**, *11*, 185–194. [CrossRef]

34. Martinez-Martinez, J.; Boere, A.; Gilg, I.C.; van Lent, J.W.M.; Witte, H.J.; van Bleijswijk, J.D.L.; Brussaard, C.P.D. New lipid envelop-containing dsDNA virus isolates infecting *Micromonas pusilla* reveal a separate phylogenetic group. *Aquat. Microb. Ecol.* **2015**, *74*, 17–28. [CrossRef]

35. Mackinder, L.C.M.; Worthy, C.A.; Biggi, G.; Hall, M.; Ryan, K.P.; Varsani, A.; Harper, G.M.; Wilson, W.H.; Brownlee, C.; Schroeder, D.C. A unicellular algal virus, *Emiliania huxleyi* virus 86, exploits an animal-like infection strategy. *J. Gen. Virol.* **2009**, *90*, 2306–2316. [CrossRef] [PubMed]

36. Frada, M.J.; Schatz, D.; Farstey, V.; Ossolinski, J.E.; Sabanay, H.; Ben-Dor, S.; Koren, I.; Vardi, A. Zooplankton may serve as transmission vectors for viruses infecting algal blooms in the ocean. *Curr. Biol.* **2014**, *24*, 2592–2597. [CrossRef] [PubMed]

37. Schwarz, K.B. Oxidative stress during viral infection: A review. *Free Radic. Biol. Med.* **1996**, *21*, 641–649. [CrossRef]

38. Kang, M.; Duncan, G.A.; Kuszynski, C.; Oyler, G.; Zheng, J.Y.; Becker, D.F.; Van Etten, J.L. Chlorovirus PBCV-1 encodes an active copper-zinc superoxide dismutase. *J. Virol.* **2014**, *88*, 12541–12550. [CrossRef] [PubMed]

39. Lartigue, A.; Burlat, B.; Coutard, B.; Chaspoul, F.; Claverie, J.M.; Abergel, C. The *Megavirus chilensis* Cu,Zn-superoxide dismutase: The first viral structure of a typical cellular copper chaperone-independent hyperstable dimeric enzyme. *J. Virol.* **2015**, *89*, 824–832. [CrossRef] [PubMed]

40. Fischer, M.G.; Kelly, I.; Foster, L.J.; Suttle, C.A. The virion of *Cafeteria roenbergensis* virus (CroV) contains a complex suite of proteins for transcription and DNA repair. *Virology* **2014**, *466*, 82–94. [CrossRef] [PubMed]

41. Hakim, M.; Ezerina, D.; Alon, A.; Vonshak, O.; Fass, D. Exploring ORFan domains in giant viruses: Structure of mimivirus sulfhydryl oxidase R596. *PLoS ONE* **2012**, *7*, e50649. [CrossRef] [PubMed]

42. Apperizeller-Herzog, C.; Ellgaard, L. The human PDI family: Versatility packed into a single fold. *Biochim. Biophys. Acta* **2008**, *1783*, 535–548. [CrossRef] [PubMed]

43. Ryser, H.J.P.; Levy, E.M.; Mandel, R.; Disciullo, G.J. Inhibition of human-immunodeficiency-virus infection by agents that interfere with thiol-disulfide interchange upon virus-receptor interaction. *Proc. Natl. Acad. Sci. USA* **1994**, *91*, 4559–4563. [CrossRef] [PubMed]

44. Schelhaas, M.; Malmstrom, J.; Pelkmans, L.; Haugstetter, J.; Ellgaard, L.; Grunewald, K.; Helenius, A. Simian virus 40 depends on ER protein folding and quality control factors for entry into host cells. *Cell* **2007**, *131*, 516–529. [CrossRef] [PubMed]

45. Moreira, D.; Brochier-Armanet, C. Giant viruses, giant chimeras: The multiple evolutionary histories of mimivirus genes. *BMC Evol. Biol.* **2008**, *8*, 12. [CrossRef] [PubMed]

46. Khan, M.; La Scola, B.; Lepidi, H.; Raoult, D. Pneumonia in mice inoculated experimentally with *Acanthamoeba polyphaga* mimivirus. *Microb. Pathog.* **2007**, *42*, 56–61. [CrossRef] [PubMed]

47. Correa, A.M.S.; Ainsworth, T.D.; Rosales, S.M.; Thurber, A.R.; Butler, C.R.; Vega Thurber, R.L. Viral outbreak in corals associated with an in situ bleaching event: Atypical herpes-like viruses and a new megavirus infecting symbiodinium. *Front. Microbiol.* **2016**, *7*, 127. [CrossRef] [PubMed]

48. Popgeorgiev, N.; Boyer, M.; Fancello, L.; Monteil, S.; Robert, C.; Rivet, R.; Nappez, C.; Azza, S.; Chiaroni, J.; Raoult, D.; et al. Marseillevirus-like virus recovered from blood donated by asymptomatic humans. *J. Infect. Dis.* **2013**, *208*, 1042–1050. [CrossRef] [PubMed]

49. Yolken, R.H.; Jones-Brando, L.; Dunigan, D.D.; Kannan, G.; Dickerson, F.; Severance, E.; Sabunciyan, S.; Talbot, C.C.; Prandovszky, E.; Gurnon, J.R.; et al. Chlorovirus ATCV-1 is part of the human oropharyngeal virome and is associated with changes in cognitive functions in humans and mice. *Proc. Natl. Acad. Sci. USA* **2014**, *111*, 16106–16111. [CrossRef] [PubMed]

50. Iyer, L.M.; Aravind, L.; Koonin, E.V. Common origin of four diverse families of large eukaryotic DNA viruses. *J. Virol.* **2001**, *75*, 11720–11734. [CrossRef] [PubMed]

51. Van Etten, J.L.; Meints, R.H. Giant viruses infecting algae. *Annu. Rev. Microbiol.* **1999**, *53*, 447–494. [CrossRef] [PubMed]

52. Garcia-Beato, R.; Salas, M.L.; Vinuela, E.; Salas, J. Role of the host cell nucleus in the replication of african swine fever virus DNA. *Virology* **1992**, *188*, 637–649. [CrossRef]

53. Goorha, R. Frog virus-3 DNA-replication occurs in 2 stages. *J. Virol.* **1982**, *43*, 519–528. [PubMed]

54. Moss, B. *Poxviridae: The Viruses and Their Replication*; Lippincott-Raven Publishers: Philadelphia, PA, USA, 1996; pp. 1163–1197.

55. Koonin, E.V.; Yutin, N. Origin and evolution of eukaryotic large nucleo-cytoplasmic DNA viruses. *Intervirology* **2010**, *53*, 284–292. [CrossRef] [PubMed]

56. Elde, N.C.; Child, S.J.; Eickbush, M.T.; Kitzman, J.O.; Rogers, K.S.; Shendure, J.; Geballe, A.P.; Malik, H.S. Poxviruses deploy genomic accordions to adapt rapidly against host antiviral defenses. *Cell* **2012**, *150*, 831–841. [CrossRef] [PubMed]

57. Filee, J. Route of NCLDV evolution: The genoic accordion. *Curr. Opin. Virol.* **2013**, *3*, 595–599. [CrossRef] [PubMed]

58. Desnues, C.; La Scola, B.; Yutin, N.; Fournous, G.; Robert, C.; Azza, S.; Jardot, P.; Monteil, S.; Campocasso, A.; Koonin, E.V.; et al. Provirophages and transpovirons as the diverse mobilome of giant viruses. *Proc. Natl. Acad. Sci. USA* **2012**, *109*, 18078–18083. [CrossRef] [PubMed]

59. La Scola, B.; Desnues, C.; Pagnier, I.; Robert, C.; Barrassi, L.; Fournous, G.; Merchat, M.; Suzan-Monti, M.; Forterre, P.; Koonin, E.; et al. The virophage as a unique parasite of the giant mimivirus. *Nature* **2008**, *455*, 100–104. [CrossRef] [PubMed]

60. Yutin, N.; Wolf, Y.I.; Raoult, D.; Koonin, E.V. Eukaryotic large nucleo-cytoplasmic DNA viruses: Clusters of orthologous genes and reconstruction of viral genome evolution. *Virol. J.* **2009**, *6*, 13. [CrossRef] [PubMed]

61. Claverie, J.M.; Abergel, C. Mimivirus and its virophage. In *Annual Review of Genetics*; Annual Reviews: Palo Alto, CA, USA, 2009; Volume 43, pp. 49–66.

62. Aherfi, S.; La Scola, B.; Pagnier, I.; Raoult, D.; Colson, P. The expanding family *Marseilleviridae*. *Virology* **2014**, *466–467*, 27–37. [CrossRef] [PubMed]

63. Colson, P.; de Lamballerie, X.; Fournous, G.; Raoult, D. Reclassification of giant viruses composing a fourth domain of life in the new order megavirales. *Intervirology* **2012**, *55*, 321–332. [CrossRef] [PubMed]

64. Colson, P.; De Lamballerie, X.; Yutin, N.; Asgari, S.; Bigot, Y.; Bideshi, D.K.; Cheng, X.W.; Federici, B.A.; Van Etten, J.L.; Koonin, E.V.; et al. "Megavirales", a proposed new order for eukaryotic nucleocytoplasmic large DNA viruses. *Arch. Virol.* **2013**, *158*, 2517–2521. [CrossRef] [PubMed]

65. Krupovic, M.; Bamford, D.H. Virus evolution: How far does the double beta-barrel viral lineage extend? *Nat. Rev. Microbiol.* **2008**, *6*, 941–948. [CrossRef] [PubMed]

66. Condit, R.C. Vaccinia, Inc.—Probing the functional substructure of poxviral replication factories. *Cell Host Microbe* **2007**, *2*, 205–207. [CrossRef] [PubMed]

67. Netherton, C.; Moffat, K.; Brooks, E.; Wileman, T. A guide to viral inclusions, membrane rearrangements, factories, and viroplasm produced during virus replication. *Adv. Virus Res.* **2007**, *70*, 101–182. [PubMed]

68. Mutsafi, Y.; Zauberman, N.; Sabanay, I.; Minsky, A. Vaccinia-like cytoplasmic replication of the giant mimivirus. *Proc. Natl. Acad. Sci. USA* **2010**, *107*, 5978–5982. [CrossRef] [PubMed]

69. Boyer, M.; Yutin, N.; Pagnier, I.; Barrassi, L.; Fournous, G.; Espinosa, L.; Robert, C.; Azza, S.; Sun, S.Y.; Rossmann, M.G.; et al. Giant marseillevirus highlights the role of amoebae as a melting pot in emergence of chimeric microorganisms. *Proc. Natl. Acad. Sci. USA* **2009**, *106*, 21848–21853. [CrossRef] [PubMed]

70. Netherton, C.L.; Wileman, T. Virus factories, double membrane vesicles and viroplasm generated in animal cells. *Curr. Opin. Virol.* **2011**, *1*, 381–387. [CrossRef] [PubMed]

71. Moniruzzaman, M.; LeCleir, G.R.; Brown, C.M.; Gobler, C.J.; Bidle, K.D.; Wilson, W.H.; Wilhelm, S.W. Genome of the brown tide virus (AaV), the little giant of the megaviridae, elucidates NCLDV genome expansion and host-virus coevolution. *Virology* **2014**, *466–467*, 60–70. [CrossRef] [PubMed]

72. Yutin, N.; Wolf, Y.I.; Koonin, E.V. Origin of giant viruses from smaller DNA viruses not from a fourth domain of cellular life. *Virology* **2014**, *466–467*, 38–52. [CrossRef] [PubMed]

73. Moniruzzaman, M.; Gann, E.R.; LeCleir, G.R.; Kang, Y.; Gobler, C.J.; Wilhelm, S.W. Diversity and dynamics of algal megaviridae members during a harmful brown tide caused by the pelagophyte, *Aureococcus anophagefferens*. *FEMS Microbiol. Ecol.* **2016**, *92*, fiw058. [CrossRef] [PubMed]

74. Yutin, N.; Koonin, E.V. Pandoraviruses are highly derived phycodnaviruses. *Biol. Direct* **2013**, *8*, 8. [CrossRef] [PubMed]

75. Lwoff, A. The concept of a virus. *J. Gen. Microbiol.* **1957**, *17*, 239–253. [CrossRef] [PubMed]

76. Twort, F.W. An investigation on the nature of ultra-microscopic viruses. *Lancet* **1915**, *2*, 1241–1243. [CrossRef]

77. D'Herelle, F. Sur un microbe invisible antagonistic des bacilles dysenterique. *C. R. Acad. Sci. Paris* **1917**, *165*, 373–375.

78. Lwoff, A.; Tournier, P. Classification of viruses. *Annu. Rev. Microbiol.* **1966**, *20*, 45–74. [CrossRef] [PubMed]

79. Raoult, D.; La Scola, B.; Birtles, R. The discovery and characterization of mimivirus, the largest known virus and putative pneumonia agent. *Clin. Infect. Dis.* **2007**, *45*, 95–102. [CrossRef] [PubMed]

80. Raoult, D.; Audic, S.; Robert, C.; Abergel, C.; Renesto, P.; Ogata, H.; La Scola, B.; Suzan, M.; Claverie, J.M. The 1.2-megabase genome sequence of mimivirus. *Science* **2004**, *306*, 1344–1350. [CrossRef] [PubMed]

81. Boyer, M.; Madoui, M.-A.; Gimenez, G.; La Scola, B.; Raoult, D. Phylogenetic and phyletic studies of informational genes in genomes highlight existence of a 4th domain of life including giant viruses. *PLoS ONE* **2010**, *5*, e15530. [CrossRef] [PubMed]

82. Felsenstein, J. *Inferring Phylogenies*; Sinauer Associates: Sunderland, MA, USA, 2004.

83. Williams, T.A.; Embley, T.M.; Heinz, E. Informational gene phylogenies do not support a fourth domain of life for nucleocytoplasmic large DNA viruses. *PLoS ONE* **2011**, *6*, e21080. [CrossRef] [PubMed]

84. Doutre, G.; Philippe, N.; Abergel, C.; Claverie, J.M. Genome analysis of the first *Marseilleviridae* representative from Australia indicates that most of its genes contribute to virus fitness. *J. Virol.* **2014**, *88*, 14340–14349. [CrossRef] [PubMed]

85. Chow, C.-E.T.; Winget, D.M.; White, R.A.; Hallam, S.J.; Suttle, C.A. Combining genomic sequencing methods to explore viral diversity and reveal potential virus-host interactions. *Front. Microbiol.* **2015**, *6*, 265. [CrossRef] [PubMed]

86. La Scola, B.; Audic, S.; Robert, C.; Jungang, L.; de Lamballerie, X.; Drancourt, M.; Birtles, R.; Claverie, J.M.; Raoult, D. A giant virus in amoebae. *Science* **2003**, *299*, 2033. [CrossRef] [PubMed]

87. Abrahão, J.S.; Dornas, F.P.; Silva, L.C.; Almeida, G.M.; Boratto, P.V.; Colson, P.; La Scola, B.; Kroon, E.G. *Acanthamoeba polyphaga* mimivirus and other giant viruses: An open field to outstanding discoveries. *Virol. J.* **2014**, *11*, 120. [CrossRef] [PubMed]

88. Kim, M.S.; Park, E.J.; Roh, S.W.; Bae, J.W. Diversity and abundance of single-stranded DNA viruses in human feces. *Appl. Environ. Microbiol.* **2011**, *77*, 8062–8070. [CrossRef] [PubMed]

89. DeLong, J.P.; Al-Ameeli, Z.; Duncan, G.; Van Etten, J.L.; Dunigan, D.D. Predators catalyze an increase in chloroviruses by foraging on the symbiotic hosts of zoochlorellae. *Proc. Natl. Acad. Sci. USA* **2016**, *113*, 13780–13784. [CrossRef] [PubMed]

90. Kristensen, D.M.; Mushegian, A.R.; Dolja, V.V.; Koonin, E.V. New dimensions of the virus world discovered through metagenomics. *Trends Microbiol.* **2010**, *18*, 11–19. [CrossRef] [PubMed]

91. Williamson, S.J.; Allen, L.Z.; Lorenzi, H.A.; Fadrosh, D.W.; Brami, D.; Thiagarajan, M.; McCrow, J.P.; Tovchigrechko, A.; Yooseph, S.; Venter, J.C. Metagenomic exploration of viruses throughout the Indian Ocean. *PLoS ONE* **2012**, *7*, e42047. [CrossRef] [PubMed]

92. Khalil, J.Y.B.; Langlois, T.; Andreani, J.; Sorraing, J.-M.; Raoult, D.; Carmoin, L.; La Scola, B. Flow cytometry sorting to separate viable giant viruses from amoeba co-culture supernatants. *Front. Cell. Infect. Microbiol.* **2016**, *6*, 202. [CrossRef] [PubMed]

93. Martinez Martinez, J.; Poulton, N.J.; Stepanauskas, R.; Sieracki, M.E.; Wilson, W.H. Targeted sorting of single virus-infected cells of the coccolithophore *Emiliania huxleyi*. *PLoS ONE* **2011**, *6*, e22520. [CrossRef] [PubMed]

94. Zablocki, O.; van Zyl, L.; Adriaenssens, E.M.; Rubagotti, E.; Tuffin, M.; Cary, S.C.; Cowana, D. High-level diversity of tailed phages, eukaryote-associated viruses, and virophage-like elements in the metaviromes of Antarctic soils. *Appl. Environ. Microbiol.* **2014**, *80*, 10. [CrossRef] [PubMed]

95. Correa, A.M.; Welsh, R.M.; Vega Thurber, R.L. Unique nucleocytoplasmic dsDNA and +ssRNA viruses are associated with the dinoflagellate endosymbionts of corals. *ISME J.* **2013**, *7*, 13–27. [CrossRef] [PubMed]

96. Lysholm, F.; Wetterbom, A.; Lindau, C.; Darban, H.; Bjerkner, A.; Fahlander, K.; Lindberg, A.M.; Persson, B.; Allander, T.; Andersson, B. Characterization of the viral microbiome in patients with severe lower respiratory tract infections, using metagenomic sequencing. *PLoS ONE* **2012**, *7*, e30875. [CrossRef] [PubMed]

97. Brussaard, C.P.D.; Kuipers, B.; Veldhuis, M.J.W. A mesocosm study of *Phaeocystis globosa* population dynamics. *Harmful Algae* **2005**, *4*, 859–874. [CrossRef]

98. Castberg, T.; Thyrhaug, R.; Larsen, A.; Sandaa, R.-A.; Heldal, M.; Van Etten, J.L.; Bratbak, G. Isolation and characterization of a virus that infects *Emiliania huxleyi* (Haptophyta). *J. Phycol.* **2002**, *38*, 767–774. [CrossRef]

99. Gobler, C.J.; Anderson, O.R.; Gastrich, M.D.; Wilhelm, S.W. Ecological aspects of viral infection and lysis in the harmful brown tide alga *Aureococcus anophagefferens*. *Aquat. Microb. Ecol.* **2007**, *47*, 25–36. [CrossRef]

100. Nagasaki, K.; Tarutani, K.; Yamaguchi, M. Growth characteristics of *Heterosigma akashiwo* virus and its possible use as a microbiological agent for red tide control. *Appl. Environ. Microbiol.* **1999**, *65*, 898–902. [PubMed]

101. Brussaard, C.P.D.; Gast, G.J.; van Duyl, F.C.; Riegmen, R. Impact of phytoplankton bloom magnitude on a pelagic microbial food web. *Mar. Ecol. Prog. Ser.* **1996**, *144*, 211–221. [CrossRef]

102. Weitz, J.S.; Wilhelm, S.W. Ocean viruses and their effects on microbial communities and biogeochemical cycles. *F1000 Biol. Rep.* **2012**, *4*, 17. [CrossRef] [PubMed]

103. Wilhelm, S.W.; Suttle, C.A. Viruses and nutrient cycles in the sea. *Bioscience* **1999**, *49*, 781–788. [CrossRef]

104. Nagasaki, K.; Bratbak, G. Isolation of viruses infecting photosynthetic and nonphotosynthetic protists. In *Manual of Aquatic Viral Ecology*; Wilhelm, S.W., Weinbauer, M.G., Suttle, C.A., Eds.; ASLO: Waco, TX, USA, 2010; pp. 92–101.

105. Brodie, J.; Zuccarello, G.C. Systematics of the species rich algae: Red algal classificiation, phylogeny and speciation. In *Reconstructing the Tree of Life. Taxonomy and Systematics of Species Rich Taxa*; Hodkinson, T.R., Parnell, J.A.N., Eds.; CRC Press: New York, NY, USA, 2006; pp. 323–336.

106. Short, S.M. The ecology of viruses that infect eukaryotic algae. *Environ. Microbiol.* **2012**, *14*, 2253–2271. [CrossRef] [PubMed]

107. Johannessen, T.V.; Bratbak, G.; Larsen, A.; Ogata, H.; Egge, E.S.; Edvardsen, B.; Eikrem, W.; Sandaa, R.A. Characterisation of three novel giant viruses reveals huge diversity among viruses infecting Prymnesiales (Haptophyta). *Virology* **2015**, *476*, 180–188. [CrossRef] [PubMed]

108. Wilhelm, S.W.; Coy, S.R.; Gann, E.R.; Moniruzzaman, M.; Stough, J.M.A. Standing on the shoulders of giant viruses: 5 lessons learned about large viruses infecting small eukaryotes and the opportunities they create. *PLoS Pathog.* **2016**, *12*, e1005752. [CrossRef] [PubMed]

109. Moniruzzaman, M.; Wurch, L.L.; Alexander, H.; Dyhrman, S.T.; Gobler, C.J.; Wilhelm, S.W. Virus-host infection dynamics of marine single-celled eukaryotes resolved from metatranscriptomics. *bioRxiv* **2016**. [CrossRef]

110. Van Etten, J.L.; Lane, L.C.; Meints, R.H. Viruses and viruslike particles of eukaryotic algae. *Microbiol. Rev.* **1991**, *55*, 586–620. [PubMed]

111. Reteno, D.G.; Benamar, S.; Kahlil, J.B.; Andreani, J.; Armstrong, N.; Klose, T.; Rossmann, M.G.; Colson, P.; Raoult, D.; La Scola, B. Faustovirus, an asfarvirus-related new lineage of giant viruses infecting amoebae. *J. Virol.* **2015**, *89*, 6585–6594. [CrossRef] [PubMed]

112. Fischer, M.G.; Allen, M.J.; Wilson, W.H.; Suttle, C.A. Giant virus with a remarkable complement of genes infects marine zooplankton. *Proc. Natl. Acad. Sci. USA* **2010**, *107*, 19508–19513. [CrossRef] [PubMed]

113. Schnepf, E.; Soeder, C.J.; Hegewald, E. Polyhedral virus-like particles lysing the aquatic phycomycete *Aphedlidium* sp., a parasite of the grean algae *Scenedesmus armatus*. *Virology* **1970**, *42*, 482–487. [CrossRef]

114. Pickett-Heaps, J.D. A possible virus infection in the green alga *Oedogonium*. *J. Phycol.* **1972**, *8*, 44–47. [CrossRef]

115. Toth, R.; Wilce, R.T. Viruslike particles in the marine alga *Chorda tomentosa* lyngye (Phaeophyceae). *J. Phycol.* **1972**, *8*, 126–130.

116. Baker, J.R.J.; Evans, L.V. The ship fouling alga *Ectocarpus*. *Protoplasma* **1973**, *77*, 1–13. [CrossRef]

117. Clitheroe, S.B.; Evans, L.V. Virus like particles in the brown algae *Ectocarpus*. *J. Ultrastruct. Res.* **1974**, *49*, 211–217. [CrossRef]

118. Swale, E.M.F.; Belcher, J.H. A light and electron microscope study of the colourless flagellate *Aulacomonas skuja*. *Arch. Microbiol.* **1973**, *92*, 91–103. [CrossRef]

119. Markey, D.R. A possivle virus infection in the brown alga *Phlaiella littoralis*. *Protoplasma* **1974**, *80*, 223–232. [CrossRef] [PubMed]

120. Moestrup, O.; Thomsen, H.A. An ultrastructural study of the flagellate *Phyramimonas orientalis* with particular emphasis on golgi apparatus activity and the flagellar apparatus. *Protoplasma* **1974**, *81*, 247–269. [CrossRef]

121. Gibbs, A.; Skotnicki, A.H.; Gardiner, J.E.; Walker, E.S.; Hollings, M. A tabamovirus of a green alga. *Virology* **1975**, *64*, 571–574. [CrossRef]

122. Oliveira, L.; Bisalputra, T. A virus infection in the brown alga *Sorocarpus uvaeformis* (Lyngbye) pringsheim (Phaeophyta, Ecocarpales). *Ann. Bot.* **1978**, *42*, 439–445. [CrossRef]

123. Sicko-Goad, L.; Walker, G. Viroplasm and large virus-like particles in the dinoflagellate *Gymnodiunium uberrimum*. *Protoplasma* **1979**, *99*, 203–210. [CrossRef]

124. Dodds, J.A.; Cole, A. Microscopy and biology of *Uronema gigas*, a filamentous eucaryotic green alga, and its associated tailed virus-like particle. *Virology* **1980**, *100*, 156–165. [CrossRef]

125. Preisig, H.R.; Hibberd, D.J. Virus-like particles and endophytic bacteria in *Paraphysomonas* and *Chromophysomonas* (Chrysophyceae). *Nord. J. Bot.* **1984**, *4*, 279–285. [CrossRef]

126. Gowing, M.M. Large virus-like particles from vacuoles of phaeodarian radiolarians and from other marine samples. *Mar. Ecol. Prog. Ser.* **1993**, *101*, 33–43. [CrossRef]

127. Smayda, T.J. Complexity in the eutrophication-harmful algal bloom relationship, with comment on the important of grazing. *Harmful Algae* **2008**, *8*, 140–151. [CrossRef]

128. Filee, J.; Pouget, N.; Chandler, M. Phylogenetic evidence for extensive lateral acquisition of cellular genes by nucleocytoplasmic large DNA viruses. *BMC Evol. Biol.* **2008**, *8*, 320. [CrossRef] [PubMed]

129. Monier, A.; Pagarete, A.; de Vargas, C.; Allen, M.J.; Read, B.; Claverie, J.-M.; Ogata, H. Horizontal gene transfer of an entire metabolic pathway between a eukaryotic alga and its DNA virus. *Genome Res.* **2009**, *19*, 1441–1449. [CrossRef] [PubMed]

130. Rosenwasser, S.; Mausz, M.A.; Schatz, D.; Sheyn, U.; Malitsky, S.; Aharoni, A.; Weinstock, E.; Tzfadia, O.; Ben-Dor, S.; Feldmesser, E.; et al. Rewiring host lipid metabolism by large viruses determines the fate of *Emiliania huxleyi*, a bloom-forming alga in the ocean. *Plant Cell Online* **2014**, *26*, 2689–2707. [CrossRef] [PubMed]

131. Vardi, A.; Van Mooy, B.A.S.; Fredricks, H.F.; Popendorf, K.J.; Ossolinski, J.E.; Haramaty, L.; Bidle, K.D. Viral glycosphingolipids induce lytic infection and cell death in marine phytoplankton. *Science* **2009**, *326*, 861–865. [CrossRef] [PubMed]

132. Filee, J.; Siguier, P.; Chandler, M. I am what I eat and I eat what I am: Acquisition of bacterial genes by giant viruses. *Trends Genet.* **2007**, *23*, 10–15. [CrossRef] [PubMed]

133. Filee, J. Multiple occurrences of giant virus core genes acquired by eukaryotic genomes: The visible part of the iceberg? *Virology* **2014**, *466–467*, 53–59. [CrossRef] [PubMed]

134. Blanc, G.; Duncan, G.; Agarkova, I.; Borodovsky, M.; Gurnon, J.; Kuo, A.; Lindquist, E.; Lucas, S.; Pangilinan, J.; Polle, J.; et al. The *Chlorella variabilis* NC64A genome reveals adaptation to photosymbiosis, coevolution with viruses, and cryptic sex. *Plant Cell* **2010**, *22*, 2943–2955. [CrossRef] [PubMed]

135. Blanc, G.; Gallot-Lavallee, L.; Maumus, F. Provirophages in the bigelowiella genome bear testimony to past encounters with giant viruses. *Proc. Natl. Acad. Sci. USA* **2015**, *112*, E5318–5326. [CrossRef] [PubMed]

136. Read, B.A.; Kegel, J.; Klute, M.J.; Kuo, A.; Lefebvre, S.C.; Maumus, F.; Mayer, C.; Miller, J.; Monier, A.; Salamov, A.; et al. Pan genome of the phytoplankton *Emiliania* underpins its global distribution. *Nature* **2013**, *499*, 209–213. [CrossRef] [PubMed]

137. Maumus, F.; Epert, A.; Nogué, F.; Blanc, G. Plant genomes enclose footprints of past infections by giant virus relatives. *Nat. Commun.* **2014**, *5*, 4268. [CrossRef] [PubMed]

138. Breitbart, M. Marine viruses: Truth or dare. *Ann Rev Mar Sci* **2012**, *4*, 425–448. [CrossRef] [PubMed]

139. Frada, M.; Probert, I.; Allen, M.J.; Wilson, W.H.; de Vargas, C. The "cheshire cat" escape strategy of the coccolithophore *Emiliania huxleyi* in response to viral infection. *Proc. Natl. Acad. Sci. USA* **2008**, *105*, 15944–15949. [CrossRef] [PubMed]

140. Yau, S.; Lauro, F.M.; DeMaere, M.Z.; Brown, M.V.; Thomas, T.; Raftery, M.J.; Andrews-Pfannkoch, C.; Lewis, M.; Hoffman, J.M.; Gibson, J.A.; et al. Virophage control of antarctic algal host-virus dynamics. *Proc. Natl. Acad. Sci. USA* **2011**, *108*, 6163–6168. [CrossRef] [PubMed]

141. Campos, R.K.; Boratto, P.V.; Assis, F.L.; Aguiar, E.R.; Silva, L.C.; Albarnaz, J.D.; Dornas, F.P.; Trindade, G.S.; Ferreira, P.P.; Marques, J.T.; et al. Samba virus: A novel mimivirus from a giant rain forest, the Brazilian amazon. *Virol. J.* **2014**, *11*, 95. [CrossRef] [PubMed]

142. Fischer, M.G.; Suttle, C.A. A virophage at the origin of large DNA transposons. *Science* **2011**, *332*, 231–234. [CrossRef] [PubMed]

143. Gaia, M.; Benamar, S.; Boughalmi, M.; Pagnier, I.; Croce, O.; Colson, P.; Raoult, D.; La Scola, B. Zamilon, a novel virophage with *Mimiviridae* host specificity. *PLoS ONE* **2014**, *9*, e94923. [CrossRef] [PubMed]

144. Levasseur, A.; Bekliz, M.; Chabrière, E.; Pontarotti, P.; La Scola, B.; Raoult, D. Mimivire is a defence system in mimivirus that confers resistance to virophage. *Nature* **2016**, *531*, 249–252. [CrossRef] [PubMed]

145. Santini, S.; Jeudy, S.; Bartoli, J.; Poirot, O.; Lescot, M.; Abergel, C.; Barbe, V.; Wommack, K.E.; Noordeloos, A.A.M.; Brussaard, C.P.D.; et al. Genome of *Phaeocystis globosa* virus PgV-16t highlights the common ancestry of the largest known DNA viruses infecting eukaryotes. *Proc. Natl. Acad. Sci. USA* **2013**, *110*, 10800–10805. [CrossRef] [PubMed]

146. Claverie, J.M.; Abergel, C. CRISPR-CAS-like system in giant viruses: Why mimivire is not likely to be an adaptive immune system. *Virol. Sin.* **2016**, *31*, 193–196. [CrossRef] [PubMed]

147. Sullivan, M.B.; Weitz, J.S.; Wilhelm, S.W. Viral ecology comes of age. *Environ. Microbiol. Repo.* **2017**, *9*, 33–35. [CrossRef] [PubMed]

8

Porcine Rotaviruses: Epidemiology, Immune Responses and Control Strategies

Anastasia N. Vlasova [1,*], Joshua O. Amimo [2,3] and Linda J. Saif [1,*]

[1] Food Animal Health Research Program, CFAES, Ohio Agricultural Research and Development Center, Department of Veterinary Preventive Medicine, The Ohio State University, Wooster, OH 44691, USA

[2] Department of Animal Production, Faculty of Veterinary Medicine, University of Nairobi, Nairobi 30197, Kenya; jamimo@uonbi.ac.ke

[3] Bioscience of Eastern and Central Africa, International Livestock Research Institute (BecA-ILRI) Hub, Nairobi 30709, Kenya

* Correspondence: vlasova.1@osu.edu (A.N.V.); saif.2@osu.edu (L.J.S.);

Academic Editors: Simon Graham and Linda Dixon

Abstract: Rotaviruses (RVs) are a major cause of acute viral gastroenteritis in young animals and children worldwide. Immunocompetent adults of different species become resistant to clinical disease due to post-infection immunity, immune system maturation and gut physiological changes. Of the 9 RV genogroups (A–I), RV A, B, and C (RVA, RVB, and RVC, respectively) are associated with diarrhea in piglets. Although discovered decades ago, porcine genogroup E RVs (RVE) are uncommon and their pathogenesis is not studied well. The presence of porcine RV H (RVH), a newly defined distinct genogroup, was recently confirmed in diarrheic pigs in Japan, Brazil, and the US. The complex epidemiology, pathogenicity and high genetic diversity of porcine RVAs are widely recognized and well-studied. More recent data show a significant genetic diversity based on the VP7 gene analysis of RVB and C strains in pigs. In this review, we will summarize previous and recent research to provide insights on historic and current prevalence and genetic diversity of porcine RVs in different geographic regions and production systems. We will also provide a brief overview of immune responses to porcine RVs, available control strategies and zoonotic potential of different RV genotypes. An improved understanding of the above parameters may lead to the development of more optimal strategies to manage RV diarrheal disease in swine and humans.

Keywords: Porcine rotavirus; group A, B, C, E and H rotaviruses; rotavirus vaccines; epidemiology; genetic variability; prevalence; active and passive immunity; swine; zoonotic potential

1. Introduction

Rotavirus (RV) is well established as a major cause of acute gastroenteritis in young children and animals, including nursing and weaned piglets [1]. The name "rotavirus" comes from the wheel-like virion appearance observed by electron microscopy. The virus is transmitted by the fecal–oral route and the infection results in destruction of mature small intestinal enterocytes [2]. RV-mediated damage is characterized by shortened villi with sparse, irregular microvilli and by mononuclear cell infiltration of the lamina propria [2]. Several mechanisms are suggested to contribute to the development of diarrhea including malabsorption due to the destruction of enterocytes, villus ischaemia, neuro-regulatory release of a vasoactive agent from infected epithelial cells. Also the RV non-structural protein 4 (NSP4) induces an age- and dose-dependent diarrheal response by acting as an enterotoxin and secretory agonist [2] (Figure 1) to: (i) stimulate Ca^{2+}-dependent cell permeability and (ii) alter the integrity of epithelial barrier.

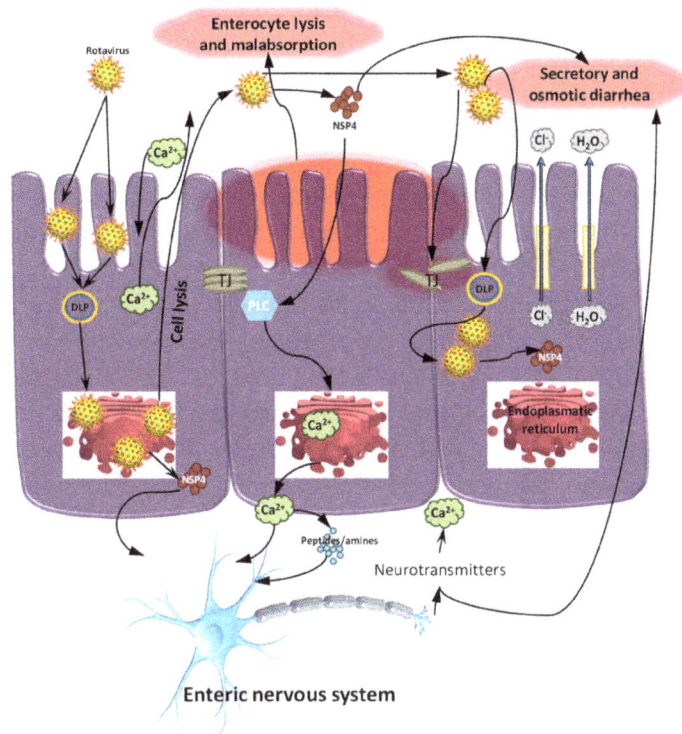

Figure 1. Potential mechanisms of rotavirus (RV) pathogenesis. RV replication inside enterocytes induces osmotic diarrhea. RV also increases the concentration of intracellular calcium (Ca^{2+}), disrupting the cytoskeleton and the tight junctions, increasing paracellular permeability. In addition, RV produces non-structural protein 4 (NSP4), an enterotoxin that induces Ca^{2+} efflux from endoplasmatic reticulum via the phospholipase C dependent (PLC) mechanism further contributing to electrolyte imbalance and secretory diarrhea. RV can also stimulate the enteric nervous system (ENS, via NSP4 dependent mechanism), further contributing to secretory diarrhea and increasing intestinal motility. Agents that can inhibit the ENS could be useful in alleviating RV diarrhea in children. Following, tryptic cleavage of viral protein 8 (VP8) from VP5, the VP8 fragment alters the localization of claudin-3, ZO-1 and occludin leading to the disruption of the barrier integrity of tight junctions (TJ) [3–6]. Late in the infectious process, RV destroys mature enterocytes, further contributing to malabsorptive or osmotic diarrhoea. RV antigens, genomic RNA and infectious particles have been found in the blood of children and blood and systemic organs in animals [7,8]. The role of systemic RV translocation in disease pathogenesis is currently unknown. DLP: double-layered particles.

RVs represent a genus in the *Reoviridae* family of double-stranded RNA (dsRNA) viruses, with a genome of 11 segments of dsRNA encoding six structural viral proteins (VP1–VP4, VP6 and VP7) and five nonstructural proteins (NSP1–NSP5/6). RVs are classified into 10 groups (A–J) based on antigenic relationships of their VP6 proteins, with provisional I and J species recently identified in sheltered dogs in Hungary and in bats in Serbia, respectively [9–12]. The outer capsid proteins, VP7 and VP4, induce neutralizing antibodies and form the basis for the G and P dual typing system [9]. The most common groups that infect humans and animals are groups A, B and C (RVA, RVB and RVC), with the highest prevalence of RVA strains that represent one of the most significant causes of acute dehydrating diarrhea from public health and veterinary health perspectives. To date, 27 different G- and 37 P-genotypes have been described in both humans and animals for RVAs [13,14]. For highly genetically diverse RVA strains, the dual (G/P) typing system was extended in 2008 to a full-genome sequence classification system, with nucleotide percent identity cut-off values established for all 11 gene segments, with the notations Gx-P[x]-Ix-Rx-Cx-Mx-Ax-Nx-Tx-Ex-Hx used for the VP7-VP4-VP6-VP1-VP2-VP3-NSP1-NSP2-NSP3-NSP4-NSP5/6 encoding genes, respectively [15].

Subsequently, a Rotavirus Classification Working Group (RCWG) was formed to set the RVA classification guidelines and maintain the proposed classification system [16] to facilitate complete classification of novel RVA strains. Currently, only RVA classification has been developed and is being maintained by the RCWG, while much less is known about the epidemiology and disease burden associated with infection by non-RVAs. However, RVB, RVC, RVE, RVH and RVI have been detected in sporadic, endemic or epidemic infections of various mammalian species, whereas RVD, RVF and RVG are found in poultry, such as chickens and turkeys [14,17–24]. RVs of groups A, B, C, E and H have been described in pigs [25–32].

In 1969, bovine RV was the first group A RV isolated in cell culture and confirmed as a cause of diarrhea in calves [33,34]. Human RV was discovered soon after, in 1973, by Bishop and colleagues [35]. Subsequent studies documented the widespread prevalence of RVA infections in young animals, including calves and pigs, and their association with diarrhea in animals <1 month of age [20,28,30,36,37]. Group C RVs were first isolated in piglets in 1980 [31] and were subsequently identified in other animals and humans [30,38–41]. Porcine RVB was first described as an RV-like agent identified in a diarrheic pig in the 1980s [29,42]. In addition to pigs, RVB strains have been also detected in cattle [43–46], lambs [47], and rats [48]. In contrast to human RVA and RVC that were described worldwide, human RVB strains have been described only in China [49–52], India [53,54], and Bangladesh [55–59]. An atypical group E porcine RV was only reported in UK swine, where a serological survey indicated a widespread distribution of antibodies to this virus in pigs older than 10 weeks [25,60]. Most recently, RVH strains were described in pigs in Japan, Brazil and in the US, where they were reportedly circulating since at least 2002 [27,61,62].

2. RV Genogroup/Genotype Classification and Prevalence in Swine

Infections by RVAs are confirmed in pigs worldwide with or without association with diarrhea [63–74]. RVA prevalence rates in pigs vary from 3.3% to 67.3% without evidence of seasonality, but with spatio-temporal fluctuations and re-emergence of certain genotypes, including G9 and G1 [67,71,75–87], with farm-level prevalence reaching 61%–74% [73,74]. Twelve G genotypes (G1 to G6, G8 to G12, and G26) and 16 P genotypes (P[1] to P[8], P[11], P[13], P[19], P[23], P[26], P[27], P[32], and P[34]) of RVA have been associated with pigs [65,67,70,72–74,84,88–91]. However, G3, G4, G5, G9 and G11 were historically considered the most common G genotypes in swine and were usually associated with P[5], P[6], P[7], P[13] and P[28] [16,89,92].

Similar to RVA, porcine RVCs are reported in most parts of the world [32,39]. Diarrhea outbreaks associated with RVCs have been documented in nursing, weaning and post-weaning pigs [31,32,93], either alone or in mixed infection with other enteric pathogens [1]. In addition, the antibody prevalence in pigs (58%–100%) shows that RVC infection may be very common and has circulated for many decades in swine herds in developed countries [32]. Recent studies on US and Canadian porcine samples demonstrated a 46% prevalence of RVC which was higher in very young (78%, ≤3 day old) and young (65%, 4–20 day old) piglets [94]. RVC genotypes G1 and G3 were initially assigned to the prototype porcine RVC Cowden and HF strains, respectively [95]. Further efforts to classify RVC strains into sequence-based genotypes resulted in identification of a total of nine G genotypes (G1–G9), seven P genotypes (P1–P7) and seven I genotypes (I1–I7) [94,96–100]. Additional attempts were made to extend RVC classification based on the sequencing of all 11 genes [101,102]; however, only limited genomic sequences are currently available. Porcine RVCs belong to G1, G3, G5–G9 genotypes and a newly described G10 genotype [103], while bovine and human RVCs are classified as G2 and G4 genotypes, respectively [94,96,97,104]. Additionally, two provisional G genotypes (G12 and G13 based on the 86% nucleotide identity cut-off value) are proposed by Niira et al. based on their recent results [105].

Rapid molecular characterization of RVB strains is hampered by the difficulty of adapting RVB strains to cell culture [32,58]. Additionally, limited and variable fecal shedding and instability in feces were shown for RVBs [44]. Complete genome sequences were obtained for several human RVB strains from Southeast Asia [55,106–108] and partial genome sequencing was done for several rat and bovine RVB strains [43–46,48,53,57,109]. Kuga and colleagues analyzed sequences of the VP7 gene of 38 porcine RVB strains from Japan (2000–2007) and the five genotypes proposed were further divided into 12 clusters, using 67% and 76% nucleotide cut-off values (66% and 79% on the amino acid level, respectively) [110]. Recent results by Marthaler et al. suggested a broader diversity of porcine RVBs based on sequencing of the VP7 gene of 68 RVB strains (collected in 2009 from 14 US states and Japan) defining 20 G genotypes based on an 80% nucleotide identity cut-off value and providing the first evidence that porcine RVB genotypes may be host species- and region-specific [111]. Therefore, porcine RVB strains of genotypes G1, G2 and G3/G5 are only found in rats, humans and bovine species, respectively, while genotypes G4, G7, G9, G13, G15 and G19 are only confirmed in pigs in Japan, and a small number of porcine RVB strains of genotypes G10 and G17 are only found in the US. An additional G genotype, G21, was detected in pigs in India [112].

Three human RVH strains from Asia (ADRV-N, J19, B219) [113–116] and a porcine RVH strain (SKA-1) were identified during 1997–2002 [27]. In 2012, three more porcine RVH strains BR63, BR60, and BR59 from Brazil were identified [62]. Surprisingly high prevalence (15%) of porcine RVH strains was recently demonstrated by Marthaler and colleagues mostly in older (21–55 days old) piglets [18]. Their data suggested that porcine RVH strains circulated in the US herds since at least 2006 and that they are evolutionarily distinct from those of humans, as well as from porcine RVH strains in Brazil and Japan [18]. Complete genome analyses of a porcine RVH identified in South Africa showed that the novel RVH strain MRC-DPRU1575 clustered together with the SKA-1 strain and known porcine RVH strains from Brazil and the USA (only for available genome segments) [117]. However, it was only distantly related to human RVH strains from Asia and an RVH-like strain recently detected in bats from Cameroon [117].

Additional data is needed to evaluate the epidemiological importance of porcine RVE strains, because porcine RVE has only been identified in the UK approximately 3 decades ago and has not been reported to expand its geographic or host range since [25].

3. Porcine RV Distribution, Genotype Prevalence and Spatio-Temporal Variations in the Americas

3.1. North America

A high prevalence of porcine RV strains of groups A, B and C among samples from diarrheic piglets collected in 2009–2011 in the US, Canada and Mexico was reported by Marthaler et al. (2014) and Homwong et al. (2016) [69,71]. The highest overall prevalence of porcine RVs of 82.1% (90%–100% for UT, PA, VA and NC, and 5%–90% for the rest of the states) was observed in the US, and similar values of 79.7% and 73.3% are reported for Canada and Mexico, respectively. In the US, the highest proportion of RVA positive samples (70.1%) was in the Southeastern states, whereas the highest detection rate of RVB and RVC samples was found in the South-central states (34.2% and 62.2%, respectively); however no genotyping results were reported. The historic prevalence of porcine RVA, RVB and RVC strains in the US was reported as 67.8%, 10.0% and 11.1%, respectively [26]. A systematic review by Papp and colleagues [72] summarized genotype prevalence and distribution for porcine historic samples collected/analysed between 1976 and 2011 from both diarrheic and non-diarrheic animals. The most prevalent G type of porcine RVA in the Americas was G5 (71.4%), followed by G4 (8.2%), G3 (3.57%), G9 (2.31%) and G11 (1.9%) [68,72,82,118] (Figure 2). The frequencies of infections by other RVA genotypes found in pigs (G1, G2, G6, G8 and G10) were ~1% or less. P[7] genotype was the most common in the Americas (77.2%), while other P-types represented less than 1% of the identified RVA strains [72]. Finally, G5P[7] was the single most prevalent combination. In contrast, the analysis of more recent US RVA strains (2004–2012) conducted by Amimo and colleagues demonstrated that the dominant G-P combination was G9P[13] found in 60.9% of positive samples [from Ohio (OH) North Carolina (NC) and Michigan (MI)], followed by G9P[7] (8.7%), G4P[13] (8.7%), G11P[13] (4.3%), and G11P[7] (4.3%), while no G5 strains were detected [67]. Additionally, despite the relatively low overall prevalence of porcine RVA strains in samples from US diarrheic and non-diarrheic animals of 9.4%, Amimo et al. reported that there was an increase in RVA detection from 5.9% in 2004 to 13.8% in 2012 [67], which may be due to the increase in the prevalence of novel or re-emerging genotypes (such as G9) because of lack of herd immunity against them.

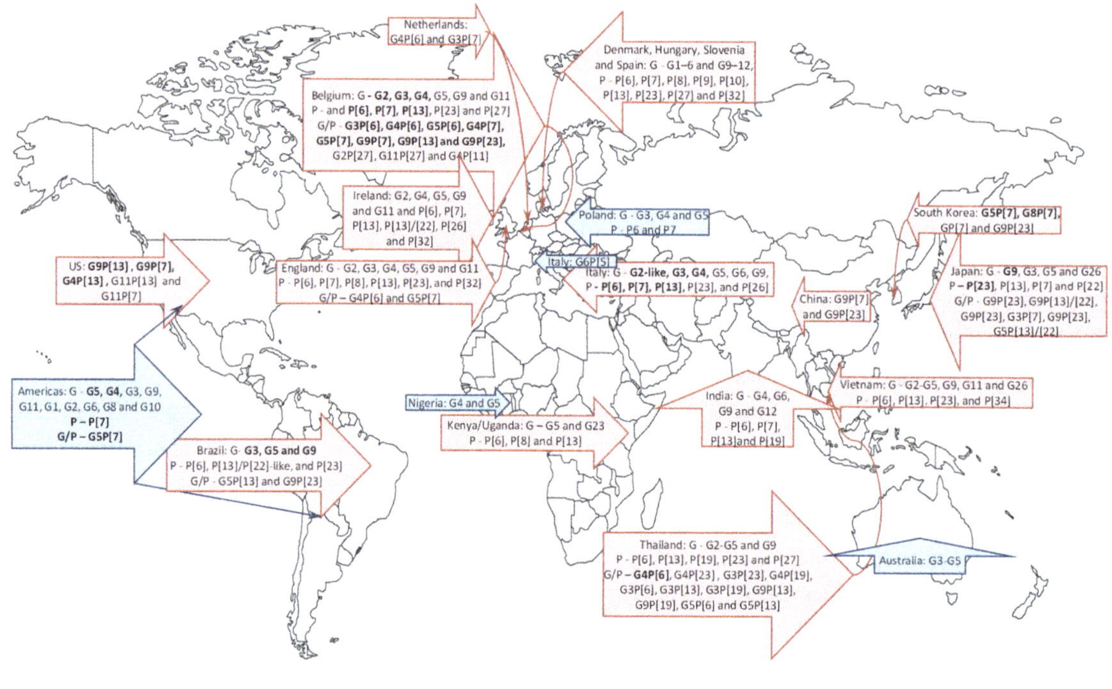

Figure 2. Global genotype distribution of porcine RVA strains reported in historic (1976–2011, blue figure arrows) and current (after 2000, pink figure arrows) studies. Porcine RVAs are also detected in Germany and Russia, but no genotyping data is available.

An earlier study by Kim et al. (1999) identified porcine RVC strains associated with diarrheal outbreaks in feeder pigs in the US [93]. Although the porcine RVC prevalence was not evaluated in this study, phylogenetic analysis demonstrated that the identified strains were more closely related to Cowden (G1) and more distant to HF (G3) strains. Recently, Amimo et al. reported a higher overall prevalence of porcine RVC strains compared to that of porcine RVA strains (19.5% versus 9.4%) in diarrheic and non-diarrheic piglets collected from several farms in the US (OH, NC and MI) in 2004–2012 [119]. In this study, the frequency of porcine RVC identification in the samples from diarrheic was higher than that in non-diarrheic piglets. The porcine RVC strains were confirmed as G3 and G6 in this study (Figure 3). Further, Marthaler analyzed 7520 porcine fecal samples (collected in 2009–2011 in the US and Canada) and identified RVC in 46% of the samples tested [94]. The porcine RVC prevalence was 16% in very young pigs (<3 days old), 21% in young pigs (4–20 days old), 42% in post-weaning pigs (21–55 days old), 13% in older pigs (455 days old), and 8% in pigs of unknown age. However, single porcine RVC infection prevalence was highest in very young (<3 days), and young pigs (4–22 days) in 78% and 65% of the RVC positive samples, respectively, whereas this percentage was much lower (6%–39%) in the older age groups. The most common VP7 genotype detected in this study was G6 (70%), followed by G5 (17%), G1 (12%), and G9 (1%); however, unlike in the study conducted by Amimo et al., no G3 strains were identified. These data suggest that despite the limited genotyping information available for porcine RVC strains, there was a possible shift in their prevalence from G1 and G3 genotypes associated with the prototype Cowden and HF strains to G6 and G5 genotypes.

The current knowledge of the genetic diversity of porcine RVB strains is mostly from two studies: Kuga et al. (2009) and Marthaler et al. (2012) from Japan and the US, respectively. They classified the existing porcine RVB strains into 20 G genotypes [110,111] (Figure 3). Due to the limited information on porcine RVB epidemiology, it is hard to provide an accurate statistics on the temporal fluctuations in porcine RVB prevalence and porcine RVB genotype distribution in the North and South Americas. However, the new findings reported by Marthaler suggest an increased porcine RVB prevalence (46.8%) in the US that was previously observed by others elsewhere and in the US [67,104,110,112], and demonstrate that remarkably diverse porcine RVB genotypes (10 G genotypes: G6, G8, G10, G11, G12, G14, G16, G17, G18 and G20 associated with various I genotypes) are currently circulating in the US, with G8, G12, G16, G18 and G20 genotypes being most prevalent [111].

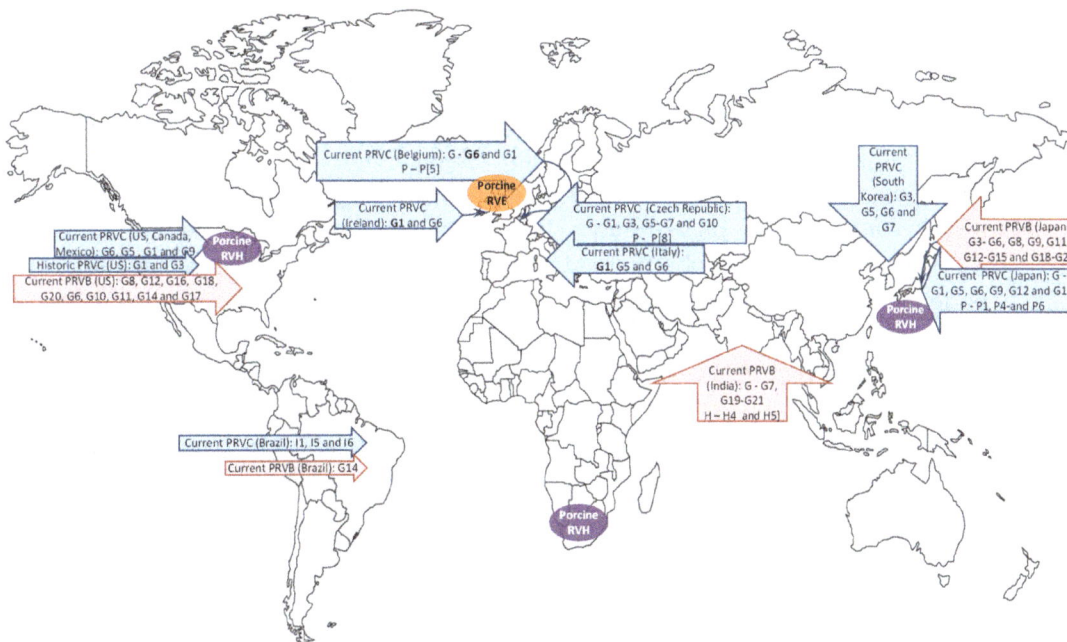

Figure 3. Global genotype distribution of porcine RVB (pink figure arrows) and RVC (blue figure arrows) strains and porcine RVE (bolded, orange circle)/RVH (bolded, purple circles) occurrence in different countries reported in historic (1976–2011) and current (after 2000) studies. Porcine RVCs are also detected in Germany and China, and porcine RVB is confirmed in Germany and Czech Republic, but no porcine RVC/RVB genotyping data is available for these countries.

3.2. South America

As reported for the US, G5, G4 and G9 genotypes of porcine RVA were most prevalent in Brazil and Argentina, with G5P[7] being the single most prevalent combination [68,72,82]. Similar to findings by Marthaler and Amimo, recent findings by Molinari (on samples collected from a single diarrheic outbreak in Brazil in a G5P[7] vaccinated herd in 2012) demonstrated that porcine RVC (78%) was the most prevalent group found in single (34%) and mixed (44%) infections, followed by porcine RVA (46%), RVB (32%), and RVH (18%) [112]. The porcine RVA genotypes detected were G5P[13] and G9P[23], that differed from the G5P[7] found in the vaccine. Another recent study from Brazil (2011–2012) demonstrated co-circulation of G3, G5, G9, and P[6], P[13]/P[22]-like, and P[23] genotypes [120], but with no indication of the historic G5P[7] genotype combination. These findings may indicate that application of the G5P[7] based porcine RVA vaccines in North and South America might have contributed to the previously reported increased prevalence of the G5P[7] strains, while subsequently developed herd immunity and selective pressure against the G5P[7] strains, resulted in their recent decline (or disappearance) and emergence of the G9 or reassortant variants. Similar to the findings by Marthaler [94], Molinari reported an increased prevalence of porcine RVC strains in Brazil in diarrheic piglets in a herd vaccinated with porcine RVA G5P[7] vaccine [112]. The VP6 gene sequence analysis demonstrated that the RVC strain possessed an I1 genotype like Cowden; however, G and P types were not determined. Another study from Brazil, confirmed the presence of three I genotypes (I1, I5, and I6) in the samples from diarrheic piglets (2004–2010) suggesting that diverse porcine RVC strains circulate in different Brazilian states [98]. Additionally, Molinari et al. reported porcine RVB genotype G14 in diarrheic pigs in Brazil, as also reported by Kuga and Marthaler [110–112].

4. Global Porcine RV Distribution and Genotype Prevalence: Africa, Europe, Asia and Australia

4.1. Africa

The presence of group A, B, C and H porcine RVs has been confirmed in several African countries [65,117,121–123]. The prevalence of porcine RVA in Kenya and Uganda reported in the recent study by Amimo et al. of 26.2% [65] was higher compared to the prevalence rates of 6.5%–25.7% reported for samples collected in 2004–2011 in the USA [65], several European countries [75,84,85], Thailand [89], and India [124], but however, lower (32.7%–38.3%) to those observed in Vietnam [74], Brazil [80] and Korea [76]. It was lower than that reported for samples collected in the US in 2009–2011 [69], in asymptomatic pigs in Italy (71.5%) [70] or previously reported for South Africa (84.6%) [123]. This study provides the first evidence that porcine RVA infections are widespread and likely endemic in East African pig herds. The 18 characterized African porcine RVA strains were classified into three different P-types including P[6], P[8] and P[13] that were associated with G5 and G23 G-types [65] (Figure 2). An increased prevalence of porcine RVA strains in diarrheic and asymptomatic suckling and weaned piglets of 41.8% was also reported in Tanzania (2014), but the identified porcine RVA positive samples were not genotyped [125]. Interestingly, although previous attempts to characterize porcine RVA strains from piglets in Nigeria by classical serotyping methods demonstrated the presence of G4 and G5 types, substantial numbers of the strains from that study was non-typeable [122]. These findings indicate that phylogenetically distinct porcine RVA genotypes/strains may circulate in African countries together with the historically common (G4 and G5) genotypes and warrants further epidemiological investigation.

Apart from some data on porcine RVB and RVC prevalence in Africa reported by Geyer et al. nearly three decades ago [123], the absence of surveillance programs and adequate diagnostic facilities have resulted in a lack of data on porcine RVB and RVC prevalence and genetic composition [65]; however, recently Amimo and colleagues demonstrated 8.3% (37/446) prevalence of porcine RVC in swine populations in Kenya (8.8%) and Uganda (7.7%) (Amimo et al., 2014, unpublished data).

A recent discovery and characterization of a porcine RVH strain from diarrheic piglets in South Africa confirmed that it was closely related to Japanese, Brazilian and the US porcine RVH, but not human RVH or bat (RVH-like) strains [117,121].

4.2. Europe

Diarrhea associated with RVA, RVB and RVC infections in pigs is an important cause of increased mortality, growth impairment, and economic losses in Europe [73,85,126,127]. porcine RVA strains of G2, G3, G4, G5, G9 and G11 and P[6], P[7], P[13], P[23] and P[27] genotypes were isolated from feces of diarrheic and non-diarrheic Belgian piglets in 2012 [128] (Figure 2). A wide range of G/P genotype combinations including; G3P[6], G4P[6], G5P[6], G4P[7], G5P[7], G9P[7], G9P[13] and G9P[23] was commonly detected in stool samples of diarrheic and non-diarrheic pigs in Belgium. Additionally, uncommon genotypes/genotype combinations were reported; G2P[27], G11P[27] and G4P[11]. During a large surveillance study in Italy (2003–2004), a total of 751 fecal samples were collected from nursing and weaned pigs involved in outbreaks of diarrhea [70]. Porcine RVA prevalence of 16.1% was identified by electron microscopy or by a commercial immunoenzyme assay. Upon either PCR genotyping or sequencing, the porcine RVA strains displayed a broad spectrum of VP7 and VP4 types, including G2-like, G3, G4, G5, G6, G9, P[6], P[7], P[13], P[23], and P[26] [70,129,130]. However, an earlier study by Martella et al. (2001) demonstrated that porcine stool samples collected in Northern Italy during a massive diarrheal outbreak in 1983–1984 contained porcine RVA strains of

G6P[5] genotype combination [127]. Furthermore, Midgley et al. (2012) analyzed a total of 1101 fecal samples from pigs collected from 134 swine farms in four European countries (Denmark, Hungary, Slovenia and Spain) in 2003–2007 [85]. The results demonstrated that porcine RVA prevalence in Danish swine was only 10% although all samples were collected from diarrheic animals. In contrast, in Slovenia where the majority of swine were asymptomatic, the porcine RVA detection rate (20%) was significantly higher than that in swine with diarrhea in Denmark. This is consistent with the results by Amimo et al. [67] showing that unlike porcine RVC [119], there was no strong association between diarrhea and porcine RVA prevalence in nursing and suckling piglets in the US. However, in Spain, porcine RVA infections were significantly more frequent in animals with diarrhea (27%) than in asymptomatic animals (7%) [75]. Among these porcine RVA positive samples, ten different G types, G1–6 and G9–12, and nine different P types, P[6], P[7], P[8], P[9], P[10], P[13], P[23], P[27], P[32], were detected. No single G type was found to be dominant across the participating countries. In Slovenia G3, G4, and G5 were all common genotypes detected in 19%–30% of the samples. In Denmark, G4 was the most common genotype (44%). G9 was only detected in Spain, where it was the most prevalent genotype (33%). Among the various P types, only P[6] was detected in all four countries, which was the most common type in both Slovenia (41%) and in Denmark (56%). Otto et al. (2015) reported a porcine RVA prevalence of 51.2%, but no genotyping data was available from this study [126]. Finally, of the three positive porcine RVA samples identified in the Netherlands in 1999–2001, two were determined to possess G4P[6] and one G3P[7] genotype constellations [131]. Collins et al. tested 292 fecal samples collected from 4–5- to 8–9-week-old asymptomatic pigs in Ireland (2005–2007) and showed that 6.5% samples were positive for porcine RVA [84]. By sequence analysis of the VP7 and VP4 (VP8*) genes, the Irish porcine RVA strains were identified as G2, G4, G5, G9 and G11 and P[6], P[7], P[13], P[13]/[22], P[26] and P[32] genotypes, respectively [84]. The G5 and G11 strains were closely related to other human and porcine G11 strains, while the G2 and G9 strains resembled porcine G2 viruses detected recently in Europe and southern Asia. However, the G4 strains were only distantly related to other G4 human and animal strains, constituting a separate G4 VP7 lineage. Winiarczyk et al. (2002) identified G3, G4 and G5 types in combination with P6 and P7 types circulating in Poland [118]. Thus, in most European countries no dominant porcine RVA genotype/genotype constellations or temporal fluctuations in their prevalence was identified; however, the findings by Martella from different years suggest some epidemiological changes over time in Italy: disappearance of G6P[5] genotype constellation in more recent compared to historic studies [70,127]. A study conducted in England between 2010 and 2012 on samples from diarrheic pigs also revealed the presence of a wide range of porcine RVA genotypes: six G types: G2, G3, G4, G5, G9 and G11 and six P types: P[6], P[7], P[8], P[13], P[23], and P[32] [132]. G4 and G5 were the most common VP7 genotypes, accounting for 25% (16/64) and 36% (23/64) of the strains, respectively, while P[6] (33%, 21/64) and P[32] (27%, 17/64) were the most common VP4 genotypes, respectively. Overall, the most common genotype combinations were G4P[6] and G5P[7], similar to those detected in the historic US samples emphasizing the current unique epidemiology of porcine RVA in England compared to other European countries.

Porcine RVC strains have been detected in feces of asymptomatically infected 4–5 week old Irish pigs (in 2005–2007) and of diarrheic piglets from the Czech Republic at low rates of 4.4% (of 292 samples) and 4.6% (of 329 samples) [104,133]. In comparison 29% and 31% of diarrheic piglets in Belgium (2014) and Germany (1999–2011), respectively, were porcine RVC positive in recent studies [73,126]. All Belgian porcine RVC strains characterized in the study belonged to genotype G6, except for one strain possessing the G1 genotype, while the VP4 genes were genetically heterogeneous, but were classified in the genotype P5 [73] (Figure 3). The majority of the Irish porcine RVC strains were identified as G1 genotype, while only two strains belonged to the genotype G6 [104] and the German porcine RVC strains were not typed [126]. A higher genetic heterogeneity was reported among Czech porcine RVC strains that were grouped into six G genotypes (G1, G3, G5–G7, and a newly described G10 genotype) based on an 85% nucleotide identity cutoff value [103]. Analysis of the VP4 gene revealed low nucleotide sequence identities between two Czech strains and other porcine (72.2%–75.3%), bovine

(74.1%–74.6%), and human (69.1%–69.3%) RVCs and was tentatively classified as a novel RVC VP4 genotype, P8 [103]. Martella et al. (2007) characterized 20 porcine RVC strains collected from distinct diarrheal outbreaks in 2003–2005 in Northern and Central Italy [97]. They belonged to G1, G5 and G6 genotypes, similar to those identified in Ireland.

A very low prevalence rate of porcine RVB was reported in Germany in samples collected between 1999 and 2013, with no genotyping data available [126]. Additionally, Smitalova et al. (2009) reported that porcine RVB was detected in 0.6% of samples from diarrheic pigs in Czech Republic [133]; but they were not genotyped. Apart from the above information, no data for porcine RVB prevalence, pathogenic potential and genetic characteristics are available for Europe. Additionally, no reports of porcine RVH are available and only one historic study confirmed circulation of porcine RVE in England [25], requiring further evaluation and verification that pigs in fact serve as natural reservoir for porcine RVE strains.

4.3. Asia

Numerous prevalence studies conducted in Asian countries demonstrated the presence of uncommon RVA genotypes in humans suspected to originate from animal sources [134–137] and reassortants of human-animal origin [83,138] including the G9 strains emerging globally or regionally in pigs and humans and the need of careful monitoring of animal RVs. Teodoroff et al. (2005) reported that genotype G9 of porcine RVA was dominant in a survey among porcine RVA strains associated with outbreaks of diarrhea in young pigs in Japan between 2000 and 2002 [139] (Figure 2). Similarly, Miyazaki et al. (2011) demonstrated that G9P[23], G9P[13]/[22], G9P[23], G3P[7], G9P[23], G5P[13]/[22], and P[7] combined with an untypeable G genotype caused four different diarrheal outbreaks in Japan in 2009–2010 that affected almost all suckling pigs born to 20% to 30% of lactating sows [90]. Further, this study provided evidence that the untypeable G genotype was a novel porcine RVA G26 genotype [90], which was confirmed by the Rotavirus Classification Working Group. A large-scale surveillance study of smallholder pig farms in the Mekong Delta, Vietnam, was conducted in 2012 and demonstrated an overall animal-level and farm-level porcine RVA prevalence of 32.7% (239/730) and 74% (77/104), respectively; however, no significant association with clinical disease was observed [74]. The study also identified six different G types and four P types in various combinations (G2, G3, G4, G5, G9, G11 and P[6], P[13], P[23], and P[34]) [74]. Additionally, one G26 strain was detected. A novel genotype P[27] in combination with G2 was identified in Thailand in samples collected in 2000–2001 [140]. Saikruang et al. (2013) reported an overall prevalence of porcine RVA of 19.8% (of 207 samples) in diarrheic samples of piglets in Thailand (2009–2010) and identified a wider variety of G-P combinations [78]. In this study, G4P[6] was identified as the most prevalent genotype (39.0%), followed by G4P[23] (12.2%), G3P[23] (7.3%), G4P[19] (7.3%), G3P[6] (4.9%), G3P[13] (4.9%), G3P[19] (4.9%), G9P[13] (4.9%), G9P[19] (4.9%), G5P[6] and G5P[13] each of 2.4%. Furthermore, G5 and G9 in combinations with P-nontypeable strains were also found as 2.4% ($n = 1$) of the collection. Among the diverse porcine RVA strains, novel genotype combinations of G4P[19] and G9P[19] were detected for the first time. Further corroborating the emergence and widespread prevalence of non-classic G and P genotypes of porcine RVA in Asia, 92.9% of porcine RVA containing stool samples collected from piglets with diarrhea in northern Thailand (2006–2008) belonged to the rare P[23] genotype combination with G9 or G3 genotypes [89]. The G9P[23] combination was reported to circulate in pigs in China as well [141]. Porcine RVA strains of the G9 genotype in combination with the P[7] and P[23] genotypes were isolated and identified as the third most important genotype in the diarrheic pigs in South Korea, after G5P[7] and G8P[7] [93]. A review by Malik and colleagues summarized the results of various surveillance studies (using ELISA-, PAGE- and PCR-based typing) suggesting the presence of G4, G6, G9, G12 and P[6], P[7], P[13] and P[19] genotypes in different regions in India [124]. Although there are no documented large-scale surveillance programs in China, the presence of porcine RVA G9P[7] in piglets with diarrhea was confirmed in Jiangsu Province, China [142], suggesting that various G9 combinations circulate in most if not all Asian countries.

Despite somewhat scarce information on porcine RVC prevalence in Asian countries, there are several reports describing different porcine RVC genotypes circulating in Japan and South Korea [99,100,105]. The genotypes described in Japan include G1, G5, G6, G9, G12 and G13 G genotypes found in combination with P1, P4–P6 P genotypes, while G3, G5, G6 and G7 G genotypes were shown to circulate in South Korea [100,105] (Figure 3). There is also a report of porcine RVC circulation in China with a prevalence rate of 16.65% among diarrheic and asymptomatic piglets (2007–2008); however, no genotyping data is available [143].

Similar to porcine RVC data, very limited information on porcine RVB prevalence and dominant genotypes circulating in most Asian countries is available. A high prevalence of porcine RVB and porcine RVB specific antibodies in porcine fecal and serum samples, respectively, are reported in several studies in Japan [110,144]. Furthermore, at least G3–G6, G8, G9, G11, G12–G15 and G18–G20 genotypes with distinct sub-clusters within the genotypes were identified in porcine samples collected in Japan between 2000 and 2007 [71,110] (Figure 3). Additional evidence of remarkable porcine RVB diversity is highlighted in a report from India that demonstrates that at least G7, G19, G20 and tentative novel G21 genotypes (associated with H4 and H5 genotypes) circulate in the Northern and Western regions of India [145].

4.4. Australia

Apart from several reports on circulation of porcine RVA G3, G4 and G5 ~3 decades ago [146–148], there is no epidemiological data for porcine RVs in this region (Figure 2).

5. Zoonotic Potential of Porcine RV Strains

Historically, RVs were believed to be host-specific; however, recent and growing evidence challenges this postulation. Diverse animal reservoirs of zoonotic RVs are suggested to include at least porcine, bovine, ovine, pteropine, rodent, avian and insectivore species [17,85,149–151]. The widely documented zoonotic potential of RVA strains is best exemplified by globally emerging human RVs, such as G9 and G12, likely originating from porcine species by gene reassortment because similar G9 and G12 VP7 specificities are often observed in piglets [139,152–154]. Additionally, numerous reports have described interspecies transmission leading to sporadic cases of human disease with RVs from different animal species origin [72,155–158]. Table 1 summarizes common (G1–G4, P[6] and P[8]) and uncommon human RV G and P genotypes (suggestive of possible emergence via re-assortment) and G/P combinations (indicating possible direct transmission) that likely originated from swine. A total of 10 G genotypes (G1–5, G9–G12 and G26) and 7 P genotypes (P[4], P[6], P[8], P[13], P[14], P[19] and P[25]) of porcine origin have been identified in humans to date, with some genotypes including G10, G11, G12, G26, P[13], P[14], P[19] and P[25] displaying regional characteristics (found only in Asian or African countries), whereas the rest were found more commonly or emerging globally (Table 1). The recent discovery showing that different P-genotypes of RVA strains interact with distinct histo-blood group antigens (HBGA, ABOH, Lewis) and sialic acids via VP4 may provide insights into regional prevalence and increased zoonotic potential of some RVAs of swine origin [159–162]. While only a few animal RVs (of P[1], P[2], P[3], and P[7]) are sialidase sensitive, cellular attachment of human and the majority of animal RVs are sialic acid independent and use HBGAs as attachment factors or (co)receptors [161]. Further, RVs bearing different P-types recognize polymorphic HBGAs in a strain-specific manner, leading to variable host-specific susceptibility among different populations. Further, a stepwise-biosynthesis of HBGAs may represent one of the mechanisms regulating age-specific susceptibility to RV infection in early life [161]. Similar polymorphic HBGAs are also observed in many animals, including pigs (A and H antigens) [163]. The latter may provide an explanation why RVA strains of the P[6] genotype (that recognize H antigen) are commonly found in and transmitted between humans and pigs in different countries, while P[19] strains in humans of potential porcine origin appear to be restricted to India, Asian and African countries coinciding with distinct polymorphisms in Lewis antigens associated with Caucasian and other populations [164].

Table 1. Human RV genotypes of suspect or confirmed porcine origin via direct transmission or multiple re-assortment events.

Porcine RV Species	G and/or P Genotype	Geographic Region	Year Samples Collected	Epidemiological Status and Medical Relevance	Reference
A	G1-G4, G9, G12 and P8	Worldwide	2000s	Commonly seen in humans *	[152,153,165]
	G3-G5, G9 and G11, as well as P[6]	Denmark, France, Hungary, Italy, Slovenia	2003-2007	G3-G5—common in humans, G5—regional in humans, P6—rare in humans	[85]
	G1 and G4	Brazil	2007	Common	[166]
	G1P[8]	China; MD, USA	2004-2009	Common	[15,167,168]
	G1, G1P[6]	Japan	2001	Common	[169]
	G1P[6]	Japan	1997	Rare	[170]
	G1P[6], G4P[6] and G12P[6]	Democratic Republic of the Congo	2007-2010	Common	[171]
	G1P[19]	India	1992	Rare	[134]
	G2	Europe	1992	Uncommon	[130]
	G3P[6], G4P[6] and G4P[8]	China, Italy, Slovenia	2003-2013	Common	[172-174]
	G3P[25]	Taiwan	2009	Rare	[175]
	G4P[6] strains, one G5P[6]	Taiwan	2006-2012	Common	[176]
	G4P[6]	Hungary, China, Argentina, Madagascar	2006-2007 2008-2009	Sporadic identification in humans worldwide	[177-181]
	G5P[6]	Japan, Bulgaria	2011 2006	Rare	[182,183]
	G5P[8]	Brazil, Argentina, Paraguay, Cameroon, China, Thailand, and Vietnam	1986-2005	Common in Asian, African and South American countries	[184]
	G9	NE, USA; India	1980s; 1190s, 1997-2000	Uncommon, emerging worldwide	[153,185,186]
	G9P[6]	India	2007	Unusual	[187]
	G9P[19]	Thailand, India	2012-2013 1989-1990	Rare	[188,189]
	G9P[19] and G9P[13]	Taiwan	2014-2015	Rare	[190]
	G9P[19] and G10P[14]	Vietnam	2007-2008	Rare	[191]
	G11P[4], G11P[6], G11P[8] G11P[25]	Nepal, Bangladesh	2001-2004	Rare	[192]
	G11P[25]	India	2005-2009	Uncommon	[193,194]
	G12P[6] and G12P[8]	Kenya, Myanmar	2010, 2011	Common	[195,196]
	G26P[19]	Vietnam	2009-2010	Atypical in humans	[197]
B	N/A	Brazil	2000s	Regional significance	[198]
C	N/A	Japan, Brazil	1982-1986, 2000-2007	Regional significance	[100,199]

* G1P[8], G2P[4], G3P[8], C4P[8], and G9P[8]) were described in ~90% of samples from humans submitted to the EuroRotaNet database (that included data for 17 European countries: Belgium, Bulgaria, Denmark, Finland, France, Germany, Greece, Hungary, Italy, Lithuania, The Netherlands, Romania, Slovenia, Spain, Sweden, UK) between 2005 and 2009 from the 16 participating countries [200–202]. Letters of different colors represent different G-genotypes for easier distinction.

Unlike porcine RVA strains which are commonly demonstrated to possess zoonotic potential [17], there is currently little evidence in support of porcine RVC interspecies transmission. Identification of porcine RVC-derived genes in human and bovine RVC strains was reported in Brazil [199]. In addition to identification of bovine RVC strain WD534tc of likely porcine origin [203], whole genome analysis of porcine RVC strains from Japan has suggested a close phylogenetic relations between the human and some of these porcine RVC strains [100]. Additionally, a possible zoonotic role of animal RVCs has also been hypothesized based on increased seroprevalence rates to RVC in human populations [7] and the high prevalence of RVC infections in some geographic areas where they may cause <5% of gastroenteritis-associated hospitalizations in childhood [204]. However, it is important to note that the limited genetic variability of RVCs in humans contrasts with the high genetic diversity currently seen in pigs [97].

More recently, RVB strains were identified from sporadic cases of infantile diarrhea in Bangladesh as opposed to adult diarrhea cases associated with RVB in China and India. These recent strains differed genetically from the Chinese strain [53,55], suggesting that diverse RVB strains are circulating in humans. Limited evidence for the zoonotic potential of some porcine RVB strains was provided by Medici and colleagues demonstrated a high nucleotide identity between the NSP2 gene sequences of human and porcine strains [198].

Overall, these data indicate that frequent surveillance of porcine RVA and additional research on porcine RVB/RVC diversity in swine are needed to control their regional and global zoonotic spread.

6. Passive and Active Immunity

Immune responses and correlates of protection against RVs in humans and different animal species (mostly against RVA) are reviewed elsewhere [205,206]. Much of the knowledge of RV immune responses has been generated using a gnotobiotic (Gn) pig model and human RV infection/vaccines. In this review, we will briefly summarize passive and active immune responses in pigs induced by human RVA strains (Figure 4), since piglets can be infected with porcine and human RV strains, and develop clinical disease [206]. In terms of innate immunity, our recent studies have demonstrated that decreased severity of human RV clinical disease and infection was associated with enhanced function and frequencies of plasmacytoid dendritic cell (pDC) and natural killer (NK) cells evident systemically and locally and systemic IL-12 responses [207], similar to observations in humans and mice [208]. Although the role of interferon (IFN)-α in protection against homologous/heterologous RV infections is debated [209–211], earlier we demonstrated that an imbalanced IFN-α production coincided with increased human RV disease/infection severity [212]. Additionally, increased expression of toll-like receptor 3 (recognizes double-stranded RNA) was associated with improved protection against human RV infection and disease in Gn piglets, suggesting it could be an attractive target for therapeutic development [213]. Finally, reduced human RV replication in Gn piglets in our recent studies was associated with increased total Ig responses in systemic and local tissues [214].

The correlates of oral human RV vaccine induced protection against challenge with human RV (G1P[8]) were the presence and concentration of RV-specific IgA antibodies or antibody-secreting-cells (ASC) in serum or intestine, and frequencies of IFN-γ producing CD4+ T cells, but not the concentration of intestinal or systemic RV-neutralizing antibodies [215–217] or VP6-specific IgA antibodies [205,206,218] (Figure 4).

Priming orally with an attenuated human RV vaccine conferred protection in piglets that was augmented by a booster with VP 2/6 virus–like particles (VLPs) [218]. This protection was correlated with immune responses to VP4 and VP7 [206]. However, systemic and intestinal immune responses to human RV NSP4 alone did not correlate with protection of Gn piglets against human RV challenge [219]. While maternally derived circulating RV-specific antibodies mediated high levels of passive protection against human RV disease, active immune responses to replicating and non-replicating human RV vaccines were suppressed, as evident by reduced numbers of ASC in the intestine which decreased protection upon experimental challenge [220,221].

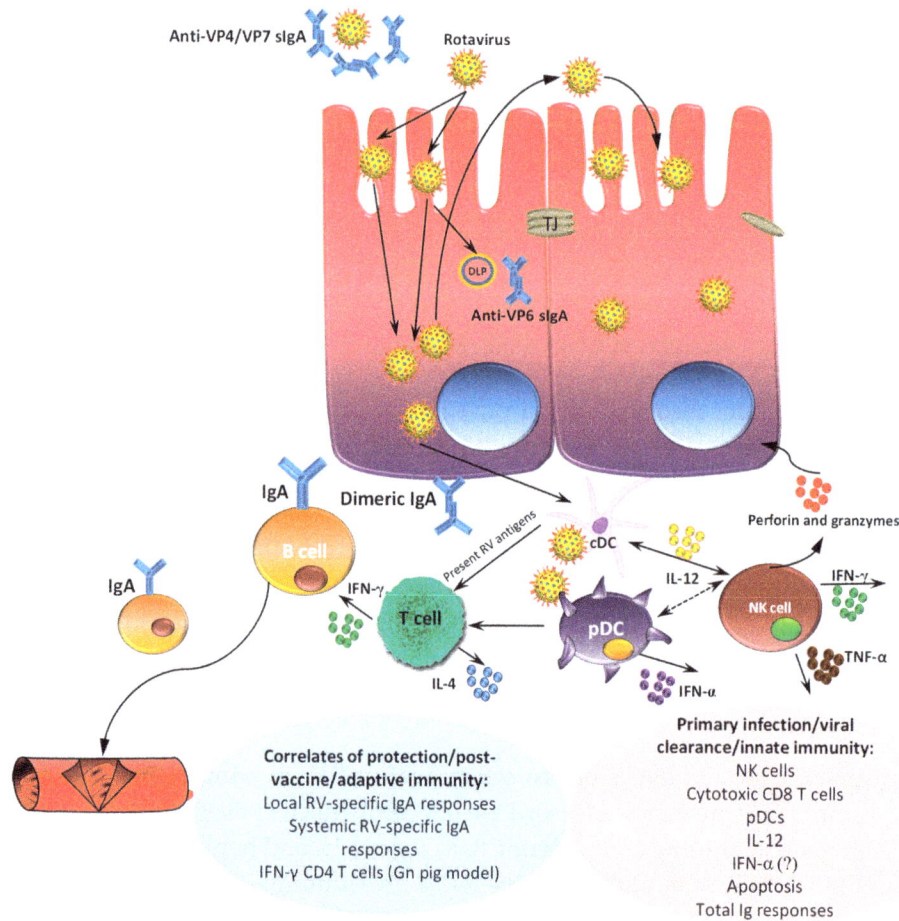

Figure 4. Immune responses to RV infection in pigs. Intestinal RV VP4/VP7 secretory immunoglobulin A (sIgA) neutralizing antibodies can prevent viral binding to enterocytes and penetration (early post-infection), while viral replication can be partially inhibited by anti-VP6 sIgA during transcytosis across enterocytes. In addition, a number of immune cells contribute to RV innate and adaptive immune responses: plasmacytoid dendritic cells (pDCs) produce antiviral (IFN-α) and pro-inflammatory (IL-12) cytokines which can inhibit RV replication or induce other immune cell subsets, including natural killer (NK) cells that produce granzymes, perforins and TNF-α and can lyse RV-infected cells. After antigen presentation by conventional dendritic cells (cDCs) to T cells, cytokine-secreting (IFN-γ in particular) RV-specific Th cells can also inhibit viral replication and activate IgA production by B cells. Additionally, RV-specific CD8 cytotoxic IFN-γ producing T cells contribute to the lysis of RV infected cells. RV induces apoptosis of intestinal epithelial (enterocytes) and immune cells; however, it is unclear whether this decreases (by eliminating infected cells) or promotes (via dissemination of the infectious particles) RV replication. Although high levels of systemic RV-neutralizing antibodies may coincide with improved protection against RV challenge, they are not correlated with protection in most studies. TJ: tight junctions. DLP: double-layered particles.

7. Porcine RVA Vaccines and Control Strategies: Potential Impact of Vaccines on Porcine RVA Genetic Diversity

Although following worldwide application of human RV vaccines, child mortality due to diarrhea declined, RV remains the most common cause of severe dehydrating diarrhea among children <5 years of age [222]. In livestock, vaccination strategies were focused on the induction of active or passive immunity, however, oral administration of attenuated RV vaccines to piglets and calves often lacked efficacy in the field [223]. The endemic porcine RV infections and the ubiquitous presence of porcine RV antibodies in swine revealed a need for strategies to boost lactogenic immunity in sows to provide passive antibodies to the neonate with colostrum and milk. The variable success of maternal RV vaccines in the field is influenced by vaccine dose, strain, inactivating agent, adjuvant, route of administration, and porcine RV exposure levels. The use of genetically engineered VLP vaccines to boost antibodies in mammary secretions showed promise because they are replication independent allowing circumvention of maternal antibody interference. However, although the immunogenicity of such VLP vaccines was high, the protective efficacy they induced was insufficient demonstrating the need for priming with live attenuated RV vaccines [224]. Nevertheless, field application of ProSystem porcine RV vaccine (which contained modified live porcine RVA strains of G4P[6] and G5P[7] genotype combinations) or G5P[7] (porcine RVA OSU) based vaccines could have resulted in the widespread circulation of porcine RVA of these genotypes for several decades and their more recent substitution by G9 and G11 genotypes or reassortant G4 and G5 variants discussed in detail in Section 3 of this review. Alternatively, they could generate herd immunity gradually decreasing the prevalence of the historic G4/G5 porcine RVA genotypes and allowing for the spread of novel emerging porcine RVAs.

8. Concluding Remarks

The remarkable diversity and genetic plasticity of porcine RVs indicate a need for further research on molecular characterization and spatio-temporal prevalence and fluctuations of endemic and emerging porcine RVs. The recent emergence of unusual G and P genotypes of porcine RVA strains worldwide, the discovery of novel porcine RV groups in different geographic regions, as well as the growing evidence of increased porcine RV prevalence and genetic diversity compared to that previously estimated suggest that porcine RV epidemiology is very complex and highly dynamic. These observations lead to at least two conclusions: (i) molecular diagnostic and characterization toolkits should be frequently updated and expanded to include novel porcine RV variants to ensure accurate epidemiological monitoring (especially for the countries where such information is lacking: African countries, Russia, Australia, etc.); (ii) a better understanding of the molecular pathogenesis and immunity to porcine RV is needed to optimize and update classical vaccine approaches to control porcine RV infections and spread. Although not highly efficacious in the field, attenuated replicating porcine RVA vaccines may be contributing directly to the genetic diversity of porcine RVs (via reassortment of vaccine strains with wild type strains and their subsequent spread) and the emergence of novel genetic variants that can evade herd immunity against the vaccine strains, as observed with human RVA vaccines, RotaTeq and Rotarix, that generate within vaccine (RotaTeq) and vaccine-wild type strain re-assortants capable of further spread in susceptible populations [225]. Alternative or additional approaches (to live attenuated vaccine use) may include wide-scale probiotic use or therapeutic applications that target the virus replication cycle to enhance innate or anamnestic immune responses, to decrease RV shedding and environmental contamination, and to alleviate porcine RV-mediated intestinal damage. Finally, although not previously well recognized, the zoonotic potential of various porcine RV genogroups/genotypes should be carefully and extensively evaluated by conducting simultaneous epidemiological studies of human and porcine RVs in the same geographic regions. Additional studies to understand the higher propensity of some genogroups/genotypes to generate re-assorted variants and cross interspecies barriers are needed, including the potential interactions of different porcine RV genotypes with HBGAs as shown for human RV strains [159–162].

Acknowledgments: This work was supported by state and federal funds appropriated to the Ohio Agricultural Research and Development Center (OARDC) of the Ohio State University Grant support was from a Ohio State University SEED Grant—#2011-077 and a National Pork Board grant (NPB 12-094) for swine health.

Author Contributions: A.N.V. and L.J.S. conceived and designed the experiments described in this review; J.O.A. performed the experiments described in this review; A.N.V. and J.O.A. analyzed the data; A.N.V. wrote this review that was edited by L.J.S.

References

1. Chang, K.; Kim, Y.; Saif, L.J. Rotavirus and reovirus. In *Diseases of Swine*, 10 ed.; Zimmerman, J.J., Karriker, L.A., Ramirez, A., Schwartz, K.J., Stevenson, G.W., Eds.; Wiley-Blackwell: West Sussex, UK, 2012; pp. 621–634.

2. Estes, M.K.; Kang, G.; Zeng, C.Q.; Crawford, S.E.; Ciarlet, M. Pathogenesis of rotavirus gastroenteritis. *Novartis Found. Symp.* **2001**, *238*, 82–96. [PubMed]

3. Nava, P.; Lopez, S.; Arias, C.F.; Islas, S.; Gonzalez-Mariscal, L. The rotavirus surface protein VP8 modulates the gate and fence function of tight junctions in epithelial cells. *J. Cell Sci.* **2004**, *117 Pt 23*, 5509–5519. [CrossRef] [PubMed]

4. Obert, G.; Peiffer, I.; Servin, A.L. Rotavirus-induced structural and functional alterations in tight junctions of polarized intestinal Caco-2 cell monolayers. *J. Virol.* **2000**, *74*, 4645–4651. [CrossRef] [PubMed]

5. Dickman, K.G.; Hempson, S.J.; Anderson, J.; Lippe, S.; Zhao, L.; Burakoff, R.; Shaw, R.D. Rotavirus alters paracellular permeability and energy metabolism in Caco-2 cells. *Am. J. Physiol. Gastrointest. Liver Physiol.* **2000**, *279*, G757–G766. [PubMed]

6. Beau, I.; Cotte-Laffitte, J.; Amsellem, R.; Servin, A.L. A protein kinase A-dependent mechanism by which rotavirus affects the distribution and mRNA level of the functional tight junction-associated protein, occludin, in human differentiated intestinal Caco-2 cells. *J. Virol.* **2007**, *81*, 8579–8586. [CrossRef]

7. Iturriza-Gomara, M.; Clarke, I.; Desselberger, U.; Brown, D.; Thomas, D.; Gray, J. Seroepidemiology of group C rotavirus infection in England and Wales. *Eur. J. Epidemiol.* **2004**, *19*, 589–595. [CrossRef] [PubMed]

8. Blutt, S.E.; Conner, M.E. Rotavirus: to the gut and beyond! *Curr. Opin. Gastroenterol.* **2007**, *23*, 39–43. [CrossRef] [PubMed]

9. Estes, M.; Greenberg, H.B. Rotaviruses. In *Fields Virology*, 5 ed.; Knipe, D.M., Howley, P., Eds.; Wolters Kluwer Health/Lippincott Williams & Wilkins: Philadelphia, PA, USA, 2013; pp. 1347–1395.

10. Matthijnssens, J.; Otto, P.H.; Ciarlet, M.; Desselberger, U.; Van Ranst, M.; Johne, R. VP6-sequence-based cutoff values as a criterion for rotavirus species demarcation. *Arch. Virol.* **2012**, *157*, 1177–1182. [CrossRef]

11. Mihalov-Kovacs, E.; Gellert, A.; Marton, S.; Farkas, S.L.; Feher, E.; Oldal, M.; Jakab, F.; Martella, V.; Banyai, K. Candidate new rotavirus species in sheltered dogs, Hungary. *Emerg. Infect. Dis.* **2015**, *21*, 660–663. [CrossRef] [PubMed]

12. Banyai, K.; Kemenesi, G.; Budinski, I.; Foldes, F.; Zana, B.; Marton, S.; Varga-Kugler, R.; Oldal, M.; Kurucz, K.; Jakab, F. Candidate new rotavirus species in Schreiber's bats, Serbia. *Infect. Genet. Evol.* **2017**, *48*, 19–26. [CrossRef] [PubMed]

13. Matthijnssens, J.; Ciarlet, M.; McDonald, S.M.; Attoui, H.; Banyai, K.; Brister, J.R.; Buesa, J.; Esona, M.D.; Estes, M.K.; Gentsch, J.R.; et al. Uniformity of rotavirus strain nomenclature proposed by the Rotavirus Classification Working Group (RCWG). *Arch. Virol.* **2011**, *156*, 1397–1413. [CrossRef] [PubMed]

14. Trojnar, E.; Sachsenroder, J.; Twardziok, S.; Reetz, J.; Otto, P.H.; Johne, R. Identification of an avian group A rotavirus containing a novel VP4 gene with a close relationship to those of mammalian rotaviruses. *J. Gen. Virol.* **2013**, *94 Pt 1*, 136–142. [CrossRef] [PubMed]

15. Matthijnssens, J.; Ciarlet, M.; Heiman, E.; Arijs, I.; Delbeke, T.; McDonald, S.M.; Palombo, E.A.; Iturriza-Gomara, M.; Maes, P.; Patton, J.T.; et al. Full genome-based classification of rotaviruses reveals a common origin between human Wa-Like and porcine rotavirus strains and human DS-1-like and bovine rotavirus strains. *J. Virol.* **2008**, *82*, 3204–3219. [CrossRef] [PubMed]

16. Matthijnssens, J.; Ciarlet, M.; Rahman, M.; Attoui, H.; Banyai, K.; Estes, M.K.; Gentsch, J.R.; Iturriza-Gomara, M.; Kirkwood, C.D.; Martella, V.; et al. Recommendations for the classification of group A rotaviruses using all 11 genomic RNA segments. *Arch. Virol.* **2008**, *153*, 1621–1629. [CrossRef] [PubMed]

17. Martella, V.; Banyai, K.; Matthijnssens, J.; Buonavoglia, C.; Ciarlet, M. Zoonotic aspects of rotaviruses. *Vet. Microbiol.* **2010**, *140*, 246–255. [CrossRef] [PubMed]

18. Marthaler, D.; Suzuki, T.; Rossow, K.; Culhane, M.; Collins, J.; Goyal, S.; Tsunemitsu, H.; Ciarlet, M.; Matthijnssens, J. VP6 genetic diversity, reassortment, intragenic recombination and classification of rotavirus B in American and Japanese pigs. *Vet. Microbiol.* **2014**, *172*, 359–366. [CrossRef] [PubMed]

19. Matthijnssens, J.; Taraporewala, Z.F.; Yang, H.; Rao, S.; Yuan, L.; Cao, D.; Hoshino, Y.; Mertens, P.P.; Carner, G.R.; McNeal, M.; et al. Simian rotaviruses possess divergent gene constellations that originated from interspecies transmission and reassortment. *J. Virol.* **2010**, *84*, 2013–2026. [CrossRef] [PubMed]

20. McNulty, M.S. Rotaviruses. *J. Gen. Virol.* **1978**, *40*, 1–18. [CrossRef] [PubMed]

21. McNulty, M.S.; Allan, G.M.; Connor, T.J.; McFerran, J.B.; McCracken, R.M. An entero-like virus associated with the runting syndrome in broiler chickens. *Avian Pathol.* **1984**, *13*, 429–439. [CrossRef] [PubMed]

22. McNulty, M.S.; Allan, G.M.; McFerran, J.B. Prevalence of antibody to conventional and atypical rotaviruses in chickens. *Vet. Rec.* **1984**, *114*, 219. [CrossRef] [PubMed]

23. McNulty, M.S.; Todd, D.; Allan, G.M.; McFerran, J.B.; Greene, J.A. Epidemiology of rotavirus infection in broiler chickens: recognition of four serogroups. *Arch. Virol.* **1984**, *81*, 113–121. [CrossRef] [PubMed]

24. Otto, P.; Liebler-Tenorio, E.M.; Elschner, M.; Reetz, J.; Lohren, U.; Diller, R. Detection of rotaviruses and intestinal lesions in broiler chicks from flocks with runting and stunting syndrome (RSS). *Avian Dis.* **2006**, *50*, 411–418. [CrossRef] [PubMed]

25. Chasey, D.; Bridger, J.C.; McCrae, M.A. A new type of atypical rotavirus in pigs. *Arch. Virol.* **1986**, *89*, 235–243. [CrossRef] [PubMed]

26. Janke, B.H.; Nelson, J.K.; Benfield, D.A.; Nelson, E.A. Relative prevalence of typical and atypical strains among rotaviruses from diarrheic pigs in conventional swine herds. *J. Vet. Diagn. Investig.* **1990**, *2*, 308–311. [CrossRef] [PubMed]

27. Wakuda, M.; Ide, T.; Sasaki, J.; Komoto, S.; Ishii, J.; Sanekata, T.; Taniguchi, K. Porcine rotavirus closely related to novel group of human rotaviruses. *Emerg. Infect. Dis.* **2011**, *17*, 1491–1493. [CrossRef]

28. Bridger, J.C.; Woode, G.N. Neonatal calf diarrhoea: identification of a reovirus-like (rotavirus) agent in faeces by immunofluorescence and immune electron microscopy. *Br. Vet. J.* **1975**, *131*, 528–535. [PubMed]

29. Bridger, J.C.; Brown, J.F. Prevalence of antibody to typical and atypical rotaviruses in pigs. *Vet. Rec.* **1985**, *116*, 50. [CrossRef] [PubMed]

30. Saif, L.J.; Rosen, B.; Parwani, A. Animal rotaviruses. In *Virus Infections of the Gastrointestinal Tract*; Kapikian, A.Z., Ed.; Marcel-Dekker: New York, NY, USA, 1994; pp. 279–367.

31. Saif, L.J.; Bohl, E.H.; Theil, K.W.; Cross, R.F.; House, J.A. Rotavirus-like, calicivirus-like, and 23-nm virus-like particles associated with diarrhea in young pigs. *J. Clin. Microbiol.* **1980**, *12*, 105–111. [PubMed]

32. Saif, L.J.; Jiang, B. Nongroup A rotaviruses of humans and animals. *Curr. Top. Microbiol. Immunol.* **1994**, *185*, 339–371.

33. Mebus, C.A.; Underdahl, N.R.; Rhodes, M.B.; Twiehaus, M.J. Calf diarrhea (scours): reproduced with a virus from field outbreak. *Neb. Agric. Exp. Stn. Res. Bull.* **1969**, *233*, 1–16.

34. Mebus, C.A.; Underdahl, N.R.; Rhodes, M.B.; Twiehaus, M.J. Further studies on neonatal calf diarrhea virus. *Proc. Annu. Meet. U. S. Anim. Health Assoc.* **1969**, *73*, 97–99. [PubMed]

35. Bishop, R.F.; Davidson, G.P.; Holmes, I.H.; Ruck, B.J. Virus particles in epithelial cells of duodenal mucosa from children with acute non-bacterial gastroenteritis. *Lancet* **1973**, *2*, 1281–1283. [CrossRef]

36. Theil, K.W. Group A rotaviruses. In *Viral Diarrheas of Man and Animals*; Saif, L.J., Theil, K.W., Eds.; CRC Press: Boca Raton, FL, USA, 1990; pp. 35–72.

37. Holmes, I.H. Rotaviruses. In *The Reoviridae*; Joklik, W.T., Ed.; Plenum Press: New York, NY, USA, 1983; pp. 359–423.

38. Rodger, S.M.; Bishop, R.F.; Holmes, I.H. Detection of a rotavirus-like agent associated with diarrhea in an infant. *J. Clin. Microbiol.* **1982**, *16*, 724–726. [PubMed]

39. Bridger, J.C.; Pedley, S.; McCrae, M.A. Group C rotaviruses in humans. *J. Clin. Microbiol.* **1986**, *23*, 760–763. [PubMed]

40. Torres-Medina, A. Isolation of an atypical rotavirus causing diarrhea in neonatal ferrets. *Lab. Anim. Sci.* **1987**, *37*, 167–171. [PubMed]

41. Tsunemitsu, H.; Saif, L.J.; Jiang, B.M.; Shimizu, M.; Hiro, M.; Yamaguchi, H.; Ishiyama, T.; Hirai, T. Isolation, characterization, and serial propagation of a bovine group C rotavirus in a monkey kidney cell line (MA104). *J. Clin. Microbiol.* **1991**, *29*, 2609–2613. [PubMed]

42. Theil, K.W.; Saif, L.J.; Moorhead, P.D.; Whitmoyer, R.E. Porcine rotavirus-like virus (group B rotavirus): characterization and pathogenicity for gnotobiotic pigs. *J. Clin. Microbiol.* **1985**, *21*, 340–345. [PubMed]

43. Barman, P.; Ghosh, S.; Das, S.; Varghese, V.; Chaudhuri, S.; Sarkar, S.; Krishnan, T.; Bhattacharya, S.K.; Chakrabarti, A.; Kobayashi, N.; et al. Sequencing and sequence analysis of VP7 and NSP5 genes reveal emergence of a new genotype of bovine group B rotaviruses in India. *J. Clin. Microbiol.* **2004**, *42*, 2816–2818. [CrossRef] [PubMed]

44. Chang, K.O.; Parwani, A.V.; Smith, D.; Saif, L.J. Detection of group B rotaviruses in fecal samples from diarrheic calves and adult cows and characterization of their VP7 genes. *J. Clin. Microbiol.* **1997**, *35*, 2107–2110. [PubMed]

45. Ghosh, S.; Varghese, V.; Sinha, M.; Kobayashi, N.; Naik, T.N. Evidence for interstate transmission and increase in prevalence of bovine group B rotavirus strains with a novel VP7 genotype among diarrhoeic calves in Eastern and Northern states of India. *Epidemiol. Infect.* **2007**, *135*, 1324–1330. [CrossRef] [PubMed]

46. Tsunemitsu, H.; Morita, D.; Takaku, H.; Nishimori, T.; Imai, K.; Saif, L.J. First detection of bovine group B rotavirus in Japan and sequence of its VP7 gene. *Arch. Virol.* **1999**, *144*, 805–815. [CrossRef] [PubMed]

47. Shen, S.; McKee, T.A.; Wang, Z.D.; Desselberger, U.; Liu, D.X. Sequence analysis and in vitro expression of genes 6 and 11 of an ovine group B rotavirus isolate, KB63: Evidence for a non-defective, C-terminally truncated NSP1 and a phosphorylated NSP5. *J. Gen. Virol.* **1999**, *80 Pt 8*, 2077–2085. [CrossRef]

48. Eiden, J.J.; Nataro, J.; Vonderfecht, S.; Petric, M. Molecular cloning, sequence analysis, in vitro expression, and immunoprecipitation of the major inner capsid protein of the IDIR strain of group B rotavirus (GBR). *Virology* **1992**, *188*, 580–589. [CrossRef]

49. Chen, G.M.; Hung, T.; Mackow, E.R. Identification of the gene encoding the group B rotavirus VP7 equivalent: primary characterization of the ADRV segment 9 RNA. *Virology* **1990**, *178*, 311–315. [CrossRef]

50. Dai, G.Z.; Sun, M.S.; Liu, S.Q.; Ding, X.F.; Chen, Y.D.; Wang, L.C.; Du, D.P.; Zhao, G.; Su, Y.; Li, J.; et al. First report of an epidemic of diarrhoea in human neonates involving the new rotavirus and biological characteristics of the epidemic virus strain (KMB/R85). *J. Med. Virol.* **1987**, *22*, 365–373. [PubMed]

51. Fang, Z.Y.; Ye, Q.; Ho, M.S.; Dong, H.; Qing, S.; Penaranda, M.E.; Hung, T.; Wen, L.; Glass, R.I. Investigation of an outbreak of adult diarrhea rotavirus in China. *J. Infect. Dis.* **1989**, *160*, 948–953. [CrossRef] [PubMed]

52. Hung, T.; Chen, G.M.; Wang, C.G.; Yao, H.L.; Fang, Z.Y.; Chao, T.X.; Chou, Z.Y.; Ye, W.; Chang, X.J.; Den, S.S.; et al. Waterborne outbreak of rotavirus diarrhoea in adults in China caused by a novel rotavirus. *Lancet* **1984**, *1*, 1139–1142.

53. Kelkar, S.D.; Zade, J.K. Group B rotaviruses similar to strain CAL-1, have been circulating in Western India since 1993. *Epidemiol. Infect.* **2004**, *132*, 745–749. [CrossRef] [PubMed]

54. Lahon, A.; Chitambar, S.D. Molecular characterization of VP4, VP6, VP7 and NSP4 genes of group B rotavirus strains from outbreaks of gastroenteritis. *Asian Pac. J. Trop. Med.* **2011**, *4*, 846–849. [CrossRef]

55. Ahmed, M.U.; Kobayashi, N.; Wakuda, M.; Sanekata, T.; Taniguchi, K.; Kader, A.; Naik, T.N.; Ishino, M.; Alam, M.M.; Kojima, K.; et al. Genetic analysis of group B human rotaviruses detected in Bangladesh in 2000 and 2001. *J. Med. Virol.* **2004**, *72*, 149–155. [CrossRef] [PubMed]

56. Saiada, F.; Rahman, H.N.; Moni, S.; Karim, M.M.; Pourkarim, M.R.; Azim, T.; Rahman, M. Clinical presentation and molecular characterization of group B rotaviruses in diarrhoea patients in Bangladesh. *J. Med. Microbiol.* **2011**, *60 Pt 4*, 529–536. [CrossRef] [PubMed]

57. Rahman, M.; Hassan, Z.M.; Zafrul, H.; Saiada, F.; Banik, S.; Faruque, A.S.; Delbeke, T.; Matthijnssens, J.; Van Ranst, M.; Azim, T. Sequence analysis and evolution of group B rotaviruses. *Virus Res.* **2007**, *125*, 219–225. [CrossRef] [PubMed]

58. Sanekata, T.; Ahmed, M.U.; Kader, A.; Taniguchi, K.; Kobayashi, N. Human group B rotavirus infections cause severe diarrhea in children and adults in Bangladesh. *J. Clin. Microbiol.* **2003**, *41*, 2187–2190. [CrossRef] [PubMed]

59. Aung, T.S.; Kobayashi, N.; Nagashima, S.; Ghosh, S.; Aung, M.S.; Oo, K.Y.; Win, N. Detection of group B rotavirus in an adult with acute gastroenteritis in Yangon, Myanmar. *J. Med. Virol.* **2009**, *81*, 1968–1974. [CrossRef] [PubMed]

60. Chasey, D.; Davies, P. Atypical rotaviruses in pigs and cattle. *Vet. Rec.* **1984**, *114*, 16–17. [CrossRef] [PubMed]

61. Marthaler, D.; Rossow, K.; Culhane, M.; Goyal, S.; Collins, J.; Matthijnssens, J.; Nelson, M.; Ciarlet, M. Widespread rotavirus H in commercially raised pigs, United States. *Emerg. Infect. Dis.* **2014**, *20*, 1195–1198. [CrossRef] [PubMed]

62. Molinari, B.L.; Lorenzetti, E.; Otonel, R.A.; Alfieri, A.F.; Alfieri, A.A. Species H rotavirus detected in piglets with diarrhea, Brazil, 2012. *Emerg. Infect. Dis.* **2014**, *20*, 1019–1022. [CrossRef] [PubMed]

63. Chinivasagam, H.N.; Thomas, R.J.; Casey, K.; McGahan, E.; Gardner, E.A.; Rafiee, M.; Blackall, P.J. Microbiological status of piggery effluent from 13 piggeries in the south east Queensland region of Australia. *J. Appl. Microbiol.* **2004**, *97*, 883–891. [CrossRef] [PubMed]

64. Amimo, J.O.; El Zowalaty, M.E.; Githae, D.; Wamalwa, M.; Djikeng, A.; Nasrallah, G.K. Metagenomic analysis demonstrates the diversity of the fecal virome in asymptomatic pigs in East Africa. *Arch. Virol.* **2016**, *161*, 887–897. [CrossRef] [PubMed]

65. Amimo, J.O.; Junga, J.O.; Ogara, W.O.; Vlasova, A.N.; Njahira, M.N.; Maina, S.; Okoth, E.A.; Bishop, R.P.; Saif, L.J.; Djikeng, A. Detection and genetic characterization of porcine group A rotaviruses in asymptomatic pigs in smallholder farms in East Africa: Predominance of P[8] genotype resembling human strains. *Vet. Microbiol.* **2015**, *175*, 195–210. [CrossRef] [PubMed]

66. Amimo, J.O.; Otieno, T.F.; Okoth, E.; Onono, J.O.; Bett, B. Risk factors for rotavirus infection in pigs in Busia and Teso subcounties, Western Kenya. *Trop. Anim. Health Prod.* **2017**, *49*, 105–112. [CrossRef] [PubMed]

67. Amimo, J.O.; Vlasova, A.N.; Saif, L.J. Detection and genetic diversity of porcine group A rotaviruses in historic (2004) and recent (2011 and 2012) swine fecal samples in Ohio: predominance of the G9P[13] genotype in nursing piglets. *J. Clin. Microbiol.* **2013**, *51*, 1142–1151. [CrossRef] [PubMed]

68. Da Silva, M.F.; Tort, L.F.; Gomez, M.M.; Assis, R.M.; de Mendonca, M.C.; Volotao Ede, M.; Leite, J.P. Phylogenetic analysis of VP1, VP2, and VP3 gene segments of genotype G5 group A rotavirus strains circulating in Brazil between 1986 and 2005. *Virus Res.* **2011**, *160*, 381–388. [CrossRef] [PubMed]

69. Homwong, N.; Diaz, A.; Rossow, S.; Ciarlet, M.; Marthaler, D. Three-Level Mixed-Effects Logistic Regression Analysis Reveals Complex Epidemiology of Swine Rotaviruses in Diagnostic Samples from North America. *PLoS ONE* **2016**, *11*, e0154734.

70. Martella, V.; Ciarlet, M.; Banyai, K.; Lorusso, E.; Arista, S.; Lavazza, A.; Pezzotti, G.; Decaro, N.; Cavalli, A.; Lucente, M.S.; et al. Identification of group A porcine rotavirus strains bearing a novel VP4 (P) Genotype in Italian swine herds. *J. Clin. Microbiol.* **2007**, *45*, 577–580. [CrossRef] [PubMed]

71. Marthaler, D.; Homwong, N.; Rossow, K.; Culhane, M.; Goyal, S.; Collins, J.; Matthijnssens, J.; Ciarlet, M. Rapid detection and high occurrence of porcine rotavirus A, B, and C by RT-qPCR in diagnostic samples. *J. Virol. Methods* **2014**, *209*, 30–34. [CrossRef] [PubMed]

72. Papp, H.; Matthijnssens, J.; Martella, V.; Ciarlet, M.; Banyai, K. Global distribution of group A rotavirus strains in horses: a systematic review. *Vaccine* **2013**, *31*, 5627–5633. [CrossRef] [PubMed]

73. Theuns, S.; Vyt, P.; Desmarets, L.M.; Roukaerts, I.D.; Heylen, E.; Zeller, M.; Matthijnssens, J.; Nauwynck, H.J. Presence and characterization of pig group A and C rotaviruses in feces of Belgian diarrheic suckling piglets. *Virus Res.* **2016**, *213*, 172–183. [CrossRef] [PubMed]

74. Pham, H.A.; Carrique-Mas, J.J.; Nguyen, V.C.; Ngo, T.H.; Nguyet, L.A.; Do, T.D.; Vo, B.H.; Phan, V.T.; Rabaa, M.A.; Farrar, J.; et al. The prevalence and genetic diversity of group A rotaviruses on pig farms in the Mekong Delta region of Vietnam. *Vet. Microbiol.* **2014**, *170*, 258–265. [PubMed]

75. Halaihel, N.; Masia, R.M.; Fernandez-Jimenez, M.; Ribes, J.M.; Montava, R.; De Blas, I.; Girones, O.; Alonso, J.L.; Buesa, J. Enteric calicivirus and rotavirus infections in domestic pigs. *Epidemiol. Infect.* **2010**, *138*, 542–548. [CrossRef] [PubMed]

76. Kim, H.J.; Park, S.I.; Ha, T.P.; Jeong, Y.J.; Kim, H.H.; Kwon, H.J.; Kang, M.I.; Cho, K.O.; Park, S.J. Detection and genotyping of Korean porcine rotaviruses. *Vet. Microbiol.* **2010**, *144*, 274–286. [CrossRef] [PubMed]

77. Katsuda, K.; Kohmoto, M.; Kawashima, K.; Tsunemitsu, H. Frequency of enteropathogen detection in suckling and weaned pigs with diarrhea in Japan. *J. Vet. Diagn. Investig.* **2006**, *18*, 350–354. [CrossRef] [PubMed]

78. Saikruang, W.; Khamrin, P.; Chaimongkol, N.; Suantai, B.; Kongkaew, A.; Kongkaew, S.; Ushijima, H.; Maneekarn, N. Genetic diversity and novel combinations of G4P[19] and G9P[19] porcine rotavirus strains in Thailand. *Vet. Microbiol.* **2013**, *161*, 255–262. [CrossRef] [PubMed]

79. Lamhoujeb, S.; Cook, A.; Pollari, F.; Bidawid, S.; Farber, J.; Mattison, K. Rotaviruses from Canadian farm samples. *Arch. Virol.* **2010**, *155*, 1127–1137. [CrossRef] [PubMed]

80. Racz, M.L.; Kroeff, S.S.; Munford, V.; Caruzo, T.A.; Durigon, E.L.; Hayashi, Y.; Gouvea, V.; Palombo, E.A. Molecular characterization of porcine rotaviruses from the southern region of Brazil: characterization of an atypical genotype G[9] strain. *J. Clin. Microbiol.* **2000**, *38*, 2443–2446. [PubMed]

81. Kusumakar, A.L.; Savita; Malik, Y.S.; Minakshi; Prasad, G. Genomic diversity among group A rotaviruses from diarrheic children, piglets, buffalo and cow calves of Madhya Pradesh. *Indian J. Microbiol.* **2010**, *50*, 83–88. [CrossRef] [PubMed]

82. Parra, G.I.; Vidales, G.; Gomez, J.A.; Fernandez, F.M.; Parreno, V.; Bok, K. Phylogenetic analysis of porcine rotavirus in Argentina: increasing diversity of G4 strains and evidence of interspecies transmission. *Vet. Microbiol.* **2008**, *126*, 243–250. [CrossRef] [PubMed]

83. Wieler, L.H.; Ilieff, A.; Herbst, W.; Bauer, C.; Vieler, E.; Bauerfeind, R.; Failing, K.; Klos, H.; Wengert, D.; Baljer, G.; et al. Prevalence of enteropathogens in suckling and weaned piglets with diarrhoea in southern Germany. *J. Vet. Med. B Infect. Dis. Vet. Public Health* **2001**, *48*, 151–159. [CrossRef] [PubMed]

84. Collins, P.J.; Martella, V.; Sleator, R.D.; Fanning, S.; O'Shea, H. Detection and characterisation of group A rotavirus in asymptomatic piglets in southern Ireland. *Arch. Virol.* **2010**, *155*, 1247–1259. [CrossRef] [PubMed]

85. Midgley, S.E.; Banyai, K.; Buesa, J.; Halaihel, N.; Hjulsager, C.K.; Jakab, F.; Kaplon, J.; Larsen, L.E.; Monini, M.; Poljsak-Prijatelj, M.; et al. Diversity and zoonotic potential of rotaviruses in swine and cattle across Europe. *Vet. Microbiol.* **2012**, *156*, 238–245. [CrossRef] [PubMed]

86. Morin, M.; Turgeon, D.; Jolette, J.; Robinson, Y.; Phaneuf, J.B.; Sauvageau, R.; Beauregard, M.; Teuscher, E.; Higgins, R.; Lariviere, S. Neonatal diarrhea of pigs in Quebec: Infectious causes of significant outbreaks. *Can. J. Comp. Med.* **1983**, *47*, 11–17. [PubMed]

87. Khamrin, P.; Peerakome, S.; Tonusin, S.; Malasao, R.; Okitsu, S.; Mizuguchi, M.; Ushijima, H.; Maneekarn, N. Changing pattern of rotavirus G genotype distribution in Chiang Mai, Thailand from 2002 to 2004: Decline of G9 and reemergence of G1 and G2. *J. Med. Virol.* **2007**, *79*, 1775–1782. [CrossRef] [PubMed]

88. Okitsu, S.; Khamrin, P.; Thongprachum, A.; Kongkaew, A.; Maneekarn, N.; Mizuguchi, M.; Hayakawa, S.; Ushijima, H. Whole-genomic analysis of G3P[23], G9P[23] and G3P[13] rotavirus strains isolated from piglets with diarrhea in Thailand, 2006–2008. *Infect. Genet. Evol.* **2013**, *18*, 74–86. [CrossRef] [PubMed]

89. Okitsu, S.; Khamrin, P.; Thongprachum, A.; Maneekarn, N.; Mizuguchi, M.; Ushijima, H. Predominance of porcine P[23] genotype rotaviruses in piglets with diarrhea in northern Thailand. *J. Clin. Microbiol.* **2011**, *49*, 442–445. [CrossRef] [PubMed]

90. Miyazaki, A.; Kuga, K.; Suzuki, T.; Kohmoto, M.; Katsuda, K.; Tsunemitsu, H. Genetic diversity of group A rotaviruses associated with repeated outbreaks of diarrhea in a farrow-to-finish farm: identification of a porcine rotavirus strain bearing a novel VP7 genotype, G26. *Vet. Res.* **2011**, *42*, 112. [CrossRef] [PubMed]

91. Collins, P.J.; Martella, V.; Buonavoglia, C.; O'Shea, H. Identification of a G2-like porcine rotavirus bearing a novel VP4 type, P[32]. *Vet. Res.* **2010**, *41*, 73. [CrossRef] [PubMed]

92. Wang, Y.H.; Kobayashi, N.; Nagashima, S.; Zhou, X.; Ghosh, S.; Peng, J.S.; Hu, Q.; Zhou, D.J.; Yang, Z.Q. Full genomic analysis of a porcine-bovine reassortant G4P[6] rotavirus strain R479 isolated from an infant in China. *J. Med. Virol.* **2010**, *82*, 1094–1102. [CrossRef] [PubMed]

93. Kim, Y.; Chang, K.O.; Straw, B.; Saif, L.J. Characterization of group C rotaviruses associated with diarrhea outbreaks in feeder pigs. *J. Clin. Microbiol.* **1999**, *37*, 1484–1488. [PubMed]

94. Marthaler, D.; Rossow, K.; Culhane, M.; Collins, J.; Goyal, S.; Ciarlet, M.; Matthijnssens, J. Identification, phylogenetic analysis and classification of porcine group C rotavirus VP7 sequences from the United States and Canada. *Virology* **2013**, *446*, 189–198. [CrossRef] [PubMed]

95. Tsunemitsu, H.; Jiang, B.; Saif, L.J. Sequence comparison of the VP7 gene encoding the outer capsid glycoprotein among animal and human group C rotaviruses. *Arch. Virol.* **1996**, *141*, 705–713. [CrossRef] [PubMed]

96. Rahman, M.; Banik, S.; Faruque, A.S.; Taniguchi, K.; Sack, D.A.; van Ranst, M.; Azim, T. Detection and characterization of human group C rotaviruses in Bangladesh. *J. Clin. Microbiol.* **2005**, *43*, 4460–4465. [CrossRef] [PubMed]

97. Martella, V.; Banyai, K.; Lorusso, E.; Decaro, N.; Bellacicco, A.; Desario, C.; Corrente, M.; Greco, G.; Moschidou, P.; Tempesta, M.; et al. Genetic heterogeneity in the VP7 of group C rotaviruses. *Virology* **2007**, *367*, 358–366. [CrossRef] [PubMed]

98. Stipp, D.T.; Alfieri, A.F.; Lorenzetti, E.; da Silva Medeiros, T.N.; Possatti, F.; Alfieri, A.A. VP6 gene diversity in 11 Brazilian strains of porcine group C rotavirus. *Virus Genes* **2015**, *50*, 142–146. [CrossRef] [PubMed]

99. Suzuki, T.; Hasebe, A.; Miyazaki, A.; Tsunemitsu, H. Phylogenetic characterization of VP6 gene (inner capsid) of porcine rotavirus C collected in Japan. *Infect. Genet. Evol.* **2014**, *26*, 223–227. [CrossRef] [PubMed]

100. Suzuki, T.; Hasebe, A.; Miyazaki, A.; Tsunemitsu, H. Analysis of genetic divergence among strains of porcine rotavirus C, with focus on VP4 and VP7 genotypes in Japan. *Virus Res.* **2015**, *197*, 26–34. [CrossRef] [PubMed]

101. Soma, J.; Tsunemitsu, H.; Miyamoto, T.; Suzuki, G.; Sasaki, T.; Suzuki, T. Whole-genome analysis of two bovine rotavirus C strains: Shintoku and Toyama. *J. Gen. Virol.* **2013**, *94 Pt 1*, 128–135. [CrossRef] [PubMed]

102. Yamamoto, D.; Ghosh, S.; Kuzuya, M.; Wang, Y.H.; Zhou, X.; Chawla-Sarkar, M.; Paul, S.K.; Ishino, M.; Kobayashi, N. Whole-genome characterization of human group C rotaviruses: identification of two lineages in the VP3 gene. *J. Gen. Virol.* **2011**, *92 Pt 2*, 361–369. [CrossRef] [PubMed]

103. Moutelikova, R.; Prodelalova, J.; Dufkova, L. Diversity of VP7, VP4, VP6, NSP2, NSP4, and NSP5 genes of porcine rotavirus C: phylogenetic analysis and description of potential new VP7, VP4, VP6, and NSP4 genotypes. *Arch. Virol.* **2015**, *160*, 1715–1727. [CrossRef] [PubMed]

104. Collins, P.J.; Martella, V.; O'Shea, H. Detection and characterization of group C rotaviruses in asymptomatic piglets in Ireland. *J. Clin. Microbiol.* **2008**, *46*, 2973–2979. [CrossRef] [PubMed]

105. Niira, K.; Ito, M.; Masuda, T.; Saitou, T.; Abe, T.; Komoto, S.; Sato, M.; Yamasato, H.; Kishimoto, M.; Naoi, Y.; et al. Whole genome sequences of Japanese porcine species C rotaviruses reveal a high diversity of genotypes of individual genes and will contribute to a comprehensive, generally accepted classification system. *Infect. Genet. Evol.* **2016**, *44*, 106–113. [CrossRef]

106. Kobayashi, N.; Naik, T.N.; Kusuhara, Y.; Krishnan, T.; Sen, A.; Bhattacharya, S.K.; Taniguchi, K.; Alam, M.M.; Urasawa, T.; Urasawa, S. Sequence analysis of genes encoding structural and nonstructural proteins of a human group B rotavirus detected in Calcutta, India. *J. Med. Virol.* **2001**, *64*, 583–588. [CrossRef] [PubMed]

107. Yamamoto, D.; Ghosh, S.; Ganesh, B.; Krishnan, T.; Chawla-Sarkar, M.; Alam, M.M.; Aung, T.S.; Kobayashi, N. Analysis of genetic diversity and molecular evolution of human group B rotaviruses based on whole genome segments. *J. Gen. Virol.* **2010**, *91 Pt 7*, 1772–1781. [CrossRef] [PubMed]

108. Yang, J.H.; Kobayashi, N.; Wang, Y.H.; Zhou, X.; Li, Y.; Zhou, D.J.; Hu, Z.H.; Ishino, M.; Alam, M.M.; Naik, T.N.; et al. Phylogenetic analysis of a human group B rotavirus WH-1 detected in China in 2002. *J. Med. Virol.* **2004**, *74*, 662–667. [CrossRef] [PubMed]

109. Petric, M.; Mayur, K.; Vonderfecht, S.; Eiden, J.J. Comparison of group B rotavirus genes 9 and 11. *J. Gen. Virol.* **1991**, *72 Pt 11*, 2801–2804. [CrossRef] [PubMed]

110. Kuga, K.; Miyazaki, A.; Suzuki, T.; Takagi, M.; Hattori, N.; Katsuda, K.; Mase, M.; Sugiyama, M.; Tsunemitsu, H. Genetic diversity and classification of the outer capsid glycoprotein VP7 of porcine group B rotaviruses. *Arch. Virol.* **2009**, *154*, 1785–1795. [CrossRef] [PubMed]

111. Marthaler, D.; Rossow, K.; Gramer, M.; Collins, J.; Goyal, S.; Tsunemitsu, H.; Kuga, K.; Suzuki, T.; Ciarlet, M.; Matthijnssens, J. Detection of substantial porcine group B rotavirus genetic diversity in the United States, resulting in a modified classification proposal for G genotypes. *Virology* **2012**, *433*, 85–96. [CrossRef] [PubMed]

112. Molinari, B.L.; Possatti, F.; Lorenzetti, E.; Alfieri, A.F.; Alfieri, A.A. Unusual outbreak of post-weaning porcine diarrhea caused by single and mixed infections of rotavirus groups A, B, C, and H. *Vet. Microbiol.* **2016**, *193*, 125–132. [CrossRef] [PubMed]

113. Yang, H.; Makeyev, E.V.; Kang, Z.; Ji, S.; Bamford, D.H.; van Dijk, A.A. Cloning and sequence analysis of dsRNA segments 5, 6 and 7 of a novel non-group A, B, C adult rotavirus that caused an outbreak of gastroenteritis in China. *Virus Res.* **2004**, *106*, 15–26. [CrossRef] [PubMed]

114. Alam, M.M.; Kobayashi, N.; Ishino, M.; Ahmed, M.S.; Ahmed, M.U.; Paul, S.K.; Muzumdar, B.K.; Hussain, Z.; Wang, Y.H.; Naik, T.N. Genetic analysis of an ADRV-N-like novel rotavirus strain B219 detected in a sporadic case of adult diarrhea in Bangladesh. *Arch. Virol.* **2007**, *152*, 199–208. [CrossRef] [PubMed]

115. Jiang, S.; Ji, S.; Tang, Q.; Cui, X.; Yang, H.; Kan, B.; Gao, S. Molecular characterization of a novel adult diarrhoea rotavirus strain J19 isolated in China and its significance for the evolution and origin of group B rotaviruses. *J. Gen. Virol.* **2008**, *89 Pt 10*, 2622–2629. [CrossRef] [PubMed]

116. Nagashima, S.; Kobayashi, N.; Ishino, M.; Alam, M.M.; Ahmed, M.U.; Paul, S.K.; Ganesh, B.; Chawla-Sarkar, M.; Krishnan, T.; Naik, T.N.; et al. Whole genomic characterization of a human rotavirus strain B219 belonging to a novel group of the genus Rotavirus. *J. Med. Virol.* **2008**, *80*, 2023–2033. [CrossRef] [PubMed]

117. Nyaga, M.M.; Peenze, I.; Potgieter, C.A.; Seheri, L.M.; Page, N.A.; Yinda, C.K.; Steele, A.D.; Matthijnssens, J.; Mphahlele, M.J. Complete genome analyses of the first porcine rotavirus group H identified from a South African pig does not provide evidence for recent interspecies transmission events. *Infect. Genet. Evol.* **2016**, *38*, 1–7. [CrossRef] [PubMed]

118. Winiarczyk, S.; Paul, P.S.; Mummidi, S.; Panek, R.; Gradzki, Z. Survey of porcine rotavirus G and P genotype in Poland and the United States using RT-PCR. *J. Vet. Med. B Infect. Dis. Vet. Public Health* **2002**, *49*, 373–378. [CrossRef] [PubMed]

119. Amimo, J.O.; Vlasova, A.N.; Saif, L.J. Prevalence and genetic heterogeneity of porcine group C rotaviruses in nursing and weaned piglets in Ohio, USA and identification of a potential new VP4 genotype. *Vet. Microbiol.* **2013**, *164*, 27–38. [PubMed]

120. Tonietti, P.O.; Hora, A.S.; Silva, F.D.; Ruiz, V.L.; Gregori, F. Phylogenetic analyses of the VP4 and VP7 genes of porcine group A rotaviruses in Sao Paulo State, Brazil: First identification of G5P[23] in piglets. *J. Clin. Microbiol.* **2013**, *51*, 2750–2753. [CrossRef] [PubMed]

121. Nyaga, M.M.; Jere, K.C.; Esona, M.D.; Seheri, M.L.; Stucker, K.M.; Halpin, R.A.; Akopov, A.; Stockwell, T.B.; Peenze, I.; Diop, A.; et al. Whole genome detection of rotavirus mixed infections in human, porcine and bovine samples co-infected with various rotavirus strains collected from sub-Saharan Africa. *Infect. Genet. Evol.* **2015**, *31*, 321–334. [CrossRef] [PubMed]

122. Atii, D.J.; Ojeh, C.K. Subgroup determination of group A rotaviruses recovered from piglets in Nigeria. *Viral Immunol.* **1995**, *8*, 151–157. [CrossRef] [PubMed]

123. Geyer, A.; Sebata, T.; Peenze, I.; Steele, A.D. Group B and C porcine rotaviruses identified for the first time in South Africa. *J. S. Afr. Vet. Assoc.* **1996**, *67*, 115–116. [PubMed]

124. Malik, Y.S.; Kumar, N.; Sharma, K.; Sircar, S.; Dhama, K.; Bora, D.P.; Dutta, T.; Prasad, M.; Tiwari, A.K. Rotavirus diarrhea in piglets: A review on epidemiology, genetic diversity and zoonotic risks. *Indian J. Anim. Sci.* **2014**, *84*, 1035–1042.

125. Gachanja, E.; Buza, J.; Petrucka, P. Prevalence of group A rotavirus in piglets in a periurban setting of Arusha, Tanzania. *J. Biosci. Med.* **2016**, *4*, 37–44.

126. Otto, P.H.; Rosenhain, S.; Elschner, M.C.; Hotzel, H.; Machnowska, P.; Trojnar, E.; Hoffmann, K.; Johne, R. Detection of rotavirus species A, B and C in domestic mammalian animals with diarrhoea and genotyping of bovine species A rotavirus strains. *Vet. Microbiol.* **2015**, *179*, 168–176. [CrossRef] [PubMed]

127. Martella, V.; Pratelli, A.; Greco, G.; Tempesta, M.; Ferrari, M.; Losio, M.N.; Buonavoglia, C. Genomic characterization of porcine rotaviruses in Italy. *Clin. Diagn. Lab. Immunol.* **2001**, *8*, 129–132. [CrossRef]

128. Theuns, S.; Desmarets, L.M.; Heylen, E.; Zeller, M.; Dedeurwaerder, A.; Roukaerts, I.D.; Van Ranst, M.; Matthijnssens, J.; Nauwynck, H.J. Porcine group A rotaviruses with heterogeneous VP7 and VP4 genotype combinations can be found together with enteric bacteria on Belgian swine farms. *Vet. Microbiol.* **2014**, *172*, 23–34. [CrossRef] [PubMed]

129. Martella, V.; Ciarlet, M.; Banyai, K.; Lorusso, E.; Cavalli, A.; Corrente, M.; Elia, G.; Arista, S.; Camero, M.; Desario, C.; et al. Identification of a novel VP4 genotype carried by a serotype G5 porcine rotavirus strain. *Virology* **2006**, *346*, 301–311. [CrossRef] [PubMed]

130. Martella, V.; Ciarlet, M.; Baselga, R.; Arista, S.; Elia, G.; Lorusso, E.; Banyai, K.; Terio, V.; Madio, A.; Ruggeri, F.M.; et al. Sequence analysis of the VP7 and VP4 genes identifies a novel VP7 gene allele of porcine rotaviruses, sharing a common evolutionary origin with human G2 rotaviruses. *Virology* **2005**, *337*, 111–123. [CrossRef] [PubMed]

131. Van der Heide, R.; Koopmans, M.P.; Shekary, N.; Houwers, D.J.; van Duynhoven, Y.T.; van der Poel, W.H. Molecular characterizations of human and animal group a rotaviruses in the Netherlands. *J. Clin. Microbiol.* **2005**, *43*, 669–675. [CrossRef] [PubMed]

132. Chandler-Bostock, R.; Hancox, L.R.; Nawaz, S.; Watts, O.; Iturriza-Gomara, M.; Mellits, K.H. Genetic diversity of porcine group A rotavirus strains in the UK. *Vet. Microbiol.* **2014**, *173*, 27–37. [CrossRef] [PubMed]

133. Smitalova, R.; Rodak, L.; Smid, B.; Psikal, I. Detection of nongroup A rotaviruses in faecal samples of pigs in the Czech Republic. *Vet. Med.* **2009**, *54*, 1–18.

134. Chitambar, S.D.; Arora, R.; Chhabra, P. Molecular characterization of a rare G1P[19] rotavirus strain from India: evidence of reassortment between human and porcine rotavirus strains. *J. Med. Microbiol.* **2009**, *58* Pt 12, 1611–1615. [CrossRef] [PubMed]

135. Nguyen, T.A.; Khamrin, P.; Trinh, Q.D.; Phan, T.G.; Pham le, D.; Hoang le, P.; Hoang, K.T.; Yagyu, F.; Okitsu, S.; Ushijima, H. Sequence analysis of Vietnamese P[6] rotavirus strains suggests evidence of interspecies transmission. *J. Med. Virol.* **2007**, *79*, 1959–1965. [CrossRef] [PubMed]

136. Duan, Z.J.; Li, D.D.; Zhang, Q.; Liu, N.; Huang, C.P.; Jiang, X.; Jiang, B.; Glass, R.; Steele, D.; Tang, J.Y.; et al. Novel human rotavirus of genotype G5P[6] identified in a stool specimen from a Chinese girl with diarrhea. *J. Clin. Microbiol.* **2007**, *45*, 1614–1617. [PubMed]

137. Matsushima, Y.; Nakajima, E.; Nguyen, T.A.; Shimizu, H.; Kano, A.; Ishimaru, Y.; Phan, T.G.; Ushijima, H. Genome sequence of an unusual human G10P[8] rotavirus detected in Vietnam. *J. Virol.* **2012**, *86*, 10236–10237. [CrossRef] [PubMed]

138. Park, S.I.; Matthijnssens, J.; Saif, L.J.; Kim, H.J.; Park, J.G.; Alfajaro, M.M.; Kim, D.S.; Son, K.Y.; Yang, D.K.; Hyun, B.H.; et al. Reassortment among bovine, porcine and human rotavirus strains results in G8P[7] and G6P[7] strains isolated from cattle in South Korea. *Vet. Microbiol.* **2011**, *152*, 55–66. [CrossRef] [PubMed]

139. Teodoroff, T.A.; Tsunemitsu, H.; Okamoto, K.; Katsuda, K.; Kohmoto, M.; Kawashima, K.; Nakagomi, T.; Nakagomi, O. Predominance of porcine rotavirus G9 in Japanese piglets with diarrhea: close relationship of their VP7 genes with those of recent human G9 strains. *J. Clin. Microbiol.* **2005**, *43*, 1377–1384. [CrossRef] [PubMed]

140. Khamrin, P.; Maneekarn, N.; Peerakome, S.; Chan-it, W.; Yagyu, F.; Okitsu, S.; Ushijima, H. Novel porcine rotavirus of genotype P[27] shares new phylogenetic lineage with G2 porcine rotavirus strain. *Virology* **2007**, *361*, 243–252. [CrossRef] [PubMed]

141. Shi, H.; Chen, J.; Li, H.; Sun, D.; Wang, C.; Feng, L. Molecular characterization of a rare G9P[23] porcine rotavirus isolate from China. *Arch. Virol.* **2012**, *157*, 1897–1903. [CrossRef] [PubMed]

142. Zhang, H.; Zhang, Z.; Wang, Y.; Wang, X.; Xia, M.; Wu, H. Isolation, molecular characterization and evaluation of the pathogenicity of a porcine rotavirus isolated from Jiangsu Province, China. *Arch. Virol.* **2015**, *160*, 1333–1338. [CrossRef] [PubMed]

143. Peng, R.; Li, D.D.; Cai, K.; Qin, J.J.; Wang, Y.X.; Lin, Q.; Guo, Y.Q.; Zhao, C.Y.; Duan, Z.J. The epidemiological characteristics of group C rotavirus in Lulong area and the analysis of diversity of VP6 gene. *Zhonghua Shi Yan He Lin Chuang Bing Du Xue Za Zhi* **2013**, *27*, 164–166. [PubMed]

144. Suzuki, T.; Soma, J.; Miyazaki, A.; Tsunemitsu, H. Phylogenetic analysis of nonstructural protein 5 (NSP5) gene sequences in porcine rotavirus B strains. *Infect. Genet. Evol.* **2012**, *12*, 1661–1668. [CrossRef] [PubMed]

145. Lahon, A.; Ingle, V.C.; Birade, H.S.; Raut, C.G.; Chitambar, S.D. Molecular characterization of group B rotavirus circulating in pigs from India: identification of a strain bearing a novel VP7 genotype, G21. *Vet. Microbiol.* **2014**, *174*, 342–352. [CrossRef] [PubMed]

146. Huang, J.; Nagesha, H.S.; Dyall-Smith, M.L.; Holmes, I.H. Comparative sequence analysis of VP7 genes from five Australian porcine rotaviruses. *Arch. Virol.* **1989**, *109*, 173–183. [CrossRef] [PubMed]

147. Huang, J.A.; Nagesha, H.S.; Holmes, I.H. Comparative sequence analysis of VP4s from five Australian porcine rotaviruses: implication of an apparent new P type. *Virology* **1993**, *196*, 319–327. [CrossRef] [PubMed]

148. Nagesha, H.S.; Huang, J.; Holmes, I.H. A variant serotype G3 rotavirus isolated from an unusually severe outbreak of diarrhoea in piglets. *J. Med. Virol.* **1992**, *38*, 79–85. [CrossRef] [PubMed]

149. Khamrin, P.; Maneekarn, N.; Peerakome, S.; Yagyu, F.; Okitsu, S.; Ushijima, H. Molecular characterization of a rare G3P[3] human rotavirus reassortant strain reveals evidence for multiple human-animal interspecies transmissions. *J. Med. Virol.* **2006**, *78*, 986–994. [CrossRef] [PubMed]

150. Marton, S.; Doro, R.; Feher, E.; Forro, B.; Ihasz, K.; Varga-Kugler, R.; Farkas, S.L.; Banyai, K. Whole genome sequencing of a rare rotavirus from archived stool sample demonstrates independent zoonotic origin of human G8P[14] strains in Hungary. *Virus Res.* **2017**, *227*, 96–103. [CrossRef] [PubMed]

151. Li, K.; Lin, X.D.; Huang, K.Y.; Zhang, B.; Shi, M.; Guo, W.P.; Wang, M.R.; Wang, W.; Xing, J.G.; Li, M.H.; et al. Identification of novel and diverse rotaviruses in rodents and insectivores, and evidence of cross-species transmission into humans. *Virology* **2016**, *494*, 168–177. [CrossRef] [PubMed]

152. Ghosh, S.; Varghese, V.; Samajdar, S.; Bhattacharya, S.K.; Kobayashi, N.; Naik, T.N. Molecular characterization of a porcine Group A rotavirus strain with G12 genotype specificity. *Arch. Virol.* **2006**, *151*, 1329–1344. [CrossRef] [PubMed]

153. Hoshino, Y.; Honma, S.; Jones, R.W.; Ross, J.; Santos, N.; Gentsch, J.R.; Kapikian, A.Z.; Hesse, R.A. A porcine G9 rotavirus strain shares neutralization and VP7 phylogenetic sequence lineage 3 characteristics with contemporary human G9 rotavirus strains. *Virology* **2005**, *332*, 177–188. [CrossRef] [PubMed]

154. Rahman, M.; Matthijnssens, J.; Yang, X.; Delbeke, T.; Arijs, I.; Taniguchi, K.; Iturriza-Gomara, M.; Iftekharuddin, N.; Azim, T.; Van Ranst, M. Evolutionary history and global spread of the emerging g12 human rotaviruses. *J. Virol.* **2007**, *81*, 2382–2390. [CrossRef] [PubMed]

155. Mukherjee, A.; Mullick, S.; Deb, A.K.; Panda, S.; Chawla-Sarkar, M. First report of human rotavirus G8P[4] gastroenteritis in India: evidence of ruminants-to-human zoonotic transmission. *J. Med. Virol.* **2013**, *85*, 537–545. [CrossRef] [PubMed]

156. Doan, Y.H.; Nakagomi, T.; Aboudy, Y.; Silberstein, I.; Behar-Novat, E.; Nakagomi, O.; Shulman, L.M. Identification by full-genome analysis of a bovine rotavirus transmitted directly to and causing diarrhea in a human child. *J. Clin. Microbiol.* **2013**, *51*, 182–189. [CrossRef] [PubMed]

157. Luchs, A.; Cilli, A.; Morillo, S.G.; Carmona Rde, C.; Timenetsky Mdo, C. Rare G3P[3] rotavirus strain detected in Brazil: possible human-canine interspecies transmission. *J. Clin. Virol.* **2012**, *54*, 89–92. [CrossRef] [PubMed]

158. Ben Hadj Fredj, M.; Heylen, E.; Zeller, M.; Fodha, I.; Benhamida-Rebai, M.; Van Ranst, M.; Matthijnssens, J.; Trabelsi, A. Feline origin of rotavirus strain, Tunisia, 2008. *Emerg. Infect. Dis.* **2013**, *19*, 630–634. [CrossRef] [PubMed]

159. Liu, Y.; Huang, P.; Tan, M.; Liu, Y.; Biesiada, J.; Meller, J.; Castello, A.A.; Jiang, B.; Jiang, X. Rotavirus VP8*: phylogeny, host range, and interaction with histo-blood group antigens. *J. Virol.* **2012**, *86*, 9899–9910. [CrossRef] [PubMed]

160. Liu, Y.; Ramelot, T.A.; Huang, P.; Liu, Y.; Li, Z.; Feizi, T.; Zhong, W.; Wu, F.T.; Tan, M.; Kennedy, M.A.; et al. Glycan Specificity of P[19] Rotavirus and Comparison with Those of Related P Genotypes. *J. Virol.* **2016**, *90*, 9983–9996. [CrossRef]

161. Huang, P.; Xia, M.; Tan, M.; Zhong, W.; Wei, C.; Wang, L.; Morrow, A.; Jiang, X. Spike protein VP8* of human rotavirus recognizes histo-blood group antigens in a type-specific manner. *J. Virol.* **2012**, *86*, 4833–4843. [CrossRef] [PubMed]

162. Van Trang, N.; Vu, H.T.; Le, N.T.; Huang, P.; Jiang, X.; Anh, D.D. Association between norovirus and rotavirus infection and histo-blood group antigen types in Vietnamese children. *J. Clin. Microbiol.* **2014**, *52*, 1366–1374. [CrossRef] [PubMed]

163. Yamamoto, F.; Yamamoto, M. Molecular genetic basis of porcine histo-blood group AO system. *Blood* **2001**, *97*, 3308–3310. [CrossRef] [PubMed]

164. Cooling, L. Blood Groups in Infection and Host Susceptibility. *Clin. Microbiol. Rev.* **2015**, *28*, 801–870. [CrossRef] [PubMed]

165. Martella, V.; Banyai, K.; Ciarlet, M.; Iturriza-Gomara, M.; Lorusso, E.; De Grazia, S.; Arista, S.; Decaro, N.; Elia, G.; Cavalli, A.; et al. Relationships among porcine and human P[6] rotaviruses: evidence that the different human P[6] lineages have originated from multiple interspecies transmission events. *Virology* **2006**, *344*, 509–519. [CrossRef] [PubMed]

166. Mascarenhas, J.D.; Leite, J.P.; Lima, J.C.; Heinemann, M.B.; Oliveira, D.S.; Araujo, I.T.; Soares, L.S.; Gusmao, R.H.; Gabbay, Y.B.; Linhares, A.C. Detection of a neonatal human rotavirus strain with VP4 and NSP4 genes of porcine origin. *J. Med. Microbiol.* **2007**, *56 Pt 4*, 524–532. [CrossRef] [PubMed]

167. Shintani, T.; Ghosh, S.; Wang, Y.H.; Zhou, X.; Zhou, D.J.; Kobayashi, N. Whole genomic analysis of human G1P[8] rotavirus strains from different age groups in China. *Viruses* **2012**, *4*, 1289–1304. [CrossRef] [PubMed]

168. Wyatt, R.G.; James, W.D.; Bohl, E.H.; Theil, K.W.; Saif, L.J.; Kalica, A.R.; Greenberg, H.B.; Kapikian, A.Z.; Chanock, R.M. Human rotavirus type 2: Cultivation in vitro. *Science* **1980**, *207*, 189–191. [CrossRef] [PubMed]

169. Do, L.P.; Nakagomi, T.; Otaki, H.; Agbemabiese, C.A.; Nakagomi, O.; Tsunemitsu, H. Phylogenetic inference of the porcine Rotavirus A origin of the human G1 VP7 gene. *Infect. Genet. Evol.* **2016**, *40*, 205–213. [CrossRef] [PubMed]

170. Do, L.P.; Nakagomi, T.; Nakagomi, O. A rare G1P[6] super-short human rotavirus strain carrying an H2 genotype on the genetic background of a porcine rotavirus. *Infect. Genet. Evol.* **2014**, *21*, 334–350. [CrossRef] [PubMed]

171. Heylen, E.; Batoko Likele, B.; Zeller, M.; Stevens, S.; De Coster, S.; Conceicao-Neto, N.; Van Geet, C.; Jacobs, J.; Ngbonda, D.; Van Ranst, M.; et al. Rotavirus surveillance in Kisangani, the Democratic Republic of the Congo, reveals a high number of unusual genotypes and gene segments of animal origin in non-vaccinated symptomatic children. *PLoS ONE* **2014**, *9*, e100953. [CrossRef] [PubMed]

172. Zhou, X.; Wang, Y.H.; Ghosh, S.; Tang, W.F.; Pang, B.B.; Liu, M.Q.; Peng, J.S.; Zhou, D.J.; Kobayashi, N. Genomic characterization of G3P[6], G4P[6] and G4P[8] human rotaviruses from Wuhan, China: Evidence for interspecies transmission and reassortment events. *Infect. Genet. Evol.* **2015**, *33*, 55–71. [CrossRef] [PubMed]

173. Martella, V.; Colombrita, D.; Lorusso, E.; Draghin, E.; Fiorentini, S.; De Grazia, S.; Banyai, K.; Ciarlet, M.; Caruso, A.; Buonavoglia, C. Detection of a porcine-like rotavirus in a child with enteritis in Italy. *J. Clin. Microbiol.* **2008**, *46*, 3501–3507. [CrossRef] [PubMed]

174. Steyer, A.; Poljsak-Prijatelj, M.; Barlic-Maganja, D.; Marin, J. Human, porcine and bovine rotaviruses in Slovenia: evidence of interspecies transmission and genome reassortment. *J. Gen. Virol.* **2008**, *89 Pt 7*, 1690–1698. [CrossRef] [PubMed]

175. Wu, F.T.; Banyai, K.; Huang, J.C.; Wu, H.S.; Chang, F.Y.; Hsiung, C.A.; Huang, Y.C.; Lin, J.S.; Hwang, K.P.; Jiang, B.; et al. Human infection with novel G3P[25] rotavirus strain in Taiwan. *Clin. Microbiol. Infec.* **2011**, *17*, 1570–1573. [CrossRef] [PubMed]

176. Hwang, K.P.; Wu, F.T.; Banyai, K.; Wu, H.S.; Yang, D.C.; Huang, Y.C.; Lin, J.S.; Hsiung, C.A.; Huang, J.C.; Jiang, B.; et al. Identification of porcine rotavirus-like genotype P[6] strains in Taiwanese children. *J. Med. Microbiol.* **2012**, *61 Pt 7*, 990–997. [CrossRef] [PubMed]

177. Papp, H.; Borzak, R.; Farkas, S.; Kisfali, P.; Lengyel, G.; Molnar, P.; Melegh, B.; Matthijnssens, J.; Jakab, F.; Martella, V.; et al. Zoonotic transmission of reassortant porcine G4P[6] rotaviruses in Hungarian pediatric patients identified sporadically over a 15 year period. *Infect. Genet. Evol.* **2013**, *19*, 71–80. [CrossRef] [PubMed]

178. Dong, H.J.; Qian, Y.; Huang, T.; Zhu, R.N.; Zhao, L.Q.; Zhang, Y.; Li, R.C.; Li, Y.P. Identification of circulating porcine-human reassortant G4P[6] rotavirus from children with acute diarrhea in China by whole genome analyses. *Infect. Genet. Evol.* **2013**, *20*, 155–162. [CrossRef] [PubMed]

179. Degiuseppe, J.I.; Beltramino, J.C.; Millan, A.; Stupka, J.A.; Parra, G.I. Complete genome analyses of G4P[6] rotavirus detected in Argentinean children with diarrhoea provides evidence of interspecies transmission from swine. *Clin. Microbiol. Infec.* **2013**, *19*, E367–E371. [CrossRef]

180. Stupka, J.A.; Carvalho, P.; Amarilla, A.A.; Massana, M.; Parra, G.I.; Argentinean National Surveillance Network for Diarrheas. National Rotavirus Surveillance in Argentina: High incidence of G9P[8] strains and detection of G4P[6] strains with porcine characteristics. *Infect. Genet. Evol.* **2009**, *9*, 1225–1231. [CrossRef] [PubMed]

181. Razafindratsimandresy, R.; Heraud, J.M.; Ramarokoto, C.E.; Rabemanantsoa, S.; Randremanana, R.; Andriamamonjy, N.S.; Richard, V.; Reynes, J.M. Rotavirus genotypes in children in the community with diarrhea in Madagascar. *J. Med. Virol.* **2013**, *85*, 1652–1660. [CrossRef] [PubMed]

182. Komoto, S.; Maeno, Y.; Tomita, M.; Matsuoka, T.; Ohfu, M.; Yodoshi, T.; Akeda, H.; Taniguchi, K. Whole genomic analysis of a porcine-like human G5P[6] rotavirus strain isolated from a child with diarrhoea and encephalopathy in Japan. *J. Gen. Virol.* **2013**, *94 Pt 7*, 1568–1575. [PubMed]

183. Mladenova, Z.; Papp, H.; Lengyel, G.; Kisfali, P.; Steyer, A.; Steyer, A.F.; Esona, M.D.; Iturriza-Gomara, M.; Banyai, K. Detection of rare reassortant G5P[6] rotavirus, Bulgaria. *Infect. Genet. Evol.* **2012**, *12*, 1676–1684. [CrossRef] [PubMed]

184. Da Silva, M.F.; Tort, L.F.; Gomez, M.M.; Assis, R.M.; Volotao Ede, M.; de Mendonca, M.C.; Bello, G.; Leite, J.P. VP7 Gene of human rotavirus A genotype G5: Phylogenetic analysis reveals the existence of three different lineages worldwide. *J. Med. Virol.* **2011**, *83*, 357–366. [CrossRef] [PubMed]

185. Mijatovic-Rustempasic, S.; Banyai, K.; Esona, M.D.; Foytich, K.; Bowen, M.D.; Gentsch, J.R. Genome sequence based molecular epidemiology of unusual US Rotavirus A G9 strains isolated from Omaha, USA between 1997 and 2000. *Infect. Genet. Evol.* **2011**, *11*, 522–527. [CrossRef] [PubMed]

186. Martinez-Laso, J.; Roman, A.; Head, J.; Cervera, I.; Rodriguez, M.; Rodriguez-Avial, I.; Picazo, J.J. Phylogeny of G9 rotavirus genotype: a possible explanation of its origin and evolution. *J. Clin. Virol.* **2009**, *44*, 52–57. [CrossRef] [PubMed]

187. Mukherjee, A.; Dutta, D.; Ghosh, S.; Bagchi, P.; Chattopadhyay, S.; Nagashima, S.; Kobayashi, N.; Dutta, P.; Krishnan, T.; Naik, T.N.; et al. Full genomic analysis of a human group A rotavirus G9P[6] strain from Eastern India provides evidence for porcine-to-human interspecies transmission. *Arch. Virol.* **2009**, *154*, 733–746. [CrossRef] [PubMed]

188. Yodmeeklin, A.; Khamrin, P.; Chuchaona, W.; Kumthip, K.; Kongkaew, A.; Vachirachewin, R.; Okitsu, S.; Ushijima, H.; Maneekarn, N. Analysis of complete genome sequences of G9P[19] rotavirus strains from human and piglet with diarrhea provides evidence for whole-genome interspecies transmission of nonreassorted porcine rotavirus. *Infect. Genet. Evol.* **2017**, *47*, 99–108. [CrossRef] [PubMed]

189. Ghosh, S.; Urushibara, N.; Taniguchi, K.; Kobayashi, N. Whole genomic analysis reveals the porcine origin of human G9P[19] rotavirus strains Mc323 and Mc345. *Infect. Genet. Evol.* **2012**, *12*, 471–477. [CrossRef] [PubMed]

190. Wu, F.T.; Banyai, K.; Jiang, B.; Liu, L.T.; Marton, S.; Huang, Y.C.; Huang, L.M.; Liao, M.H.; Hsiung, C.A. Novel G9 rotavirus strains co-circulate in children and pigs, Taiwan. *Sci. Rep.* **2017**, *7*, 40731. [CrossRef] [PubMed]

191. Do, L.P.; Kaneko, M.; Nakagomi, T.; Gauchan, P.; Agbemabiese, C.A.; Dang, A.D.; Nakagomi, O. Molecular epidemiology of Rotavirus A, causing acute gastroenteritis hospitalizations among children in Nha Trang, Vietnam, 2007–2008: Identification of rare G9P[19] and G10P[14] strains. *J. Med. Virol.* **2017**, *89*, 621–631. [CrossRef] [PubMed]

192. Matthijnssens, J.; Rahman, M.; Ciarlet, M.; Zeller, M.; Heylen, E.; Nakagomi, T.; Uchida, R.; Hassan, Z.; Azim, T.; Nakagomi, O.; et al. Reassortment of human rotavirus gene segments into G11 rotavirus strains. *Emerg. Infect. Dis.* **2010**, *16*, 625–630. [CrossRef] [PubMed]

193. Shetty, S.A.; Mathur, M.; Deshpande, J.M. Complete genome analysis of a rare group A rotavirus, G11P[25], isolated from a child in Mumbai, India, reveals interspecies transmission and reassortment with human rotavirus strains. *J. Med. Microbiol.* **2014**, *63 Pt 9*, 1220–1227. [CrossRef] [PubMed]

194. Mullick, S.; Mukherjee, A.; Ghosh, S.; Pazhani, G.P.; Sur, D.; Manna, B.; Nataro, J.P.; Levine, M.M.; Ramamurthy, T.; Chawla-Sarkar, M. Genomic analysis of human rotavirus strains G6P[14] and G11P[25] isolated from Kolkata in 2009 reveals interspecies transmission and complex reassortment events. *Infect. Genet. Evol.* **2013**, *14*, 15–21. [CrossRef] [PubMed]

195. Komoto, S.; Wandera Apondi, E.; Shah, M.; Odoyo, E.; Nyangao, J.; Tomita, M.; Wakuda, M.; Maeno, Y.; Shirato, H.; Tsuji, T.; et al. Whole genomic analysis of human G12P[6] and G12P[8] rotavirus strains that have emerged in Kenya: identification of porcine-like NSP4 genes. *Infect. Genet. Evol.* **2014**, *27*, 277–293. [CrossRef] [PubMed]

196. Ide, T.; Komoto, S.; Higo-Moriguchi, K.; Htun, K.W.; Myint, Y.Y.; Myat, T.W.; Thant, K.Z.; Thu, H.M.; Win, M.M.; Oo, H.N.; et al. Whole Genomic Analysis of Human G12P[6] and G12P[8] Rotavirus Strains that Have Emerged in Myanmar. *PLoS ONE* **2015**, *10*, e0124965. [CrossRef] [PubMed]

197. My, P.V.; Rabaa, M.A.; Donato, C.; Cowley, D.; Phat, V.V.; Dung, T.T.; Anh, P.H.; Vinh, H.; Bryant, J.E.; Kellam, P.; et al. Novel porcine-like human G26P[19] rotavirus identified in hospitalized paediatric diarrhoea patients in Ho Chi Minh City, Vietnam. *J. Gen. Virol.* **2014**, *95 Pt 12*, 2727–2733. [CrossRef] [PubMed]

198. Medici, K.C.; Barry, A.F.; Alfieri, A.F.; Alfieri, A.A. Genetic analysis of the porcine group B rotavirus NSP2 gene from wild-type Brazilian strains. *Braz. J. Med. Biol. Res.* **2010**, *43*, 13–16. [CrossRef] [PubMed]

199. Gabbay, Y.B.; Borges, A.A.; Oliveira, D.S.; Linhares, A.C.; Mascarenhas, J.D.; Barardi, C.R.; Simoes, C.M.; Wang, Y.; Glass, R.I.; Jiang, B. Evidence for zoonotic transmission of group C rotaviruses among children in Belem, Brazil. *J. Med. Virol.* **2008**, *80*, 1666–1674. [CrossRef] [PubMed]

200. Iturriza-Gomara, M.; Dallman, T.; Banyai, K.; Bottiger, B.; Buesa, J.; Diedrich, S.; Fiore, L.; Johansen, K.; Koopmans, M.; Korsun, N.; et al. Rotavirus genotypes co-circulating in Europe between 2006 and 2009 as determined by EuroRotaNet, a pan-European collaborative strain surveillance network. *Epidemiol. Infect.* **2011**, *139*, 895–909. [CrossRef] [PubMed]

201. Iturriza-Gomara, M.; Dallman, T.; Banyai, K.; Bottiger, B.; Buesa, J.; Diedrich, S.; Fiore, L.; Johansen, K.; Korsun, N.; Kroneman, A.; et al. Rotavirus surveillance in europe, 2005–2008: web-enabled reporting and real-time analysis of genotyping and epidemiological data. *J. Infect. Dis.* **2009**, *200* (Suppl. S1), S215–S221. [CrossRef] [PubMed]

202. Iturriza-Gomara, M.; Green, J.; Brown, D.W.; Ramsay, M.; Desselberger, U.; Gray, J.J. Molecular epidemiology of human group A rotavirus infections in the United Kingdom between 1995 and 1998. *J. Clin. Microbiol.* **2000**, *38*, 4394–4401. [PubMed]

203. Chang, K.O.; Nielsen, P.R.; Ward, L.A.; Saif, L.J. Dual infection of gnotobiotic calves with bovine strains of group A and porcine-like group C rotaviruses influences pathogenesis of the group C rotavirus. *J. Virol.* **1999**, *73*, 9284–9293. [PubMed]

204. Banyai, K.; Jiang, B.; Bogdan, A.; Horvath, B.; Jakab, F.; Meleg, E.; Martella, V.; Magyari, L.; Melegh, B.; Szucs, G. Prevalence and molecular characterization of human group C rotaviruses in Hungary. *J. Clin. Virol.* **2006**, *37*, 317–322. [CrossRef] [PubMed]

205. Desselberger, U.; Huppertz, H.I. Immune responses to rotavirus infection and vaccination and associated correlates of protection. *J. Infect. Dis.* **2011**, *203*, 188–195. [CrossRef] [PubMed]

206. Saif, L.J.; Ward, L.A.; Yuan, L.; Rosen, B.I.; To, T.L. The gnotobiotic piglet as a model for studies of disease pathogenesis and immunity to human rotaviruses. *Arch. Virol. Suppl.* **1996**, *12*, 153–161. [PubMed]

207. Vlasova, A.N.; Shao, L.; Kandasamy, S.; Fischer, D.D.; Rauf, A.; Langel, S.N.; Chattha, K.S.; Kumar, A.; Huang, H.C.; Rajashekara, G.; et al. Escherichia coli Nissle 1917 protects gnotobiotic pigs against human rotavirus by modulating pDC and NK-cell responses. *Eur. J. Immunol.* **2016**, *46*, 2426–2437. [CrossRef] [PubMed]

208. Narvaez, C.F.; Angel, J.; Franco, M.A. Interaction of rotavirus with human myeloid dendritic cells. *J. Virol.* **2005**, *79*, 14526–14535. [CrossRef] [PubMed]

209. Feng, N.; Kim, B.; Fenaux, M.; Nguyen, H.; Vo, P.; Omary, M.B.; Greenberg, H.B. Role of interferon in homologous and heterologous rotavirus infection in the intestines and extraintestinal organs of suckling mice. *J. Virol.* **2008**, *82*, 7578–7590. [CrossRef] [PubMed]

210. Angel, J.; Franco, M.A.; Greenberg, H.B.; Bass, D. Lack of a role for type I and type II interferons in the resolution of rotavirus-induced diarrhea and infection in mice. *J. Interferon Cytokine Res.* **1999**, *19*, 655–659. [CrossRef] [PubMed]

211. Vancott, J.L.; McNeal, M.M.; Choi, A.H.; Ward, R.L. The role of interferons in rotavirus infections and protection. *J. Interferon Cytokine Res.* **2003**, *23*, 163–170. [CrossRef] [PubMed]

212. Vlasova, A.N.; Chattha, K.S.; Kandasamy, S.; Siegismund, C.S.; Saif, L.J. Prenatally acquired vitamin A deficiency alters innate immune responses to human rotavirus in a gnotobiotic pig model. *J. Immunol.* **2013**, *190*, 4742–4753. [CrossRef] [PubMed]

213. Vlasova, A.N.; Chattha, K.S.; Kandasamy, S.; Liu, Z.; Esseili, M.; Shao, L.; Rajashekara, G.; Saif, L.J. Lactobacilli and bifidobacteria promote immune homeostasis by modulating innate immune responses to human rotavirus in neonatal gnotobiotic pigs. *PLoS ONE* **2013**, *8*, e76962. [CrossRef] [PubMed]

214. Kandasamy, S.; Vlasova, A.N.; Fischer, D.; Kumar, A.; Chattha, K.S.; Rauf, A.; Shao, L.; Langel, S.N.; Rajashekara, G.; Saif, L.J. Differential Effects of Escherichia coli Nissle and Lactobacillus rhamnosus Strain GG on Human Rotavirus Binding, Infection, and B Cell Immunity. *J. Immunol.* **2016**, *196*, 1780–1789. [CrossRef] [PubMed]

215. Azevedo, M.S.; Yuan, L.; Iosef, C.; Chang, K.O.; Kim, Y.; Nguyen, T.V.; Saif, L.J. Magnitude of serum and intestinal antibody responses induced by sequential replicating and nonreplicating rotavirus vaccines in gnotobiotic pigs and correlation with protection. *Clin. Diagn. Lab. Immunol.* **2004**, *11*, 12–20. [CrossRef] [PubMed]

216. Yuan, L.; Kang, S.Y.; Ward, L.A.; To, T.L.; Saif, L.J. Antibody-secreting cell responses and protective immunity assessed in gnotobiotic pigs inoculated orally or intramuscularly with inactivated human rotavirus. *J. Virol.* **1998**, *72*, 330–338. [PubMed]

217. Yuan, L.; Wen, K.; Azevedo, M.S.; Gonzalez, A.M.; Zhang, W.; Saif, L.J. Virus-specific intestinal IFN-gamma producing T cell responses induced by human rotavirus infection and vaccines are correlated with protection against rotavirus diarrhea in gnotobiotic pigs. *Vaccine* **2008**, *26*, 3322–3331. [CrossRef] [PubMed]

218. Yuan, L.; Iosef, C.; Azevedo, M.S.; Kim, Y.; Qian, Y.; Geyer, A.; Nguyen, T.V.; Chang, K.O.; Saif, L.J. Protective immunity and antibody-secreting cell responses elicited by combined oral attenuated Wa human rotavirus and intranasal Wa 2/6-VLPs with mutant Escherichia coli heat-labile toxin in gnotobiotic pigs. *J. Virol.* **2001**, *75*, 9229–9238. [CrossRef] [PubMed]

219. Iosef, C.; Chang, K.O.; Azevedo, M.S.; Saif, L.J. Systemic and intestinal antibody responses to NSP4 enterotoxin of Wa human rotavirus in a gnotobiotic pig model of human rotavirus disease. *J. Med. Virol.* **2002**, *68*, 119–128. [CrossRef] [PubMed]

220. Hodgins, D.C.; Kang, S.Y.; deArriba, L.; Parreno, V.; Ward, L.A.; Yuan, L.; To, T.; Saif, L.J. Effects of maternal antibodies on protection and development of antibody responses to human rotavirus in gnotobiotic pigs. *J. Virol.* **1999**, *73*, 186–197. [PubMed]

221. Nguyen, T.V.; Yuan, L.; Azevedo, M.S.; Jeong, K.I.; Gonzalez, A.M.; Iosef, C.; Lovgren-Bengtsson, K.; Morein, B.; Lewis, P.; Saif, L.J. High titers of circulating maternal antibodies suppress effector and memory B-cell responses induced by an attenuated rotavirus priming and rotavirus-like particle-immunostimulating complex boosting vaccine regimen. *Clin. Vaccine Immunol. CVI* **2006**, *13*, 475–485. [CrossRef] [PubMed]

222. Tate, J.E.; Parashar, U.D. Rotavirus vaccines in routine use. *Clin. Infect. Dis.* **2014**, *59*, 1291–1301. [CrossRef] [PubMed]

223. Saif, L.J.; Fernandez, F.M. Group A rotavirus veterinary vaccines. *J. Infect. Dis.* **1996**, *174* (Suppl. S1), S98–S106. [CrossRef] [PubMed]

224. Azevedo, M.P.; Vlasova, A.N.; Saif, L.J. Human rotavirus virus-like particle vaccines evaluated in a neonatal gnotobiotic pig model of human rotavirus disease. *Expert Rev. Vaccines* **2013**, *12*, 169–181. [CrossRef] [PubMed]

225. Gautam, R.; Mijatovic-Rustempasic, S.; Esona, M.D.; Tam, K.I.; Quaye, O.; Bowen, M.D. One-step multiplex real-time RT-PCR assay for detecting and genotyping wild-type group A rotavirus strains and vaccine strains (Rotarix(R) and RotaTeq(R)) in stool samples. *PeerJ* **2016**, *4*, e1560. [CrossRef] [PubMed]

9

Structure of Ty1 Internally Initiated RNA Influences Restriction Factor Expression

Leszek Błaszczyk [1], Marcin Biesiada [1], Agniva Saha [2], David J. Garfinkel [2] and Katarzyna J. Purzycka [1,*]

[1] Institute of Bioorganic Chemistry, Polish Academy of Sciences, Poznan 61-704, Poland; blaszcz@ibch.poznan.pl (L.B.); biesiada@ibch.poznan.pl (M.B.)

[2] Department of Biochemistry & Molecular Biology, University of Georgia, Athens, GA 30602, USA; agniva.saha@gmail.com (A.S.); djgarf@uga.edu (D.J.G.)

* Correspondence: purzycka@ibch.poznan.pl

Academic Editor: Eric O. Freed

Abstract: The long-terminal repeat retrotransposon Ty1 is the most abundant mobile genetic element in many *Saccharomyces cerevisiae* isolates. Ty1 retrotransposons contribute to the genetic diversity of host cells, but they can also act as an insertional mutagen and cause genetic instability. Interestingly, retrotransposition occurs at a low level despite a high level of Ty1 RNA, even though *S. cerevisiae* lacks the intrinsic defense mechanisms that other eukaryotes use to prevent transposon movement. p22 is a recently discovered Ty1 protein that inhibits retrotransposition in a dose-dependent manner. p22 is a truncated form of Gag encoded by internally initiated Ty1i RNA that contains two closely-spaced AUG codons. Mutations of either AUG codon compromise p22 translation. We found that both AUG codons were utilized and that translation efficiency depended on the Ty1i RNA structure. Structural features that stimulated p22 translation were context dependent and present only in Ty1i RNA. Destabilization of the 5′ untranslated region (5′ UTR) of Ty1i RNA decreased the p22 level, both in vitro and in vivo. Our data suggest that protein factors such as Gag could contribute to the stability and translational activity of Ty1i RNA through specific interactions with structural motifs in the RNA.

Keywords: RNA structure; Ty1 retrotransposon; Gag; translation regulation

1. Introduction

Ty1 is a long-terminal repeat (LTR) retrotransposon in the *Pseudoviridae* family and the most abundant mobile genetic element in the *Saccharomyces cerevisiae* reference strain [1]. Ty1 contains *GAG* and *POL* genes bracketed by LTRs and proliferates in the yeast genome by integrating new copies through an RNA-mediated mechanism [2]. Dimeric Ty1 RNA is present in virus-like particles (VLPs) [3] that are comprised of the capsid protein Gag and Gag-Pol; the latter being synthesized by a programmed +1 frameshift event that occurs at overlapping leucine codons in *GAG* and *POL* [4]. *POL* encodes protease (PR), reverse transcriptase (RT) and integrase (IN), which are required for protein maturation, reverse transcription and integration, respectively. Gag is a VLP structural component and is expressed as a 441-amino acid precursor (p49) that undergoes a C-terminal cleavage by PR to produce the mature 401-residue protein (p45). Ty1 Gag binds RNA in vitro [5,6] and serves as a multifunctional regulator that orchestrates retrotransposon replication [7].

Ty1 contributes to the genetic diversity of *S. cerevisiae* and closely related species, however, these elements can also act as insertional mutagens and cause genetic instability by recombination-mediated gene rearrangements. Overloading the genome with retrotransposon insertions is another scenario that could be lethal to the cell. Paradoxically, Ty1 retrotransposition occurs at low rate, despite a high level of Ty1 RNA [2]. *S. cerevisiae* also lack the intrinsic defense mechanisms to prevent retrotransposition that are typically active in other eukaryotes, including DNA methylation [8,9], and the expression of several host proteins, such as apolipoprotein B mRNA-editing enzyme catalytic polypeptide-like 3 (APOBEC3) family members [10] or RNAi components [11,12]. Early on, a region of Ty1 required for copy number control (CNC) was identified but the mechanism underlying CNC remained puzzling [13]. Recent genetic analysis of the CNC region identified mutations abrogating CNC that map within *GAG* downstream of two internal AUG codons [14,15]. The separation of function phenotype displayed by one of the *GAG* mutations suggests that Ty1 encodes a protein that restricts its movement. Indeed, the recently discovered protein p22 inhibits retrotransposition in a dose-dependent manner and mediates CNC. p22 is encoded by the C-terminal half of Ty1 *GAG*, and similar to Gag-p49, undergoes maturation by Ty1 protease to form p18. However, p22 is encoded by internally initiated Ty1i RNA that contains two closely spaced AUG codons. Ribosomal profiling analyses show preferential usage of AUG1, but mutational analysis of Ty1i RNA initiation codons AUG1 and AUG2 suggests that both have the potential to be utilized for p22 translation. p18 expressed from either AUG1 or AUG2 confers strong inhibition of Ty1 mobility that correlates with their level of expression. Also, p22/p18 target Gag and inhibit several steps in the process of retrotransposition prior to reverse transcription [14–16].

Like programmed Ty1 frameshifting, employing multiple start codons to initiate the synthesis of p22 is reminiscent of the non-canonical translation strategies that viruses use to maximize their coding potential [17]. Canonical 5′-end-dependent translation initiation generally permits only one protein to be synthesized from a particular mRNA. However, the leaky scanning mechanism allows the production of functionally distinct proteins from a single transcript containing multiple initiation codons. In these cases, a suboptimal sequence surrounding the first AUG codon limits its recognition, which allows ribosomal scanning and translation from downstream initiation codons [17]. This strategy is commonly employed by RNA viruses, including retroviruses [18].

We have shown that p22 translation is a cap-dependent event, however, our results suggest that the structure of 5′ UTR of Ty1i mRNA may contribute to the efficiency of translation [14]. Secondary and tertiary structures of 5′ UTRs play important roles in the regulation of translation by affecting the recruitment, positioning and movement of ribosomes [19]. Folding of the 5′ UTR into an ensemble of secondary structures may influence the initiation of translation either positively or negatively. The nature of this effect is attributed, at least in part, to the thermodynamic stability of the structural elements formed in the 5′ UTR, their guanine-cytosine (GC) content, and positioning in relation to the 5′ cap and AUG initiation codon. Hairpin structures of even moderate thermodynamic stability located close to the 5′-end of the mRNA prevent cap-dependent formation of the preinitiation complexes and can lead to translation inhibition [20–23]. On the other hand, secondary structures present in the coding region may stimulate translation if placed at particular distances downstream of the initiation codon [24,25]. This stimulatory effect may be caused by a hairpin structure that pauses migration of the preinitiation complexes. Hairpin structures can be important for mRNAs containing AUG codons located in suboptimal sequence contexts, and thus undergo translation via leaky scanning. Structure-dependent pausing of the preinitiation complexes provides more time for the recognition of AUG codons in an unfavorable context. Whether this is a general mechanism remains to be determined, however, analysis of the predicted secondary structures downstream of initiation codons suggests that this may be the case [26]. The structural context of the AUG codon can modulate translation efficiency [27]. Coding sequences can also participate in the folding of the 5′ untranslated regions that modulate RNA stability [28,29]. However, coding sequence contributions to translation initiation remain understudied since functional and structural characterization is usually conducted on isolated 5′ UTR sequences.

We set out to characterize how p22 translation is initiated. Our work suggests that both AUG codons can be utilized but AUG1 is used preferentially and translation efficiency strongly depends on the Ty1i RNA structure. Features stimulating p22 translation are context dependent as revealed by specific structures in Ty1i mRNA that are absent in full length genomic mRNA. The 5′ UTR of p22 mRNA interacts with the coding region and destabilization at the secondary or 3D structural levels results in a decrease in p22 translation. Also, our data supports the idea that protein factors such as Gag interact with a structural motif in Ty1i RNA to modulate its stability and translation.

2. Materials and Methods

2.1. Preparation of the RNA Constructs for Structure Probing Experiments and In Vitro Translation Assays

All DNA templates for secondary structure probing experiments and in vitro translation were amplified from plasmid pBDG433, which contains transcribed sequences of Ty1-H3 subcloned into the riboprobe vector pSP64 (Promega, Madison, WI, USA). Forward and reverse primers are listed in Table S1. Each construct was confirmed by DNA sequencing. In vitro transcription reactions were performed using MEGAscript or MEGAshortscript T7 transcription kits (ThermoFisher, Waltham, MA, USA), as recommended by the manufacturer. RNA transcripts were purified using Direct-zol RNA MiniPrep Kit (Zymo Research, Irvine, CA, USA) and their integrity was monitored by formaldehyde agarose gel electrophoresis. Capped transcripts were synthesised in the presence of the ARCA Cap Analog (ThermoFisher). RNA used for native gel electrophoresis was [^{32}P]-labelled at their 3′-ends with T4 RNA ligase (ThermoFisher) according to standard procedures.

2.2. Selective Acylation Analysed by Primer Extension (SHAPE)

The reaction mixture (100 μL) containing 20 pmol of RNA in SHAPE renaturation buffer (10 mM Tris-HCl pH 8.0, 100 mM KCl, 0.1 mM ethylenediaminetetraacetic acid (EDTA), pH 8.0) was heated at 95 °C for 3 min and placed on ice for 5 min. Fifty microliters of 3× SHAPE folding buffer (120 mM Tris-HCl pH 8.0, 600 mM KCl, 1.5 mM EDTA pH 8.0, 15 mM MgCl$_2$) was added and samples were incubated for 30 min at 37 °C. Folded RNA was separated equally into two reactions and mixed with the 20 mM N-methylisatoic anhydride (NMIA) in dimethyl sulfoxide (DMSO) (2 mM final concentration of NMIA) or DMSO alone. Both reactions were incubated for 45 min at 37 °C followed by purification of RNA using Direct-zol RNA MiniPrep Kit.

2.3. DMS Modification

RNA (20 pmol in 50 μL) was refolded using the same conditions as those employed in the SHAPE experiments, then divided equally into two 24 μL reactions. Refolded RNA samples were mixed with 1 μL of dimethyl sulphate (DMS) in ethanol (0.5% final concentration) or ethanol alone. Both reactions were incubated 1 min at room temperature and mixed with 475 μL of stop solution (200 mM sodium acetate, 4.8 M β-mercaptoethanol). RNA was purified using Direct-zol RNA MiniPrep Kit immediately after stopping the reaction.

2.4. Hydroxyl Radical Probing

RNA samples (10 pmol) were refolded by heating at 95 °C for 2 min in water followed by incubation at 25 °C for 5 min. Next, 3× SHAPE folding buffer was added and the reaction was incubated for 25 min at 37 °C, then diluted 20× with 20 mM Tris-HCl pH 8.0. To initiate the production of hydroxyl radicals, 1.5 μL of 2.5 mM (NH$_4$)Fe(SO$_4$)$_2$, 50 mM sodium ascorbate, 1.5% H$_2$O$_2$ and 2.75 mM EDTA were applied separately to the wall of the tube followed by centrifugation. Six microliters of water were added to the control reaction. Reactions were incubated for 10 s at room temperature, then quenched by the addition of thiourea and EDTA to final concentrations of 20 mM and 40 mM, respectively. RNA was recovered using Direct-zol RNA MiniPrep Kit.

2.5. Reverse Transcription and Data Processing

A reaction containing 2–5 pmol RNA, 10 pmol of fluorescently labelled primer PR5 or PR6 (Table S1) (Cy5 (+reagent) or Cy5.5 (control reaction)) and 0.1 mM EDTA pH 8.0 was incubated at 95 °C for 3 min, 37 °C for 10 min and 55 °C for 2 min, and then reverse transcribed at 50 °C for 45 min using Superscript III Reverse Transcriptase (ThermoFisher) as described previously [30]. Sequencing reactions were carried out using primers fluorescently labelled with LicorIR-800 (ddT) or WellRed D2 (ddA) and a Thermo Sequenase Cycle Sequencing Kit, according to the manufacturer's protocol (Affymetrix, Santa Clara, CA, USA). Reverse transcription reactions and sequencing ladders were purified using ZR DNA Sequencing Clean-up Kit (ZymoResearch). cDNA samples were analysed on a GenomeLab GeXP Analysis System (Beckman–Coulter, Brea, CA, USA). Raw data were processed as described [31]. At least four repetitions were obtained for each reaction.

2.6. In Vitro Translation

In vitro translation experiments were carried out using wheat germ extract (WGE) as recommended by the manufacturer (Promega). The reaction mixture containing 12.5 μL of WGE lysate, 80 μM amino acid mixture minus methionine, 1.25 μL of [^{35}S]-labelled methionine (1000 Ci/mmol) (Hartmann Analytic, Braunschweig, Germany), 79 mM potassium acetate, 20 units of ribonuclease inhibitor (ThermoFisher) and 1 pmol of refolded capped or uncapped RNA in the final volume of 25 μL was incubated for 1 hour at 25 °C. Translation products were resolved on sodium dodecyl sulphate (SDS)-polyacrylamide gels followed by radioisotope imaging using a FLA 5100 image analyser (Fuji, Minato, Tokyo, Japan). Bands intensities were analysed using MultiGauge software (Fuji). At least three repetitions were obtained for each in vitro translation reaction.

2.7. Native Gel Electrophoresis

[^{32}P]-labelled RNA was refolded in SHAPE renaturation buffer by heating at 95 °C for 5 min and 4 °C for 5 min. SHAPE folding buffer contained increasing $MgCl_2$ concentrations ranging from 0.1 to 10 mM. The reaction mixture (15 μL) was incubated at 37 °C for 25 min following the addition of 1.5 μL of 25% ficoll. Samples were analysed by native polyacrylamide gel electrophoresis using 12% gels in 0.5× TB at 4 °C. Electrophoresis was carried out at a gel temperature of 4 °C (DNApointer, Biovectis, Warsaw, Poland) [32]. Gels were dried, exposed to a phosphorimager screen, and scanned using FLA 5100 image analyser.

2.8. Ty1 Gag Expression and Purification

A Ty1 Gag-p45-GST fusion protein was expressed in Escherichia coli (E. coli) strain BL21(DE3)pLysS (Invitrogen, Carlsbad, CA, USA). Six liters of cells were grown in Luria-Bertani (LB) medium containing 50 μg/mL ampicillin and 34 μg/mL chloramphenicol at 28 °C to an OD_{600} of 0.7. Prior to isopropyl β-D-1-thiogalactopyranoside (IPTG) induction, cells were incubated for 30 min at 18 °C. Following the addition of IPTG (0.8 mM), the culture was induced at 18 °C overnight. Cells were pelleted by centrifugation at 4000 g for 10 min at 4 °C and resuspended in lysis buffer (50 mM Tris-HCl pH 8.0, 1 M NaCl, 10 mM β-mercaptoethanol, 2.5 mM DTT, 0.1 mM $ZnCl_2$, 0.5 mg/mL lysozyme, and protease inhibitor (Roche, Basel, Switzerland)). The cell suspension was sonicated 40 × 2 s on ice with a 30 s pause after each pulse. Debris was removed by centrifugation at 20,000 g for 20 min at 4 °C. Nucleic acids were precipitated using 0.45% polyethyleneimine and pelleted by centrifugation at 30,000 g for 30 min at 4 °C. The supernatant was mixed with 1.5–2 mL of Glutathione Sepharose 4B (GE Healthcare, Little Chalfont, UK) and incubated for 1 h at 4 °C with gentle agitation followed by centrifugation at 700 g for 5 min. The Glutathione Sepharose beads were loaded onto a column and washed with 10 column volumes (10 mL/wash) of wash buffer (50 mM Tris-HCl pH 8.0, 1 M NaCl, 10 mM β-mercaptoethanol, 2.5 mM DTT, 0.1 mM $ZnCl_2$). The glutathione S-transferase (GST) tag was removed by thrombin cleavage (GE Healthcare) at 4 °C for 12 h with gentle agitation. Ty1 Gag p45 was eluted using wash buffer, concentrated with centrifugal filtration (Millipore, Billerica, MA, USA), aliquoted and stored at −80 °C.

2.9. Filter Binding Assay

Reactions were performed in binding buffer (50 mM Tris-HCl pH 7.5, 40 mM KCl, 2 mM $MgCl_2$, 0.01% Triton X-100) containing different concentrations of NaCl (50, 100, 150, 200, 250, 500 mM). [^{32}P]-labeled domain I of Ty1i RNA (0.2 nM) was incubated for 4 min at 95 °C without magnesium ions and Triton X-100, and slowly cooled to 37 °C. $MgCl_2$ and Triton X-100 were added following incubation for 10 min at 37 °C. Ty1 Gag protein solutions were prepared by sequential two-fold dilution of Gag in binding buffer. The binding reaction was initiated by mixing equal volumes of RNA and Gag protein in a microplate (final concentration of RNA was 0.1 nM). The reactions were incubated for 15 min at 24 °C, filtered and washed with 2×200 μL binding buffer containing 50 mM NaCl. A 96-well dot-blot (Minifold, Whatman, Maidstone, UK) was used with nitrocellulose (Protran, Whatman, Maidstone, UK) on top and charged nylon (Hybond N+, GE Healthcare) membranes on the bottom. Prior to use, both membranes were soaked in binding buffer containing 50 mM NaCl. After filtration, membranes were dried and exposed to a phosphoimager screen. Data were fitted to the Hill equation using Origin 8.5 software (OriginLab, Northampton, MA, USA).

2.10. H1Δ Plasmid and Yeast Strains

The H1Δ deletion (T1015 - A1035) was generated by overlap PCR using flanking oligonucleotides Ty335F (5′-TGGTAGCGCCTGTGCTTCGGTTAC-3′) and RP1 (5′-ATAGTCAAT AGCACTAGACC-3′), and overlapping oligonucleotides B (5′-GAAAGAATTTTCATGATAGGA TGTCTTTGACCCAGGTAGGTAG-3′) and C (5′-GGTCAAAGACATCCTATCATGAAAATTCTT TCCAAAAGTATTGAAAAAA-3′). Wild-type pGPOLΔ (pBDG1130) [14] was used as the template for PCR. Nucleotide sequences correspond to the reference Ty1-H3 element (GenBank M10876.1). The H1Δ PCR product was cloned into pGPOLΔ using XhoI and BglII. The resulting plasmid pBAS47 is denoted as H1Δ. The H1Δ insert in pBAS47 was verified by DNA sequencing. Plasmids pBDG1130 and pBAS47 were transformed into the following strains: DG2196 (1 Ty1) [13] to generate DG2374 and YAS89, and DG3582 (0 Ty1) [14] to generate YAS85 and YAS87, respectively.

2.11. Northern and Western Blotting

Yeast cultures for total cellular RNA and protein extraction were grown in SC-Ura + 2% glucose medium at 22 °C for 24 h. RNA was extracted using the MasterPure Yeast RNA purification kit (Epicenter Biotechnologies, Madison, WI, USA) [14]. For each strain, 8 μg total RNA was separated on a 1.2% formaldehyde-agarose gel and subjected to Northern blot analysis using [^{32}P]-labeled riboprobes corresponding to Ty1 nucleotides 1266–1601 and *ACT1*, followed by phosphorimaging using a STORM 840 phosphorimager and ImageQuant software (GE Healthcare) [13]. Protein isolation and Western blot analysis to detect p22 was performed as described previously [14]. A rabbit polyclonal antisera against Pgk1 (kindly provided by Jeremy Thorner) was used at a 1:100,000 dilution. Immune complexes were detected with enhanced chemiluminescence (ECL) reagent (GE Healthcare). The amount of p22 relative to Pgk1 was estimated by densitometry using Quantity One software (Bio-Rad). Northern and Western analyses using the 0 Ty1 and 1 Ty1 strains containing pGPOLΔ or pH1Δ were repeated twice and representative results are presented. Also, independent Western analyses using the 0 Ty1 strain containing pGPOLΔ or pH1Δ were repeated three more times.

Ty1*his3-AI* mobility frequencies were determined as described previously [13,33]. Briefly, a single colony was resuspended in 1 mL water and four; 1 mL SC-Ura cultures were inoculated with 5 μL of cell suspension. Quadruplicate cultures for each strain were grown at 22 °C for three days. Cells were pelleted, resuspended in 1 mL water, and dilutions spread on SC-Ura and SC-Ura-His plates were incubated at 30 °C for 4 days. The frequency of Ty1*his3-AI* mobility was calculated by the number of His$^+$ Ura$^+$ colonies/the number of Ura$^+$ colonies per mL of culture.

2.12. RNA 3D Structure Prediction

Structure prediction experiments were performed by RNAComposer [34] webserver [35]. The AUG1AUG2 RNA domain I sequence: GGGUCAAAGACAUCCUAUCCGUUGAUUA UACGGAUAUCAUGAAAAUUCUUCCAAAAGUAUUGAAAAAAUGCAAUCUGAUACCC and secondary structure topology in dot bracket notation: (((((...(((((((..(((((((.......))))))))............ (((.((.......)).)))).)))...)))...)))) were used as input data. The 3-way junction of domain I of AUG1AUG2 RNA was generated by RNAComposer, therefore, it was substituted by the elements introduced by the user. This element was chosen from RNA structures deposited in Research Collaboratory for Structural Bioinformatics (RCSB) Protein Data Bank (PDB) database following the criteria of the highest homology of secondary structure topology and sequence. More than 10 batches with different three-way junction structures were run. Ten models were generated for every batch. The resulting models were clustered based on the agreement with the hydroxyl radical cleavage data and the energy. Hydroxyl radical cleavage reactivity indexes from experiments were compared with indexes denoting atomic crowding around phosphorus at the corresponding nucleotide residue. The models with correct energy [36] and the best similarity were accepted.

3. Results and Discussion

3.1. Both AUG Codons in Ty1i RNA Can Be Recognized for Translation Initiation

Our previous results demonstrated that p22 translation can be initiated from AUG1 and AUG2 codons and is strictly cap-dependent. Also, either AUG1 or AUG2 can function to initiate translation when the other is mutated [14]. However, a number of questions remain unanswered: (i) Are both AUGs active for translation when present in the same RNA? (ii) Or is one codon translated preferentially? (iii) Does leaky scanning account for p22 synthesis from AUG2? Moreover, deleting the 5' UTR or mutating AUG1 or AUG2 decreases the level of p22 in vivo. For AUG1 and AUG2 codon mutants, the decrease in the p22 level is significantly larger than expected considering that one AUG codon is still present. These results suggest that the structure of the 5' terminal part of Ty1i RNA may influence p22 translation.

Translational activity of both AUG codons could be beneficial and contribute to the evolutionary diversification of p22. To gain insights into translation from AUG1 and AUG2 in Ty1i RNA, we performed in vitro translation assays using three derivatives of AUG1AUG2 RNA [14]. AUG1AUG2 RNA started at nt 1000 of Ty1, comprised the 5' UTR and p22 open reading frame (ORF), and ended with a natural stop codon (Figure 1). The difference between p22 proteins translated from AUG1 and AUG2 is only 10 amino acid residues (30 nt). Such a small size difference makes the two proteins difficult to separate by gel electrophoresis and obscures simultaneous analysis of the translation levels from both AUGs. To overcome this difficulty, we synthesized AUG1AUG2* RNA in which AUG2 (including its Kozak context) is 30 nucleotides downstream of the original AUG2, and introduced a GCG alanine codon in place of AUG2 (Figure 2). This modification increased the distance between AUG1 and AUG2* to 60 nt (20 amino acids), which allowed separation of the two translation products. A frameshift mutation (insertion of AU between U1050 and C1051) was introduced in AUG1frsAUG2 RNA (Figure 2). In this case, translation from AUG1 occurred out of frame in relation to AUG2 and resulted in the synthesis of a 49-amino acid peptide. Translation of the AUG1AUG2* and AUG1frsAUG2 RNAs allowed us to determine if both AUGs were recognized for translation. The third RNA, AUG1stopAUG2, contained an insertion of a single U between U1060 and U1061, which introduced a premature stop codon following translation from AUG1 (Figure 2). This RNA mutation was designed to help determine the level of p22 translated from AUG2. Each construct was also designed to avoid the introduction of rare codons that could obscure translation.

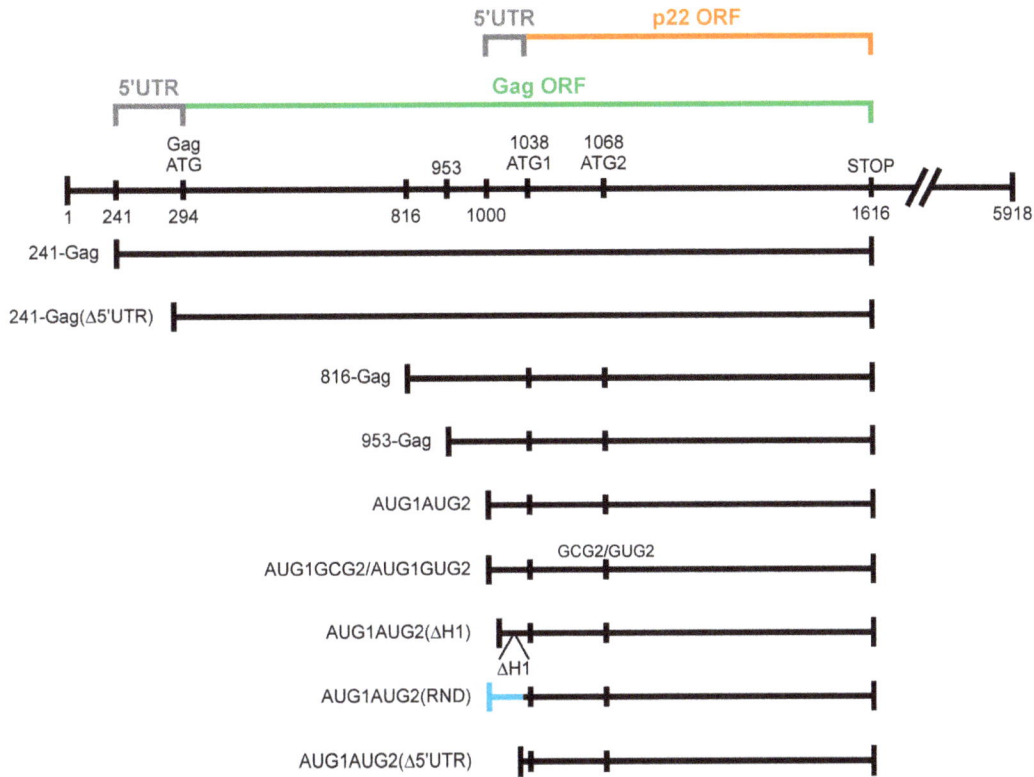

Figure 1. RNA constructs used in this study. Nucleotide positions correspond to the Ty1H3 DNA sequence (GenBank accession M18706.1) [15]. 5′ UTR: 5′ untranslated region, ORF: open reading frame.

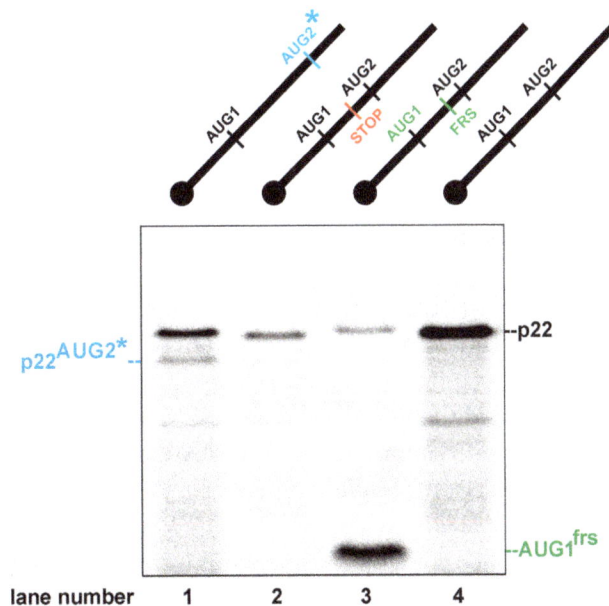

Figure 2. In vitro translation of Ty1i RNA and its derivatives in wheat germ extract. In vitro transcribed, capped RNA AUG1AUG2*, AUG1^{stop}AUG2, AUG1^{frs}AUG2 and AUG1AUG2 were translated in the presence of ^{35}S-methionine followed by electrophoresis and autoradiography. Schematic representation of RNA molecules is shown above the gel (see text for details).

AUG1AUG2* RNA was translated into two products: p22^{AUG1} synthesized from the natural AUG1 and the shorter protein p22^{AUG2*} (Figure 2, lane 1). p22^{AUG1} / p22^{AUG2*} were synthesized in a ratio of 5:1, which indicates that AUG1 is the main site of p22 translation initiation in AUG1AUG2*

RNA. However, the translational activity of AUG1AUG2* RNA decreased 75% when compared with wild-type AUG1AUG2 RNA. Two proteins were also translated from the AUG1^{frs}AUG2 RNA: a faster migrating out of frame AUG1^{frs} peptide, and p22^{AUG2}, which originated from the natural AUG2 triplet (Figure 2, lane 3). AUG1^{frs}/p22^{AUG2} were synthesized in a ratio of 6:1, which is similar to AUG1AUG2*, and confirms that AUG1 is utilized preferentially for p22 initiation in these two RNAs. As expected, p22^{AUG2} that initiated from AUG2 was detected with AUG1^{STOP}AUG2 RNA (Figure 2, lane 2). The level of AUG2-initiated p22 was low but comparable between different constructs.

Taken together, the results of in vitro translation show that both AUG codons present in Ty1i RNA can be actively translated and AUG1 is preferentially utilized to initiate p22 synthesis. Our results also suggest that leaky scanning is the most likely mechanism for p22 translation from AUG2. Experimental support for leaky scanning is illustrated by the decrease of AUG2 translation levels from AUG1AUG2* and AUG1^{frs}AUG2 RNAs (having both p22 AUG codons) in comparison to GCG1AUG2 RNA mutant where only AUG2 is present [14]. Moreover, the translational activity of AUG1AUG2* and AUG1^{frs}AUG2 RNAs was significantly lower when compared to wild-type AUG1AUG2 RNA. These results raise the possibility that AUG1AUG2* and AUG1^{frs}AUG2 RNAs affect the structure of the 5' UTR of Ty1i RNA, leading to translation inhibition, and that the 5' UTR may also regulate the production of p22.

3.2. The 5' UTR of mRNA Interacts with the p22 Coding Region

Significant loss of translational activity from AUG1 in AUG1GCG2 [14] (Figure 1), AUG1AUG2* and AUG1^{frs}AUG2 RNAs suggests that the structure of the region containing AUG1 and AUG2 is important for p22 translation. Therefore, we performed selective 2'-hydroxyl acylation analyzed by primer extension (SHAPE) [37] on the 5' terminal region of Ty1i RNA to examine its secondary structure. N-methylisatoic anhydride (NMIA) preferentially modifies 2'OH groups of single-stranded and flexible nucleotides in RNA. Primer extension of fluorescently labeled primers by reverse transcriptase is blocked at modified positions in RNA, and these truncated DNA products can be identified using capillary electrophoresis. Secondary RNA structures were obtained by computational analysis of the reverse transcription products. Secondary structure probing experiments were carried out on AUG1AUG2 RNA that was used in the in vitro translation studies. This ~630 nt long RNA contained the 5' UTR of Ty1i RNA (37 nt) and coding sequence of p22 (Figure 1).

Figure 3 shows a secondary structure model of the 5' terminal part of the Ty1i RNA [15] predicted using the *RNAstructure* software [38,39] which incorporates experimental constraints from SHAPE mapping.

Our results suggest that Ty1i RNA folds into two major domains. The smaller domain I (G1000–1083) and larger domain II (A1096–U1501) were connected by a 12nt-long single-stranded region (A1084–G1095).

Interestingly, domain I included the Ty1i 5' UTR and p22 coding sequence, and contained both p22 initiation codons (Figure 3). This structure is organized by the interaction of the proximal part of the 5' UTR (G1000–U1012) with a stretch of coding sequence (A1068–C1083; stems S1–S3). Also, two hairpin structures were present. Hairpin H1 (U1015–A1035) was composed of residues from the 5' UTR while hairpin H2 (U1048–A1066) contained nucleotides from the coding sequence. A three-way junction connected hairpins H1, H2 and stem S1.

The data from SHAPE probing support the predicted structure of domain I. Nucleotides within single-stranded regions were reactive towards the SHAPE reagent, including apical loops of both hairpins, internal loops, bulges and mismatches. The presented structure was also supported by dimethyl sulfate (DMS) probing. DMS methylates N1 of adenosines and N3 of cytidines that have an accessible Watson–Crick edge of the base rings [40]. In our structure, almost every A and C residue predicted to be single-stranded was susceptible to DMS methylation. However, some nucleotides in the hairpin H2 stem were methylated moderately by DMS but remained unreactive towards NMIA.

These results support the idea that the C1052 and A1064-A1066 hairpin region is constrained by non-standard base pairing.

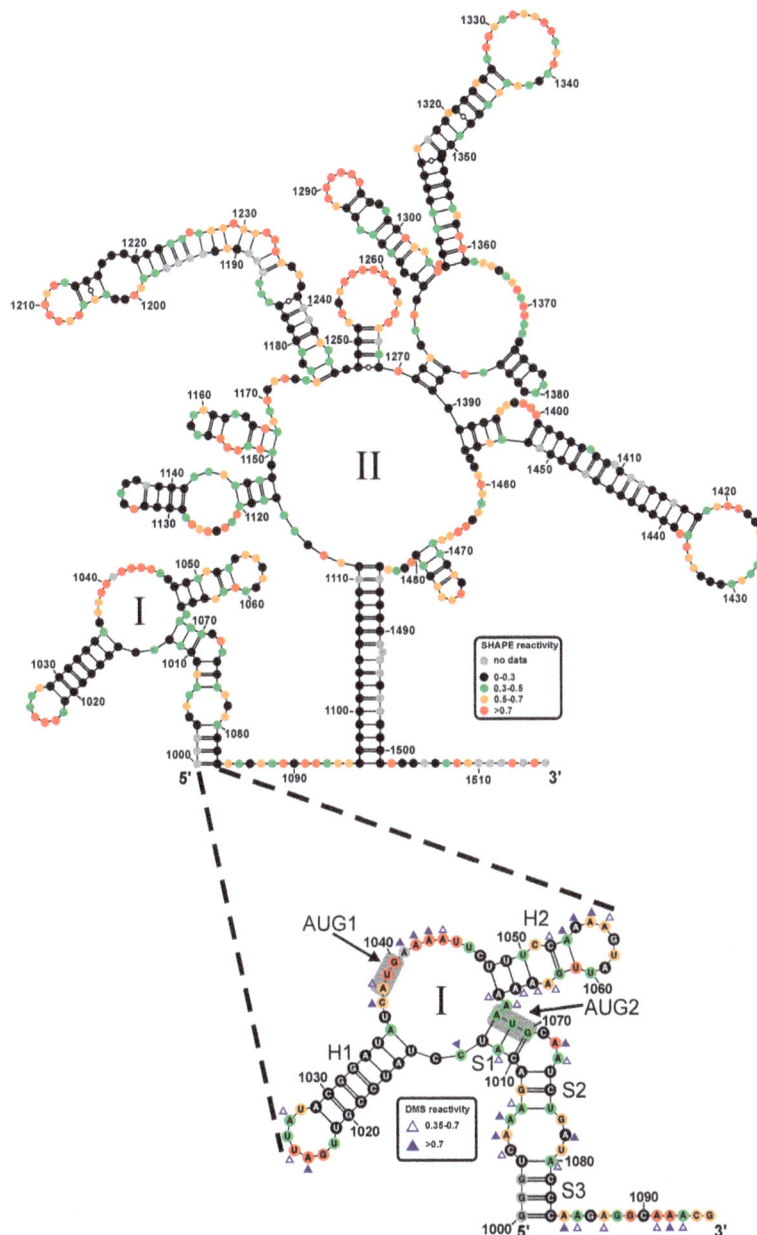

Figure 3. Secondary structure model of the 5′ terminal segment of Ty1i RNA (upper panel) and a detailed view of domain I (bottom panel) predicted by the *RNAstructure* software with experimental constraints [38]. Nucleotides are coloured according to their selective 2′-hydroxyl acylation analyzed by primer extension (SHAPE) reactivity (black, green, orange, red). The blue triangles (filled and open) represent dimethyl sulfate (DMS) modifications.

Interestingly, domain I contained both p22 initiation codons localized in different structural contexts (Figure 3). AUG1 constituted part of the 12nt-long single-stranded region U1036–C1047 while AUG2 was embedded in the double-stranded S1 stem that was formed by interactions of nts 1068–1070 with the residues of the 5′ UTR (C1010–U1012). The S1 stem may be thermodynamically unstable since the AUG2 triplet was somewhat reactive against NMIA.

Domain II folded into a large multibranched structure (Figure 3) organized by extensive pairing between A1096–C1111 and G1485–U1501. As a result, a 16 bp duplex region was formed. Domain II contained a complex junction that connected six simple hairpin structures and one branched region in a three-way junction motif. The majority of the single stranded regions were well mapped by NMIA. Importantly, the NMIA modification pattern of nucleotides spanning domain II in AUG1AUG2 RNA was very similar to the same region mapped inside VLPs using in virio SHAPE [30] (please note that the numbering herein corresponds to the complete Ty1H3 element while the numbering in reference [30] corresponds to Ty1 genomic RNA [30]). This result suggests that our in vitro folding conditions recapitulate the native structure of Ty1 RNA.

3.3. The 3D Structural Integrity of Domain I Affects p22 Translation

We reported that the combined level of p22 synthesized from AUG1GCG2 and GCG1AUG2 RNA constitutes only 30% of that obtained from wild-type AUG1AUG2 RNA [14]. Secondary structure probing of AUG1AUG2 RNA revealed that both p22 initiation codons were located within the same domain. Thus, mutation of AUG1 or AUG2 could cause structural perturbations that inhibit p22 translation. Since the in vitro translation results (Figure 2) identified AUG1 as a main translation initiation site for p22 synthesis, we hypothesized that mutating AUG2 to GCG strongly inhibited translation from AUG1 due to changing the structural context of AUG1 in domain I. The AUG2 to GCG mutation also introduced a U–G wobble pair as well as A–C mismatch that could affect the double-stranded character of the S1 stem.

To determine if the GCG mutation altered the structure of domain I, we performed secondary structure probing of AUG1GCG2 RNA using SHAPE. Although the overall reactivity pattern of the AUG1GCG2 RNA was preserved (Figure 4A), the region of domain I containing the GCG mutation (A1066–A1071) became highly reactive. This alteration suggests that the mutant RNA residues in the S1 stem are single-stranded or this region is highly unstable. Additionally, several nucleotides in hairpin H2 displayed a different pattern of reactivity: G1057–A1059, U1061 and G1062 exhibited higher reactivity while A1055 had decreased reactivity. Surprisingly, the structural motifs in the neighborhood of AUG1 remained essentially the same in wild type and mutant AUG1AUG2 RNA. Moreover, the GCG mutation did not change the secondary structure of domain II (data not shown). Overall, our data suggests that the GCG mutation disrupts the three-dimensional structure of domain I, which in turn inhibits the translation of p22 from AUG1.

Our model suggests that a three-way junction element (Figure 3) governs the special organization of domain I. By disrupting the S1 stem, the GCG mutation might change the topology and relative positioning of the H1 and H2 hairpins. Changes in the three-dimensional structure of RNA molecules can be monitored by native polyacrylamide gel electrophoresis [41]. Therefore, we subjected the isolated domain I (nts G1000–C1083) containing the GCG mutation (domain I^{GCG2}) along with the wild-type domain I to native gel electrophoresis (Figure 4B). We observed a slower mobility of domain I^{GCG2} RNA, which may reflect a change in the three-dimensional structure of domain I when compared with wild type. Migration of both wild-type and GCG mutated domain I remained unchanged at a higher concentration of Mg^{2+} ions, suggesting that this part of Ty1i RNA undergoes unimolecular folding [42].

The results obtained by native gel electrophoresis suggest that the double-stranded character of the S1 stem is an important factor stabilizing the three-dimensional structure of domain I. To help preserve the double-stranded character of stem S1, we mutated AUG2 to a GUG valine codon that changed only the first U–A pair to a U–G wobble pair (Figure 1). Secondary structure probing of AUG1GUG2 mutant RNA indicated that the S1 stem was slightly destabilized (Figure S1). Moreover, two residues directly upstream of the S1 stem (A1066 and A1067) were more reactive, suggesting an enhancement of local flexibility. A1066 and A1067 were also strongly modified in AUG1GCG2 mutant RNA. Some of the nucleotides in the H2 hairpin that changed their reactivity in AUG1GCG2 RNA behaved in a similar manner in AUG1GUG2 RNA. Higher reactivity of U1058 and A1062 as well as lack of reactivity of A1055 was detected. A1063 was also less reactive in AUG1GUG2 RNA when compared to wild type AUG1AUG2. Importantly, the structural context of AUG1 was preserved, which is similar to the AUG1GCG2 and AUG1GUG2 mutants. Taken together, our data suggest that the GUG2 mutation destabilized the S1 stem much less than the GCG2 mutation, and the structural integrity of the S1 stem and hairpin H2 are important determinants for the proper three-dimensional structure of domain I.

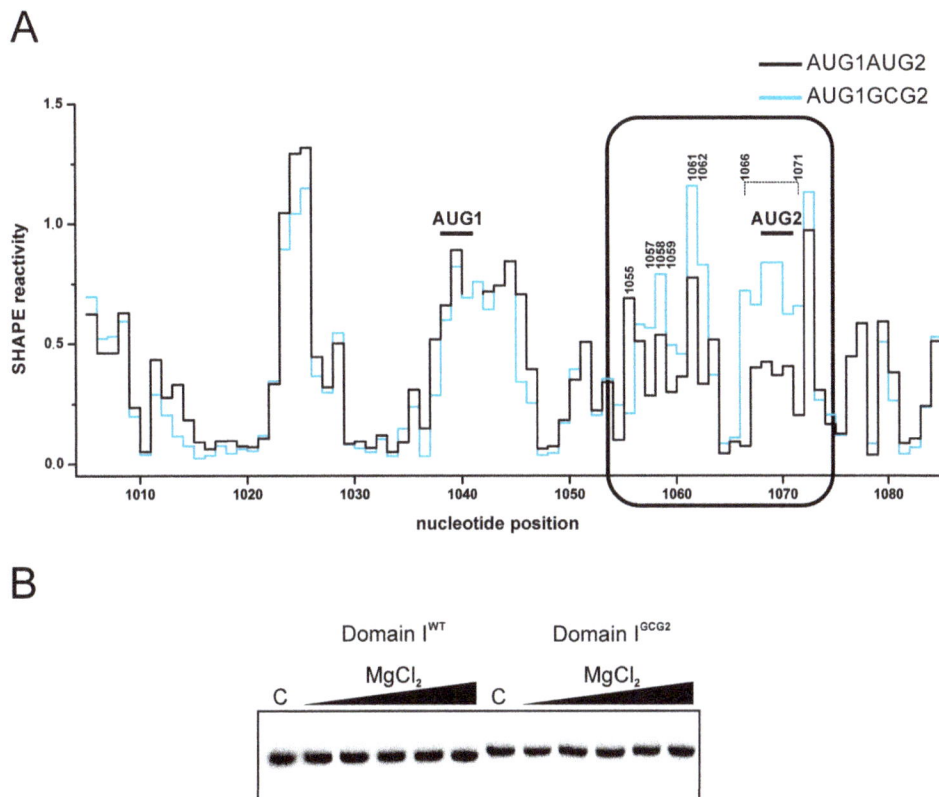

Figure 4. (**A**) SHAPE reactivity profile of the AUG1AUG2 (black) and AUG1GCG2 (blue) domain I as a function of nucleotide position. Nucleotides that changed their reactivity in domain I^{GCG2} are indicated. (**B**) Native gel electrophoresis of the [^{32}P]-labeled wild-type and mutated domain I of Ty1i RNA at increasing concentrations of MgCl$_2$. C: control reaction without MgCl$_2$. WT: wild type.

Mutation of AUG2 to GCG2 markedly inhibits p22 translation (Figure 2) [14]. Since we determined that the GUG2 mutation had a less profound effect on the domain I secondary structure, we analyzed the translational activity of capped and uncapped AUG1GUG2 RNA along with AUG1GCG2 and AUG1AUG2 RNA in vitro (Figure 5A). In agreement with our previous study [14], p22 translation from AUG1GCG2 RNA was inhibited to ~15% of the initial value calculated for AUG1AUG2 RNA. Interestingly, the translation of p22 from AUG1GUG2 RNA was also inhibited to ~20% when compared with wild type RNA. These results further extend our finding that the structural integrity of the domain I of Ty1i RNA contributes significantly to the efficient translation of the p22 from AUG1, and even small structural changes impair translation in vitro.

Placement of the initiation codon in thermodynamically stable secondary structures can decrease its translational activity [43]. However, the calculated thermodynamic stability [44] of domain I in wild-type Ty1i RNA was only −25.2 kcal/mol, and AUG1 was predicted to reside in a long single-stranded region (Figure 3). To assess the thermodynamic stability of the 5′ terminal segment of Ty1i RNA, we determined the reactivity profile of AUG1AUG2 RNA by SHAPE mapping at different temperatures (Figure 5B). SHAPE analysis at 37 °C and 60 °C identified residues within domain I that changed their reactivity at 60 °C. Interestingly, the most pronounced effects were observed in the regions prone to destabilization in RNA mutants AUG1GCG2 and AUG1GUG2 (Figure 4 and Supplementary Figure S1). At 60 °C, the nucleotide stretch A1067–G1077 (including AUG2) as well as the opposite strand A1005–C1013 became highly reactive, suggesting that the strands dissociate. Also, several residues located in the hairpin H2 stem (U1049–C1052) and in the apical loop (A1059–U1061) were altered, suggesting that the region containing AUG2 and hairpin H2 is less stable than other parts of domain I.

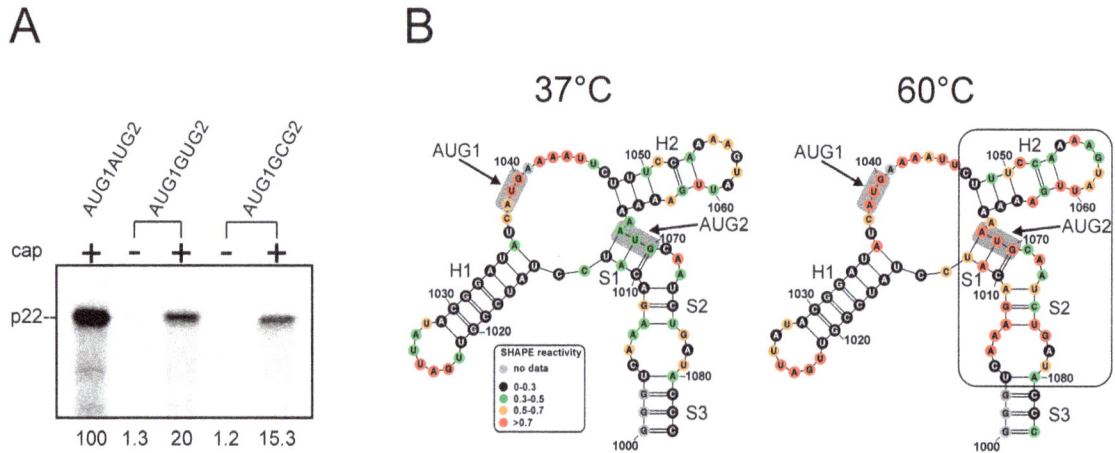

Figure 5. In vitro translation of AUG2 mutational variants of Ty1i RNA and melting profile of AUG1AUG2 RNA. (**A**) In vitro transcribed capped or uncapped transcripts were translated using wheat germ extract in the presence of ^{35}S-methionine. Calculated translation activity (in relation to the capped AUG1AUG2 RNA) is shown below the gel. (**B**) Melting of AUG1AUG2 RNA followed by SHAPE at 37 °C and 60 °C. Nucleotides are coloured according to their reactivity (black, green, orange, red). The segment of domain I with the strongest changes at 60 °C is boxed.

3.4. Structure of Domain I Specific for Ty1i RNA Stimulates p22 Translation

In vitro translation and secondary structure probing of the 5′ terminal part of wild-type and mutant Ty1i transcripts suggest that domain I plays an important role in the efficient translation of p22 from AUG1. Previous results show that p22 is not translated from the full-length genomic RNA [15]. These findings motivated us to ask whether the structure of domain I was stable in the context of a larger RNA that more closely resembles Ty1 genomic RNA. To this end, we analyzed a ~1400 nt RNA (nts 241–999 using the coordinates of the complete Ty1H3 element), termed 241-Gag RNA, that began from the first nucleotide of the genomic Ty1 RNA, and included the structured 5′ UTR [30,45] and Gag coding sequence (Figure 1). Comparison of SHAPE reactivity profiles of 241-Gag and AUG1AUG2 RNAs revealed different modification patterns of domain I (Figure 6A).

The reactivity of the region encompassing AUG2 (A1067–A1072) increased in 241-Gag RNA while the proximal part of the single-stranded region connecting domains I and II (A1084–G1089) lost accessibility to NMIA modification. The observed alterations suggest that domain I and the neighboring regions fold differently when the 5′-terminal sequence of genomic RNA is present in the transcript.

The secondary structure of the full-length Ty1 RNA has been determined inside virus-like particles (VLPs) by in virio SHAPE analysis [30]. In the proposed structure for Ty1 genomic RNA, the sequence encompassing domain I is folded differently than in Ty1i RNA (Figure 6B). Interactions between C979–U983 and A1085–G1089 extended domain I in the full-length transcript. Moreover, the structural context of the p22 initiation codons differed significantly. Unlike their context in Ty1i RNA, AUG1 was fully paired with the C1010–U1012 in full-length Ty1 RNA. Interestingly, the C1010–U1012 region was also paired but with the AUG2 codon forming the S1 stem in Ty1i RNA (Figure 3). AUG2 was localized in the stem of a predicted unstable hairpin G1057–C1071. The only common structural element within the region encompassing domain I in the full-length Ty1 and Ty1i RNAs was hairpin H1, suggesting that hairpin H1 folds independently of the structural elements present in its vicinity.

Importantly, comparing the reactivity profiles of 241-Gag and full-length Ty1 RNA [30] revealed that domain I folding was similar (Figure 6B). The main difference was AUG1 reactivity, which was high in 241-Gag RNA and low in full-length Ty1 RNA. This difference suggests that the cellular environment in this region, such as the presence of the Gag chaperone, folds the RNA into a more stable structure.

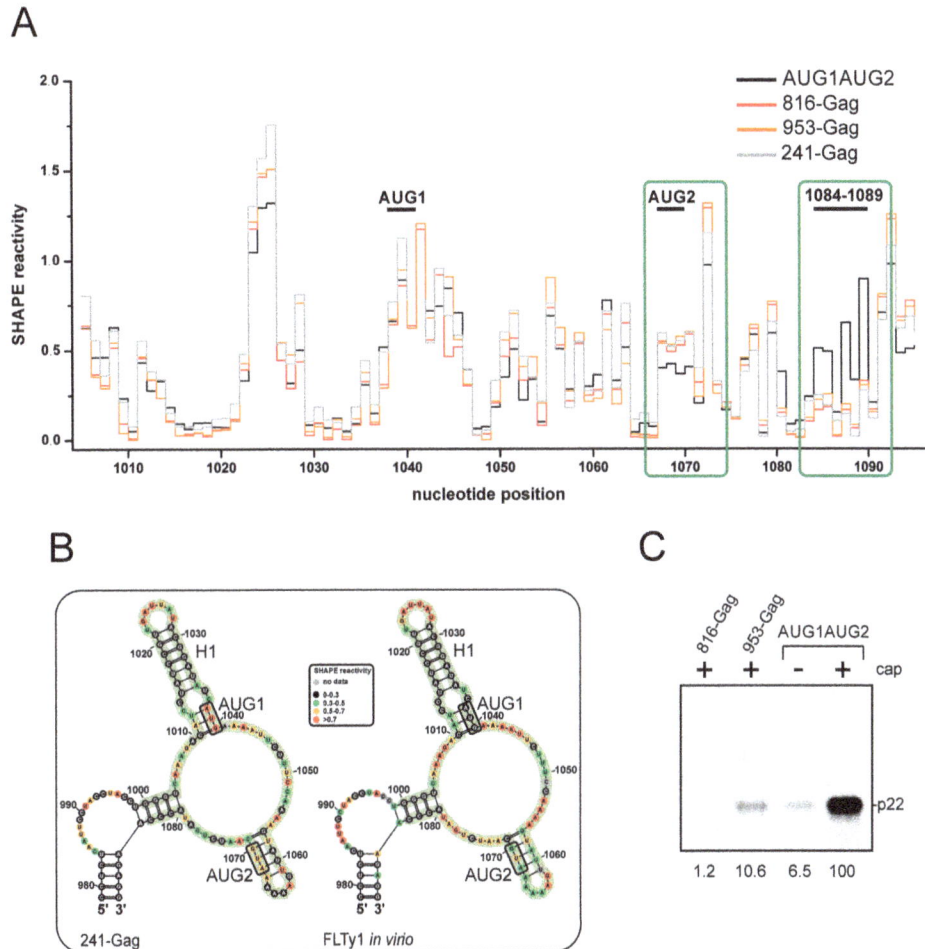

Figure 6. Secondary structure probing and the in vitro translation of 241-Gag RNA and its derivatives. (**A**) Reactivity plot of nucleotides spanning domain I in AUG1AUG2 RNA (black), 816-Gag RNA (red), 953-Gag RNA (orange) and 241-Gag RNA (grey). Regions showing consistent differences in reactivity are boxed (green). (**B**) Comparison of the secondary structure models of domain I obtained in vitro for 241-Gag RNA (left) and full-length genomic Ty1 RNA within virus-like particles (VLPs) (in virio conditions; right). Nucleotides that cover domain I in Ty1i RNA are marked (green background). p22 initiation codons and the H1 hairpin are also highlighted. (**C**) In vitro translation of sequential variants of Ty1 genomic RNA. Capped or uncapped transcripts were translated in wheat germ extract in the presence of the ^{35}S-methionine. Quantitation of the translation products is shown below the gel.

The distinct structure of the region encompassing domain I in the full-length Ty1 RNA raised a question concerning how domain I might influence p22 translation. The initiation of p22 synthesis from the 241-Gag RNA is unlikely to occur, which raises the possibility that p22 synthesis requires a specific structure of domain I in Ty1i RNA [14]. The presence of the Gag AUG initiation codon as well as seven internal in-frame AUG codons before encountering AUG1 would preclude migration of the preinitiation complexes downstream of the AUG1 and AUG2 initiation codons. Additionally, the 5′ UTR of Ty1i RNA in the 241-Gag RNA would be extended to over 700 nucleotides, which could greatly affect the scanning mechanism. To address whether a specific structure of the domain I of Ty1i RNA is necessary for the efficient translation of p22, we synthesized 816-Gag and 953-Gag RNAs (Figure 1). Both RNA molecules were designed to possess full-length folding of domain I, which is supported by their similar reactivity profile when compared to 241-Gag RNA (Figure 6A). The 816-Gag and 953-Gag RNAs were translated in vitro in wheat germ extract (Figure 6C). We observed that p22 protein was poorly translated from both RNA molecules and could be detected only when capped transcripts were used. Low levels of translation from extended Ty1 transcripts with the full-length-like folding of the region 1000–1083 suggests that the structure of the domain I observed in Ty1i RNA specifically stimulates p22 translation from AUG1.

3.5. The Ty1i RNA 5′ UTR Stimulates p22 Translation

To further understand the role of the Ty1i 5′ UTR in p22 translation, we analyzed in vitro several mutant RNA constructs (Figure 1). In AUG1AUG2(Δ5′ UTR), 32 of 37 nucleotides of the 5′ UTR have been deleted while in AUG1AUG2(RND) the same sequence was replaced by 32 random nucleotides. In AUG1AUG2(ΔH1), the common structural element of full-length Ty1 and Ty1i RNA (hairpin H1) was deleted (nts 1015–1031). Also, all transcripts maintained an intact Kozak context adjacent to the AUG1 initiation codon.

We observed significant inhibition of p22 translation from all three RNA constructs (Figure 7A). Deleting the 5′ UTR inhibited p22 translation by 40% when compared to wild-type AUG1AUG2 RNA. These results suggest that the Ty1i 5′ UTR is required for efficient p22 synthesis. Since shortening the 5′ UTR to only six nucleotides could interfere with ribosome scanning [46–48], we analyzed 241-Gag(Δ5′ UTR) RNA possessing 5′ UTR that was also reduced to six nucleotides. However, the translation of Gag was unaffected (Figure 7B). This result suggests that the inhibitory effect observed for AUG1AUG2(Δ5′ UTR) may impair the structure of domain I. The important role of the 5′ UTR in p22 translation was also supported by the translation of AUG1AUG2(RND) and AUG1AUG2(ΔH1) RNAs. Despite having a 5′ UTR of the same length as wild-type, AUG1AUG2(RND) RNA displayed >70% inhibition in p22 translation. A 55% inhibition of p22 synthesis was also observed with AUG1AUG2(ΔH1) RNA. Taken together, our data suggest a stimulatory role for the Ty1i 5′ UTR in the translation of p22 due to its involvement in the folding of domain I.

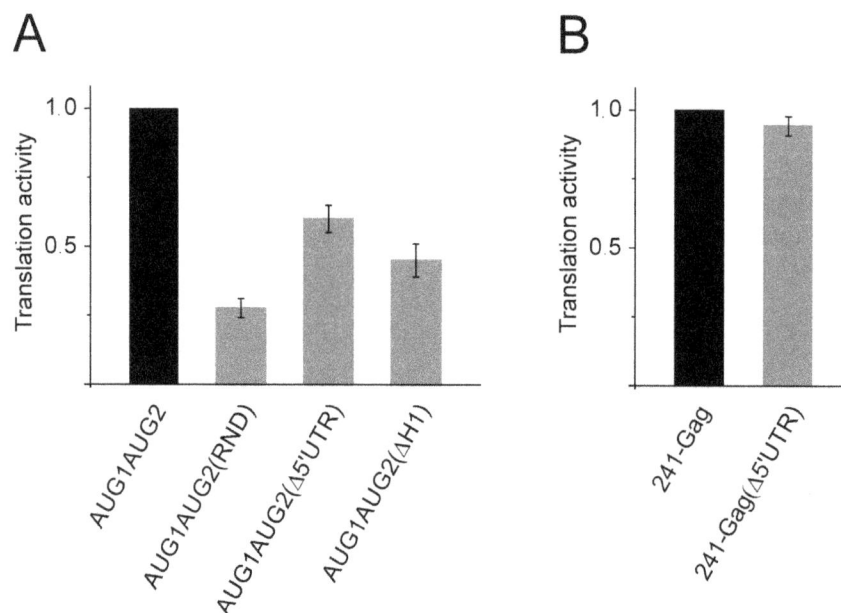

Figure 7. In vitro translation of the capped variants of the AUG1AUG2 RNA and 241-Gag RNA. Translational efficiency was normalized to the amount of the protein product synthesized from AUG1AUG2 RNA (**A**) or 241-Gag RNA (**B**).

3.6. Gag Interacts Specifically with Ty1i Domain I In Vitro

Translation initiation can be regulated not only by RNA structure but also by protein factors that interact with structural elements in mRNAs [19]. Since the amount of Gag and p22 determines the level of inhibition of Ty1 mobility [49], perhaps Gag modulates the efficiency and/or timing of p22 translation. Potential Gag binding sites in the 5′ terminal part of Ty1i RNA were detected by hydroxyl radical footprinting of AUG1AUG2 RNA complexed with recombinant Gag-p45 (Figure 8A). The protected sequences were identified by comparing the reactivity profiles of AUG1AUG2 RNA in the presence and absence of Gag. Only regions in domain I displayed decreased susceptibility to hydroxyl radical cleavage in the presence of Gag, including residues A1011–C1019 that comprise part of the S1 stem and the hairpin H1 stem. Another potential Gag binding site was localized in the p22 coding region (nts A1084–G1095) connecting domains I and II. In particular, C1081–C1090 was protected from the cleavage in the presence of Gag (Figure 8A,B).

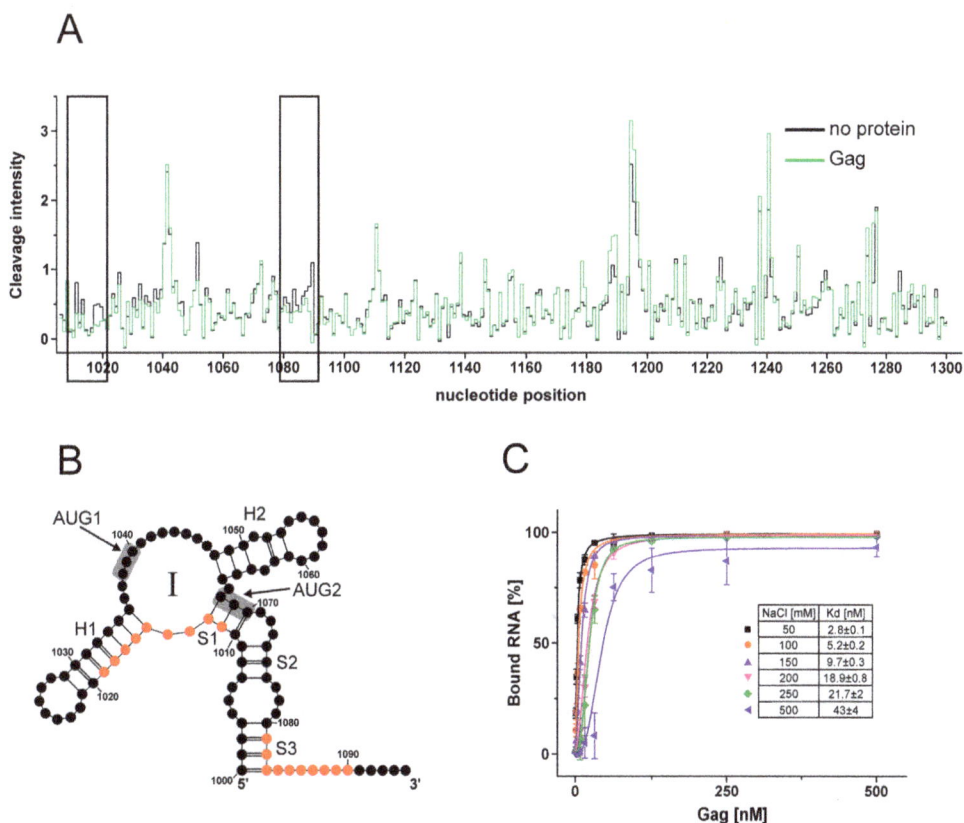

Figure 8. RNA binding properties of recombinant Ty1 Gag-p45. (**A**) Hydroxyl radical reactivity plots of protein free AUG1AUG2 RNA (black) in comparison with RNA probed in the presence of Gag (green). Regions showing consistent decreased reactivity over several nucleotides in the presence of Gag are boxed. (**B**) 2D structure model of Ty1i domain I with the positions protected from hydroxyl radical cleavage in the presence of the Ty1 Gag are indicated (red). (**C**) Filter-binding assay performed with Ty1i domain I RNA and Gag at different concentrations of NaCl (50–500 mM). The lines correspond to the best fit of the data. The error bars represent standard deviations. Kd: dissociation constant.

To further investigate the interaction between Gag and domain I, we calculated dissociation constants of RNA/protein complex formation using a double filter binding assay (Figure 8C). We used isolated domain I that was extended by the single-stranded stretch connecting domain I and II (RNA $I^{1000-1095}$) to encompass both Gag binding sites. The calculated dissociation constant (Kd ~3 nM) suggests that there is a high affinity binding site for Gag in domain I. To examine whether Gag binding is specific, we determined the Kd with increasing concentrations of NaCl, which is often used to compete out non-specific RNA/protein interactions [31]. The Gag/domain I interaction was slightly affected in the 100–250 mM NaCl range and persisted even at 500 mM NaCl (Kd ~43 nM). Taken together, the results from chemical footprinting and filter binding suggest that the interaction between Gag and domain I is strong and highly specific.

3.7. Deleting the Hairpin H1 Sequence Decreases Stability of Ty1i RNA In Vivo

To investigate the effects of the H1 hairpin on Ty1i RNA and p22 expression in vivo as well as on Ty1 transposition, a mutated pGPOLΔ plasmid was constructed (pBAS47, termed H1Δ) that expresses Ty1i RNA lacking the H1 sequence (U1015–A1035) from the 5′ UTR (Figure 9). Wild type pGPOLΔ is a multicopy expression plasmid containing most of the Ty1 5′LTR and *GAG* that is driven by the *GAL1* promoter [15]. When yeast cells containing pGPOLΔ are grown in glucose media, *GAL1* promoted transcription of Ty1 is repressed. However, Ty1i RNA and p22 are still expressed from pGPOLΔ under glucose repression since Ty1i RNA is transcribed from internal initiation sites.

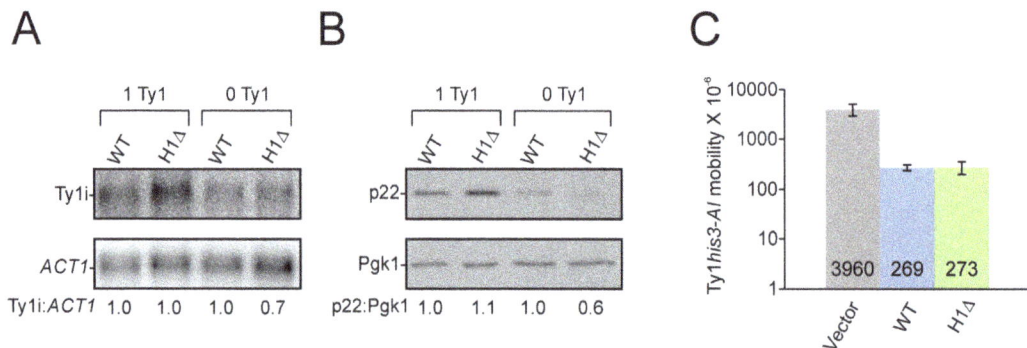

Figure 9. Effect of the hairpin H1 deletion on Ty1i RNA, p22 protein expression and Ty1*his3-AI* mobility. (**A**) Northern blotting of total RNA from the 1 Ty1 strain (DG2196) and 0 Ty1 strain (DG3582) containing either wild type (WT) pGPOLΔ or mutant pH1Δ plasmids. A [^{32}P]-labeled Ty1 riboprobe (nt 1266 to 1601) was used to detect Ty1i RNA. *ACT1* mRNA served as a loading control. Below are Ty1i:*ACT1* ratios as determined by phosphorimaging. (**B**) Whole cell extracts from strains used in (A) were immunoblotted with p18 antiserum to detect p22. Pgk1 served as a loading control. p22:Pgk1 ratios were determined by densitometry. (**C**) Quantitative Ty1*his3-AI* mobility assayed in the 1 Ty1 strain containing one genomic Ty1*his3-AI* element and empty vector, WT, or H1Δ plasmids. All strains were grown in glucose containing medium to repress *GAL1*-promoted Ty1 expression. Bars denote standard deviation.

We investigated the effect of H1Δ on Ty1i RNA level in a *S. paradoxus* strain with 1 chromosomal Ty1 element (DG2196; 1 Ty1) and the isogenic Ty1-less parent (DG3582; 0 Ty1) that contain WT pGPOLΔ or pH1Δ plasmids (Figure 9A). Northern blotting of total RNA from these strains showed no change in Ty1i RNA levels in the H1Δ mutant compared to the wild type (WT) plasmid in the 1 Ty1 strain. However, Ty1i H1Δ RNA levels decreased about 30% compared to WT Ty1i RNA in the 0 Ty1 strain (refer to Materials and Methods). These results suggest that the H1 hairpin may affect the stability of Ty1i RNA. In the 1 Ty1 strain, however, the defect in Ty1i H1Δ RNA stability was not evident. This may be due to additional Gag binding sites on Ty1i RNA that stabilize the transcript in the 1 Ty1 strain, as suggested by hydroxyl radical footprinting (Figure 8). Note that Gag binding sites C1081–C1090 remain intact in Ty1i H1Δ RNA and could function in vivo.

Total cell extracts from the same strains were subjected to Western analysis using an antiserum that detects p22 [14] (Figure 9B). The level of p22 remained about the same in the 1 Ty1 strain containing WT or H1Δ plasmids. In the 0 Ty1 strain, p22 decreased 43% (±12%) in the mutant pH1Δ when compared to WT pGPOLΔ. These results suggest that there is a correlation between p22 and Ty1i RNA levels (Figure 9A) in both strain backgrounds containing WT or H1Δ plasmids.

Finally, we asked if deleting the H1 hairpin from the Ty1i RNA affected Ty1 mobility (Figure 9C). A quantitative Ty1 mobility assay was performed in the 1 Ty1 yeast strain containing empty vector (Vector), WT or H1Δ plasmids. The single element in the 1 Ty1 strain is marked with the retrotransposition indicator gene *his3-AI* [33]. A Ty1*HIS3* genomic insertion that occurs following splicing of the *AI* (artificial intron) will complement the *HIS3* deletion mutation present in the strain. Therefore, the number of His+ colonies generally reflect the level of Ty1 mobility. As expected for cells undergoing Ty1 CNC, the level of Ty1*his3-AI* mobility decreased about 15-fold from plasmid-based expression of p22 [13,14]. However, H1Δ and WT displayed similar levels of Ty1 mobility, suggesting that deleting the H1 hairpin does not affect Ty1 CNC despite the modest decrease in p22 observed in the 0 Ty1 strain (Figure 9B). Perhaps removing only one of the Gag binding sites in domain I of Ty1i RNA is not enough to affect CNC because Gag produced in the 1 Ty1 strain stabilizes Ty1 RNA through binding to other sites.

3.8. AUG1 is Exposed in a 3D Structural Model of Domain I RNA

Our Ty1i RNA structural and functional studies indicate that the 3D structure of domain I is important for efficient p22 translation. However, determining the 3D structure of RNA in solution is challenging. Therefore, we combined chemical probing experiments to map RNA secondary (Figure 3) and tertiary structures using RNAComposer [34]. To reveal the tertiary fold of domain I of AUG1AUG2 RNA and support RNAComposer predictions [36], we also used hydroxyl radicals to produce strand breaks. This approach allows one to map solvent exposed regions of the nucleic acid backbone. This analysis predicted >100 different 3D structures of domain I and clustered them based on their agreement with the hydroxyl radical cleavage data and the energy of the final RNA 3D structure. The structures that best-fit the hydroxyl radical cleavage data allowed us to explain the gain in SHAPE reactivity of H2 apical loop nucleotides upon S1 stem destabilization in the AUG1GCG2 and AUG1GUG2 RNA mutants. Our models suggest that the H2 hairpin stem bends due to the presence of an internal loop containing unpaired C1051 and A1063, which causes an apical loop of H2 to be positioned close to the 3-way junction. Thus, disruption of junction geometry due to S1 unwinding is likely to affect H2 apical loop reactivity. The best models shared the common feature of coaxial positioning of the S1 stem and H1 hairpin. Such an organization of the 3-way junction places AUG1 on the surface of the molecule between hairpins H1 and H2, and may contribute to AUG1's preferential use for initiating the translation of p22 (Figure 10).

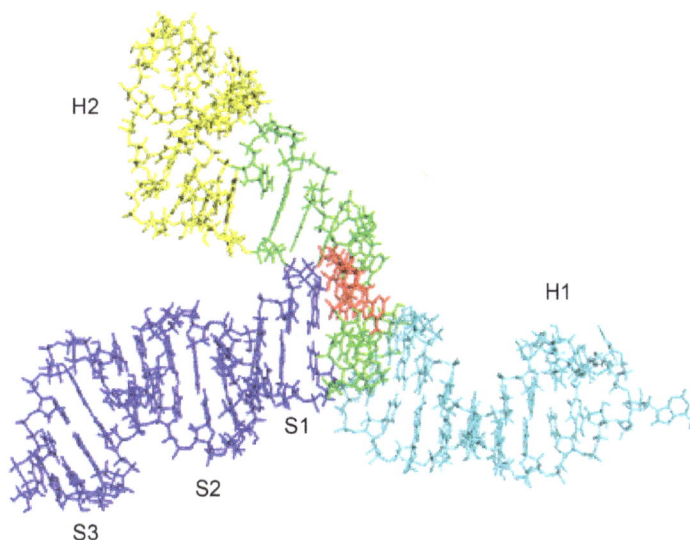

Figure 10. A 3D structure model of Ty1i RNA domain I. Structural elements are annotated: hairpin H1 (cyan), hairpin H2 (yellow), stem S1–3 (blue) and 3-way junction (green). AUG1 sequence is marked in red.

4. Conclusions

Translation initiation is the rate-limiting step of protein synthesis and is highly regulated by RNA binding factors and structural properties of the messenger RNA. This coordinated action allows cells to rapidly adapt to their environment without the need of de novo mRNA synthesis and transport from the nucleus to the cytoplasm [50]. In addition, a wide variety of viruses exploit variations in translation initiation to expand their coding capacity from a limited set of transcripts, including the use of alternative initiation codons and internal ribosome entry sites [17]. In the present work, we address how the Ty1 restriction factor p22 is translated from Ty1i RNA using a combination of structural and functional approaches. We show that two p22 initiation codons on Ty1i RNA are embedded in structural domain I, which is formed by an interaction between the 5′ UTR and the coding sequence. Our in vitro translation experiments show that both p22 initiation codons can be utilized but that AUG1 is used preferentially. We demonstrate that the structural integrity of Ty1i RNA is critical for the efficient expression of p22 from AUG1. Even small changes in the domain I sequence that disrupt its secondary and tertiary structure result in strong inhibition of p22 synthesis. Our studies have mapped two high affinity Ty1 Gag binding sites located in domain I of Ty1i RNA. Deletion of one of the binding sites leads to a decrease in the p22 level in vivo by destabilizing Ty1i RNA. Our work supports the hypothesis that structural motifs of domain I are not only important for the efficient translation of p22 protein but may also contribute to the stability of Ty1i RNA via interactions with Gag. Such interactions raise the possibility of an autogenous control loop where Gag positively controls the synthesis of p22, which in turn inhibits Gag function and mediates Ty1 CNC. However, more work will be required to understand how Gag binding to Ty1i RNA contributes to its stability.

Acknowledgments: We thank Agnieszka Kiliszek and Katarzyna Pachulska-Wieczorek for valuable discussions, and Jeremy Thorner for providing the Pgk1 antiserum. This work was supported by the Ministry of Science and Higher Education Poland [0492/IP1/2013/72], Foundation for Polish Science [HOMING PLUS/2012-6/12] (KJP), NIH grant GM095622 (DJG) and funds from UGARF (DJG). KJP also acknowledges support from the Ministry of Science and Higher Education Poland (MNiSW, fellowship for outstanding young scientists).

Author Contributions: L.B., M.B., A.S., D.J.G. and K.J.P. conceived and designed the experiments; L.B., M.B., A.S. performed the experiments; L.B., M.B., A.S., D.J.G. and K.J.P. analyzed the data; L.B., M.B., A.S., D.J.G. and K.J.P. wrote the paper.

References

1. Kim, J.M.; Vanguri, S.; Boeke, J.D.; Gabriel, A.; Voytas, D.F. Transposable elements and genome organization: A comprehensive survey of retrotransposons revealed by the complete *saccharomyces cerevisiae* genome sequence. *Genome Res.* **1998**, *8*, 464–478. [PubMed]

2. Curcio, M.J.; Lutz, S.; Lesage, P. The Ty1 ltr-retrotransposon of budding yeast. *Microbiol. Spectr.* **2015**, *3*, 1–35. [PubMed]

3. Feng, Y.X.; Moore, S.P.; Garfinkel, D.J.; Rein, A. The genomic RNA in Ty1 virus-like particles is dimeric. *J. Virol.* **2000**, *74*, 10819–10821. [CrossRef] [PubMed]

4. Belcourt, M.F.; Farabaugh, P.J. Ribosomal frameshifting in the yeast retrotransposon Ty: tRNAs induce slippage on a 7 nucleotide minimal site. *Cell* **1990**, *62*, 339–352. [CrossRef]

5. Roth, J.F.; Kingsman, S.M.; Kingsman, A.J.; Martin-Rendon, E. Possible regulatory function of the *saccharomyces cerevisiae* Ty1 retrotransposon core protein. *Yeast* **2000**, *16*, 921–932. [CrossRef]

6. Mellor, J.; Fulton, A.M.; Dobson, M.J.; Roberts, N.A.; Wilson, W.; Kingsman, A.J.; Kingsman, S.M. The Ty transposon of *saccharomyces cerevisiae* determines the synthesis of at least three proteins. *Nucleic Acids Res.* **1985**, *13*, 6249–6263. [CrossRef] [PubMed]

7.   Pachulska-Wieczorek, K.; Le Grice, S.F.; Purzycka, K.J. Determinants of genomic RNA encapsidation in the *saccharomyces cerevisiae* long terminal repeat retrotransposons Ty1 and Ty3. *Viruses* **2016**, *8*. [CrossRef] [PubMed]

8.   Bourc'his, D.; Bestor, T.H. Meiotic catastrophe and retrotransposon reactivation in male germ cells lacking Dnmt3L. *Nature* **2004**, *431*, 96–99. [CrossRef] [PubMed]

9.   Yoder, J.A.; Walsh, C.P.; Bestor, T.H. Cytosine methylation and the ecology of intragenomic parasites. *Trends Genet.* **1997**, *13*, 335–340. [CrossRef]

10. Harris, R.S.; Dudley, J.P. Apobecs and virus restriction. *Virology* **2015**, *479–480*, 131–145. [CrossRef] [PubMed]

11. Drinnenberg, I.A.; Fink, G.R.; Bartel, D.P. Compatibility with killer explains the rise of RNAi-deficient fungi. *Science* **2011**, *333*, 1592. [CrossRef] [PubMed]

12. Drinnenberg, I.A.; Weinberg, D.E.; Xie, K.T.; Mower, J.P.; Wolfe, K.H.; Fink, G.R.; Bartel, D.P. RNAi in budding yeast. *Science* **2009**, *326*, 544–550. [CrossRef] [PubMed]

13. Garfinkel, D.J.; Nyswaner, K.; Wang, J.; Cho, J.Y. Post-transcriptional cosuppression of Ty1 retrotransposition. *Genetics* **2003**, *165*, 83–99. [PubMed]

14. Nishida, Y.; Pachulska-Wieczorek, K.; Blaszczyk, L.; Saha, A.; Gumna, J.; Garfinkel, D.J.; Purzycka, K.J. Ty1 retrovirus-like element gag contains overlapping restriction factor and nucleic acid chaperone functions. *Nucleic Acids Res.* **2015**, *43*, 7414–7431. [CrossRef] [PubMed]

15. Saha, A.; Mitchell, J.A.; Nishida, Y.; Hildreth, J.E.; Ariberre, J.A.; Gilbert, W.A.; Garfinkel, D.J. A trans-dominant form of gag restricts Ty1 retrotransposition and mediates copy number control. *J. Virol.* **2015**, *89*, 3922–3938. [CrossRef] [PubMed]

16. Tucker, J.M.; Larango, M.E.; Wachsmuth, L.P.; Kannan, N.; Garfinkel, D.J. The Ty1 retrotransposon restriction factor p22 targets gag. *PLoS Genet.* **2015**, *11*, e1005571. [CrossRef] [PubMed]

17. Firth, A.E.; Brierley, I. Non-canonical translation in RNA viruses. *J Gen. Virol.* **2012**, *93*, 1385–1409. [CrossRef] [PubMed]

18. Bolinger, C.; Boris-Lawrie, K. Mechanisms employed by retroviruses to exploit host factors for translational control of a complicated proteome. *Retrovirology* **2009**, *6*, 8. [CrossRef] [PubMed]

19. Pfingsten, J.S.; Kieft, J.S. RNA structure-based ribosome recruitment: Lessons from the dicistroviridae intergenic region ireses. *RNA* **2008**, *14*, 1255–1263. [CrossRef] [PubMed]

20. Kozak, M. Circumstances and mechanisms of inhibition of translation by secondary structure in eucaryotic mRNArs. *Mol. Cell. Biol.* **1989**, *9*, 5134–5142. [CrossRef] [PubMed]

21. Sagliocco, F.A.; Vega Laso, M.R.; Zhu, D.; Tuite, M.F.; McCarthy, J.E.; Brown, A.J. The influence of 5'-secondary structures upon ribosome binding to mRNA during translation in yeast. *J Biol. Chem.* **1993**, *268*, 26522–26530. [PubMed]

22. Vega Laso, M.R.; Zhu, D.; Sagliocco, F.; Brown, A.J.; Tuite, M.F.; McCarthy, J.E. Inhibition of translational initiation in the yeast *saccharomyces cerevisiae* as a function of the stability and position of hairpin structures in the mRNA leader. *J. Biol. Chem.* **1993**, *268*, 6453–6462. [PubMed]

23. Babendure, J.R.; Babendure, J.L.; Ding, J.H.; Tsien, R.Y. Control of mammalian translation by mRNA structure near caps. *RNA* **2006**, *12*, 851–861. [CrossRef] [PubMed]

24. Kozak, M. Context effects and inefficient initiation at non-AUG codons in eucaryotic cell-free translation systems. *Mol. Cell. Biol.* **1989**, *9*, 5073–5080. [CrossRef] [PubMed]

25. Kozak, M. Downstream secondary structure facilitates recognition of initiator codons by eukaryotic ribosomes. *Proc. Natl. Acad. Sci. USA* **1990**, *87*, 8301–8305. [CrossRef] [PubMed]

26. Kochetov, A.V.; Palyanov, A.; Titov, I.I.; Grigorovich, D.; Sarai, A.; Kolchanov, N.A. AUG_hairpin: Prediction of a downstream secondary structure influencing the recognition of a translation start site. *BMC Bioinform.* **2007**, *8*, 318. [CrossRef] [PubMed]

27. Kozak, M. Influences of mRNA secondary structure on initiation by eukaryotic ribosomes. *Proc. Natl. Acad. Sci. USA* **1986**, *83*, 2850–2854. [CrossRef] [PubMed]

28. Blaszczyk, L.; Ciesiolka, J. Secondary structure and the role in translation initiation of the 5'-terminal region of p53 mRNA. *Biochemistry* **2011**, *50*, 7080–7092. [CrossRef] [PubMed]

29. Gorska, A.; Blaszczyk, L.; Dutkiewicz, M.; Ciesiolka, J. Length variants of the 5' untranslated region of p53 mRNA and their impact on the efficiency of translation initiation of p53 and its n-truncated isoform deltanp53. *RNA Biol.* **2013**, *10*, 1726–1740. [CrossRef] [PubMed]

30. Purzycka, K.J.; Legiewicz, M.; Matsuda, E.; Eizentstat, L.D.; Lusvarghi, S.; Saha, A.; Le Grice, S.F.; Garfinkel, D.J. Exploring Ty1 retrotransposon RNA structure within virus-like particles. *Nucleic Acids Res.* **2013**, *41*, 463–473. [CrossRef] [PubMed]

31. Pachulska-Wieczorek, K.; Blaszczyk, L.; Biesiada, M.; Adamiak, R.W.; Purzycka, K.J. The matrix domain contributes to the nucleic acid chaperone activity of HIV-2 Gag. *Retrovirology* **2016**, *13*, 18. [CrossRef] [PubMed]

32. Purzycka, K.J.; Pachulska-Wieczorek, K.; Adamiak, R.W. The in vitro loose dimer structure and rearrangements of the HIV-2 leader RNA. *Nucleic Acids Res.* **2011**, *39*, 7234–7248. [CrossRef] [PubMed]

33. Curcio, M.J.; Garfinkel, D.J. Single-step selection for Ty1 element retrotransposition. *Proc. Natl. Acad. Sci. USA* **1991**, *88*, 936–940. [CrossRef] [PubMed]

34. Popenda, M.; Szachniuk, M.; Antczak, M.; Purzycka, K.J.; Lukasiak, P.; Bartol, N.; Blazewicz, J.; Adamiak, R.W. Automated 3D structure composition for large RNAs. *Nucleic Acids Res.* **2012**, *40*, e112. [CrossRef] [PubMed]

35. RNAComposer. Automated RNA Structure 3D Modeling Server. Available online: http://rnacomposer.ibch.poznan.pl/ (accessed on 27 October 2016).

36. Biesiada, M.; Purzycka, K.J.; Szachniuk, M.; Blazewicz, J.; Adamiak, R.W. Automated RNA 3D structure prediction with RNAcomposer. *Methods Mol. Biol.* **2016**, *1490*, 199–215. [PubMed]

37. Wilkinson, K.A.; Merino, E.J.; Weeks, K.M. Selective 2′-hydroxyl acylation analyzed by primer extension (shape): Quantitative RNA structure analysis at single nucleotide resolution. *Nat. Protoc.* **2006**, *1*, 1610–1616. [CrossRef] [PubMed]

38. Deigan, K.E.; Li, T.W.; Mathews, D.H.; Weeks, K.M. Accurate shape-directed RNA structure determination. *Proc. Natl. Acad. Sci. USA* **2009**, *106*, 97–102. [CrossRef] [PubMed]

39. Reuter, J.S.; Mathews, D.H. RNAstructure: Software for RNA secondary structure prediction and analysis. *BMC Bioinform.* **2010**, *11*, 129. [CrossRef] [PubMed]

40. Tijerina, P.; Mohr, S.; Russell, R. DMS footprinting of structured RNAs and RNA-protein complexes. *Nat. Protoc.* **2007**, *2*, 2608–2623. [CrossRef] [PubMed]

41. Pachulska-Wieczorek, K.; Purzycka, K.J.; Adamiak, R.W. New, extended hairpin form of the TAR-2 RNA domain points to the structural polymorphism at the 5′ end of the HIV-2 leader RNA. *Nucleic Acids Res.* **2006**, *34*, 2984–2997. [CrossRef] [PubMed]

42. Woodson, S.A.; Koculi, E. Analysis of RNA folding by native polyacrylamide gel electrophoresis. *Methods Enzymol.* **2009**, *469*, 189–208. [PubMed]

43. Araujo, P.R.; Yoon, K.; Ko, D.; Smith, A.D.; Qiao, M.; Suresh, U.; Burns, S.C.; Penalva, L.O. Before it gets started: Regulating translation at the 5′ UTR. *Comp. Funct. Genom.* **2012**, *2012*, 475731. [CrossRef] [PubMed]

44. Mathews, D.H. RNA secondary structure analysis using RNAstructure. *Curr. Protoc. Bioinform.* **2014**, *46*. [CrossRef]

45. Huang, Q.; Purzycka, K.J.; Lusvarghi, S.; Li, D.; Legrice, S.F.; Boeke, J.D. Retrotransposon Ty1 RNA contains a 5′-terminal long-range pseudoknot required for efficient reverse transcription. *RNA* **2013**, *19*, 320–332. [CrossRef] [PubMed]

46. Dikstein, R. Transcription and translation in a package deal: The tisu paradigm. *Gene* **2012**, *491*, 1–4. [CrossRef] [PubMed]

47. Kozak, M. A short leader sequence impairs the fidelity of initiation by eukaryotic ribosomes. *Gene Expr.* **1991**, *1*, 111–115. [PubMed]

48. Kozak, M. Pushing the limits of the scanning mechanism for initiation of translation. *Gene* **2002**, *299*, 1–34. [CrossRef]

49. Garfinkel, D.J.; Tucker, J.M.; Saha, A.; Nishida, Y.; Pachulska-Wieczorek, K.; Błaszczyk, L.; Purzycka, K.J. A self-encoded capsid derivative restricts Ty1 retrotransposition in Saccharomyces. *Curr. Genet.* **2015**, 1–9. [CrossRef] [PubMed]

50. Sonenberg, N.; Hinnebusch, A.G. Regulation of translation initiation in eukaryotes: Mechanisms and biological targets. *Cell* **2009**, *136*, 731–745. [CrossRef] [PubMed]

10

Envelope Protein Mutations L107F and E138K Are Important for Neurovirulence Attenuation for Japanese Encephalitis Virus SA14-14-2 Strain

Jian Yang [1,2,†], Huiqiang Yang [1,†], Zhushi Li [1], Wei Wang [1], Hua Lin [1], Lina Liu [1], Qianzhi Ni [1], Xinyu Liu [3], Xianwu Zeng [1], Yonglin Wu [4] and Yuhua Li [1,3,5,*]

[1] Department of Viral Vaccine, Chengdu Institute of Biological Products Co., Ltd., China National Biotech Group, Chengdu 610023, China; jiany74@163.com (J.Y.); yang-anan@163.com (H.Y.); changdc123@sina.com (Z.L.); suntina926@163.com (W.W.); scciqlh@126.com (H.L.); linaliu@163.com (L.L.); nqz1986@126.com (Q.N.); zengxw64@163.com (X.Z.)

[2] Department of Microbiology and Immunology, North Sichuan Medical College, Nanchong 637007, China

[3] Department of Arbovirus Vaccine, National Institutes for Food and Drug Control, Beijing 100050, China; xinyuliu@hotmail.com

[4] China National Biotech Group, Beijing 100029, China; wuyonglin@sinopharm.com

[5] State Key Laboratory of Biotherapy and Cancer Center, West China Hospital, Sichuan University and Collaborative Innovation Center for Biotherapy, Chengdu 610000, China

* Correspondence: liyuhua@nifdc.org.cn

† These authors contributed equally to this work.

Academic Editor: Michael Holbrook

Abstract: The attenuated Japanese encephalitis virus (JEV) strain SA14-14-2 has been successfully utilized to prevent JEV infection; however, the attenuation determinants have not been fully elucidated. The envelope (E) protein of the attenuated JEV SA14-14-2 strain differs from that of the virulent parental SA14 strain at eight amino acid positions (E107, E138, E176, E177, E264, E279, E315, and E439). Here, we investigated the SA14-14-2-attenuation determinants by mutating E107, E138, E176, E177, and E279 in SA14-14-2 to their status in the parental virulent strain and tested the replication capacity, neurovirulence, neuroinvasiveness, and mortality associated with the mutated viruses in mice, as compared with those of JEV SA14-14-2 and SA14. Our findings indicated that revertant mutations at the E138 or E107 position significantly increased SA14-14-2 virulence, whereas other revertant mutations exhibited significant increases in neurovirulence only when combined with E138, E107, and other mutations. Revertant mutations at all eight positions in the E protein resulted in the highest degree of SA14-14-2 virulence, although this was still lower than that observed in SA14. These results demonstrated the critical role of the viral E protein in controlling JEV virulence and identified the amino acids at the E107 and E138 positions as the key determinants of SA14-14-2 neurovirulence.

Keywords: attenuation mechanism; Japanese encephalitis virus; SA14-14-2; neuroinvasiveness; neurovirulence

1. Introduction

The Japanese encephalitis virus (JEV) belongs to the *Flavivirus* genus and causes frequent endemic and epidemic infections in Asia, with JEV infection leading to acute encephalitis in humans and resulting in high mortality rates. The wild-type JEV SA14 strain was isolated from mosquitoes in Xi'An, China in 1954, and the attenuated JEV SA14-14-2 strain was obtained by serial passages of the JEV SA14 strain in mouse brain and primary hamster kidney (PHK) cells, followed by purification by plaque screening [1]. The purified SA14-14-2 strain was used to produce the attenuated live Japanese encephalitis (JE) vaccine for humans, with >600 million doses of this vaccine being administered in China and other countries in Southeast Asia, including Korea, Nepal, India, and Thailand, since 1989. The safety and efficacy of this vaccine have been well demonstrated by clinical data [2], and on 10 September 2013, it passed World Health Organization prequalification and was entered into the list of vaccines available for international purchase. As with all attenuated live viral vaccines, its reversion to virulent status remains a concern. This study explored the molecular mechanisms underpinning the attenuated neurovirulence of the live JE vaccine (SA14-14-2) by reverting specific amino acids in the SA14-14-2 envelope (E) protein to their counterparts in the parental virulent strain (SA14) and testing the virulence of the revertant viruses.

Our findings indicated no neurovirulence observed in adult mice inoculated intracerebrally (i.c.) with attenuated JEV SA14-14-2 at 10^6 plaque-forming units (PFUs), as compared with mice inoculated with the parental strain, which caused 100% mortality in mice within 1 week. The marked virulence attenuation of JEV SA14-14-2 is believed to result from specific substitutions at 24 amino acid positions, including eight amino acid mutations in the E protein, throughout the viral genome [3–5], as well as mutations in nonstructural proteins [6]. However, the specific mutations that determine the attenuated SA14-14-2 phenotype remain unknown.

The attenuated yellow fever virus (YFV) 17D strain differs from its parental Asibi strain by 32 amino acid substitutions. Among these, 12 mutations are located in the E protein. Remarkably, as few as one mutation (E303) in the E protein can change the attenuated phenotype of the Asibi strain [7]. The crucial role of amino acid mutations in the E protein, associated with attenuation, was reported in other attenuated viral vaccines, including the chimeric yellow fever-dengue 1 vaccine virus [8]. That study hypothesized that the attenuated phenotype of the JEV SA14-14-2 strain might also be attributed to specific mutations in the E protein. Here, we investigated the roles of five amino acid residues (E107, E138, E176/177, and E279) in the E protein in the attenuated strains, as compared with the virulent parental strain (Table 1), followed by an assessment of the neurovirulence and neuroinvasiveness of these revertants in mice. Our results demonstrated that amino acids at the E138 and E107 positions played key roles in neurovirulence attenuation in the JEV SA14-14-2 strain.

Table 1. Amino acid differences in the viral E protein between Japanese encephalitis virus (JEV) strains SA14, SA14-14-2, and SA14-5-3 [3,4].

Positions in E Protein [a]	Virulent Strain			Attenuated Strain		
	SA14/USA	SA14/CDC	SA14/JAP	SA14-14-2/PHK	SA14-14-2/PDK	SA14-5-3
E107	L	L	L	F	F	F
E138	E	E	E	K	K	K
E176	I	I	I	V	V	V
E177	T	T	T	A	T	T
E264	Q	Q	Q	H	Q	Q
E279	K	K	K	M	M	M
E315	A	V	A	V	V	V
E439	K	R	K	R	R	R

[a] Amino acids that differ between the virulent JEV strain SA14 and the attenuated SA14-14-2 and SA14-5-3 strains are highlighted in bold letters. E177 was studied along with E176 due to their close proximity.

2. Materials and Methods

2.1. Cells, Plasmids, and Viruses

BHK-21 cells (CCL-10; American Type Culture Collection, Manassas, VA, USA) were cultured in an Eagle minimum essential medium (MEM; Gibco; Thermo Fisher Scientific, Waltham, MA, USA) supplemented with 10% heat-inactivated fetal bovine serum. The multiple-cloning site of the low-copy plasmid pACNR was modified to contain the restriction sites *Asc*I, *Kas*I, *Bgl*II, *BspE*I,

*Bam*HI, *Bcl*I, *Xba*I, and *Xho*I. The JEV SA14-14-2 strain was generated in PHK cells isolated from 9- to 10-day-old specific pathogen-free (SPF) hamsters at the Chengdu Institute of Biological Products Co., Ltd. (Chengdu, China).

2.2. DNA Cloning

The RNA of the JEV SA14-14-2 strain was extracted using a High Pure viral RNA kit (Roche, Basel, Switzerland), and cDNA was synthesized by reverse transcription (RT) using SuperScript III reverse transcriptase (Invitrogen, Carlsbad, CA, USA). Briefly, 20 ng RNA was mixed with 10 pmol 3′-terminal primers, heated for 5 min at 65 °C, cooled on ice for 1 min, and then incubated with SuperScript III in the recommended buffer for 1 h at 55 °C, followed by heating to 70 °C for 15 min. cDNA amplification was performed with the phusion polymerase (New England Biolabs, Ipswich, MA, USA) using a touchdown polymerase chain reaction (PCR) program: one cycle at 98 °C for 1 min, 10 cycles at 98 °C for 15 s, 58.5 °C to 53.5 °C for 15 s, and 72 °C for 3 min, followed by 20 cycles at an annealing temperature of 53.5 °C and elongation for 10 min at 72 °C. PCR products were purified using a DNA purification kit (Qiagen, Hilden, Germany) and cloned into the pGEM-T easy vector (Promega, Durham, NC, USA). The correct clones were identified by DNA sequencing.

All plasmids were constructed using two-plasmid systems as described previously [9,10]. One plasmid contained the 5′ terminal 3.4-kb cDNA and the other contained the 3′ terminal 7.6-kb fragment of the SA14-14-2 strain. The first fragment (1–476 nt) contained *Asc*I and *Kas*I restriction sites [11] and was cloned into the low-copy plasmid pACNR. The second fragment, from position 476 to 2654, and the third fragment, from position 2640 to 3446, were inserted into the *Kas*I/*Bgl*II and *Bgl*II/*Bsp*EI sites, respectively, to generate the plasmid pACNR-5′JEV (harboring the 5′ terminal 3.4-kb fragment). The fourth fragment, from position 3444 to 5581; the fifth fragment, from position 5575 to 7092; the sixth fragment, from position 7086 to 9136; and the seventh fragment, from position 9130 to 10977, were cloned into the pACNR to create the plasmid pACNR-3′JEV (containing the 3′ terminal 7.6-kb fragment). This 7.6-kb fragment of JEV was then inserted into the plasmid pACNR-5′JEV to create the plasmid pACNR-JEV containing the full-length cDNA of JEV SA14-14-2. Mutations in the E protein gene were generated by PCR-based site-directed mutagenesis, and all plasmids were sequenced to verify the engineered mutations.

2.3. In Vitro RNA Transcription, Transfection, and Viral Recovery

The pACNR-JEV plasmid was linearized by restriction digest using *Xho*I and used as a template for in vitro transcription. The RNA used for transfection was synthesized using the RiboMAX large-scale RNA production system Sp6 kit (Promega) in the presence of Ribo m7G cap analog (Promega). Reaction products were treated with DNase I (RQ1 RNase-free DNase; Promega), followed by purification with the RNeasy mini kit (Qiagen). BHK-21 cells were washed twice with cold phosphate-buffered saline, then 4×10^6 cells in 200 μL were mixed with the synthesized RNA in vitro (1 μg) and pulsed at 140 V for 25 ms using a Gene Pulser II apparatus (Bio-Rad, Hercules, CA, USA). Transfected BHK-21 cells were cultivated at 37 °C in a 5% CO_2 incubator, and the viruses were harvested at day 5 post-transfection upon observation of the cytopathic effect. The harvested viruses were passaged two additional times in BHK-21 cells, titrated for the plaque assay, and stored at −80 °C until further use.

2.4. Nucleotide Sequencing of the Revertant Viruses

Briefly, viral RNA was extracted from the recovered viruses using the High Pure viral RNA kit (Roche). cDNA from position 468 to 2667 containing the prM/E protein gene was synthesized by RT, followed by the amplification of the prM/E fragment using the phusion polymerase (New England Biolabs). The PCR products were purified using the QIAquick gel extraction kit (Qiagen) and sequenced to determine the consensus sequence (Invitrogen).

2.5. Growth Analysis of Revertants and Control Viruses

BHK-21 cells were infected with the revertants or control viruses at a multiplicity of infection of 0.5. After 1 h of absorption at 37 °C, viral inocula were removed, and 20 mL MEM containing 2% inactivated newborn calf serum was added. Culture supernatant (1 mL) was collected every 24-h post-infection for 96 h. Titers of the collected viruses were determined as described for the plaque assay.

2.6. Mouse Experiments

To assess and compare neurovirulence, groups ($n = 4$) of 3-week-old SPF Kunming mice were inoculated with 0.03 mL of 10-fold dilutions of the revertants or the control viruses by the i.c. route, and inoculated mice were monitored for 14 days. All moribund mice were euthanized, and the median lethal dose (LD_{50}) was determined by the Reed and Muench calculation. Neurovirulence results for each virus were recorded as LD_{50} ($log_{10}PFU$; the viral dose capable of inducing 50% mortality). Neuroinvasiveness was measured by inoculating 3-week-old SPF Kunming mice with 0.1 mL of 10-fold dilutions of the revertants or the control viruses by the subcutaneous (s.c.) route, and the neuroinvasive results were also recorded as LD_{50} ($log_{10}PFU$). The average survival time (AST) was determined by inoculating 0.03 mL of viruses containing equal plaque titers (5.18 $log_{10}PFU$) in another group of mice ($n = 6$) by the i.c. route. Mice in a moribund condition were euthanized and scored as deaths.

2.7. Statistical Analysis

Statistical analysis of the AST was performed using analysis of variance, and a $p < 0.05$ was considered statistically significant. All analyses were performed using SPSS software version 17.0 (SPSS, Inc., Chicago, IL, USA).

2.8. Ethical Approval

The experimental protocols involving mice were approved by the Experimental Animal Welfare and Ethical Committee of the National Institutes for Food and Drug Control, China.

3. Results

3.1. Construction of Infectious JEV Full-Length cDNA Clones Containing Specific Reverse Mutations in the E Protein

All pACNR-JEV plasmids containing specific mutations were verified by sequencing, and the viruses used for testing were amplified by three passages in the BHK-21 cells. The E protein-coding region of each virus was sequenced an additional time, with the results confirming that the sequences of all engineered plasmids and revertant viruses were correct and that no new mutations had been introduced.

3.2. Growth Analysis of Revertants and Control Viruses

One mechanism of viral attenuation involves crippled viral replication [12]; therefore, the effects of reverse mutations on JEV replication were measured by infecting BHK-21 cells, followed by a determination of the production of revertants and control viruses at different time intervals following infection. Growth-curve results showed that all viruses exhibited similar replication capacities, although the SA14 virus replicated at a modestly faster rate, with 5.7 $log_{10}PFU/mL$ at 24-h post-infection, which was higher than the other viruses tested. However, the peak SA14 titer was not the highest among all viruses, which was likely due to the highest SA14 titer not being collected at the denoted time points (Figure 1).

Additionally, analysis of the plaque sizes of all viruses revealed that those of SA14 (2–3 mm) were larger than those of SA14-14-2 (1–2 mm) and the other viruses (1–2 mm). We observed no significant difference in plaque size between SA14-14-2 and all the revertant viruses, except for that of rJEV4 (E279) (0.5–1 mm), which was significantly smaller than those of the other viruses (Figure 2).

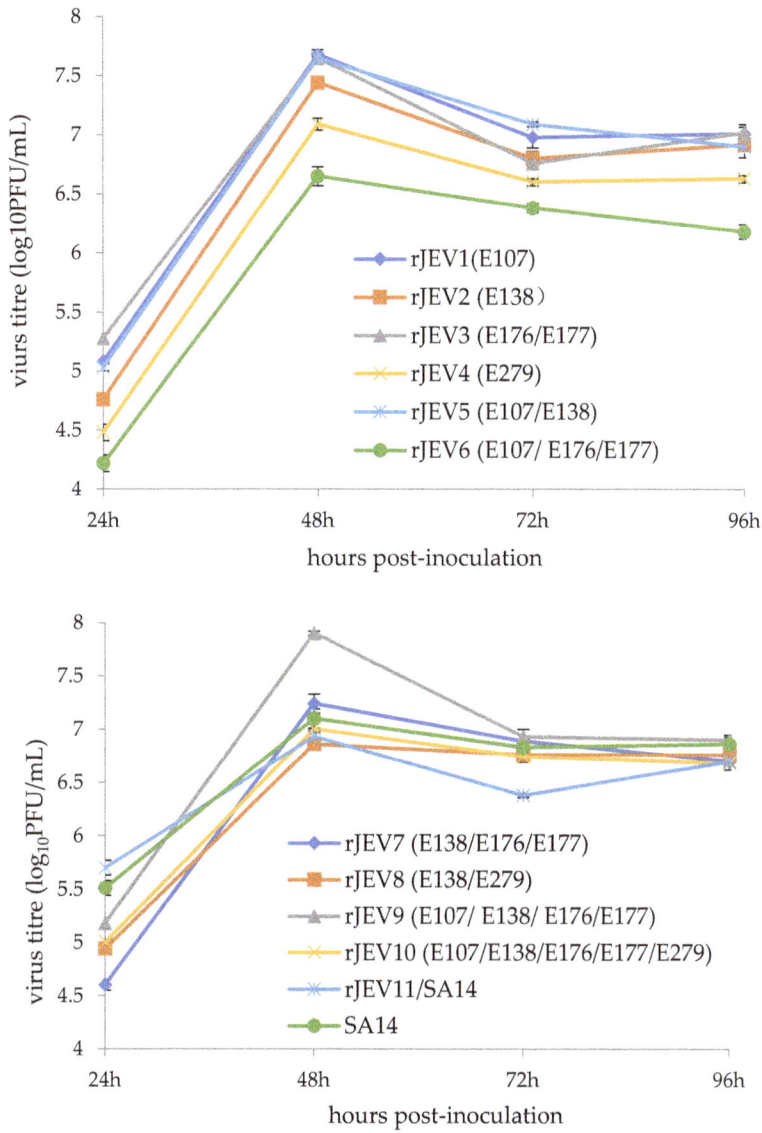

Figure 1. Growth curves of the revertants and the control viruses.

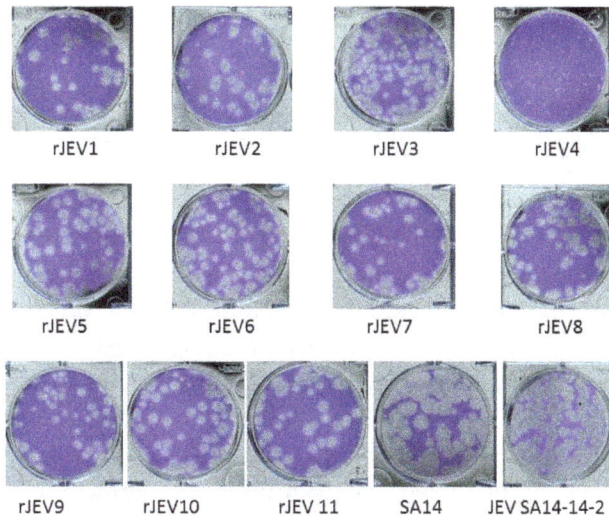

Figure 2. Plaque sizes of the revertants and the control viruses.

3.3. Mutation at Residue E138 in Combination with E107 Is Critical to the Attenuated Neurovirulence of JEV SA14-14-2

To determine the amino acids in the E protein that attenuate JEV SA14-14-2 neurovirulence, we measured the LD_{50} ($log_{10}PFU$) values of all the revertant viruses (Table 2), with low LD_{50} ($log_{10}PFU$) values indicating high degrees of neurovirulence. Among revertants containing a single amino acid substitution, rJEV1 (E107) and rJEV2 (E138) exhibited lower LD_{50} ($log_{10}PFU$) values as compared with that of the SA14-14-2 virus, whereas the revertant viruses rJEV3 (E176/E177) and rJEV4 (E279) exhibited similar LD_{50} ($log_{10}PFU$) values to that of SA14-14-2. Among these four revertants, rJEV2 (E138) exhibited the lowest LD_{50} ($log_{10}PFU$) value, indicating the highest degree of neurovirulence, followed by rJEV1 (E107). The reverse mutation of E138 in combination with E107 significantly decreased the LD_{50} ($log_{10}PFU$) value as compared with those of rJEV1 (E107) and rJEV2 (E138), and the LD_{50} ($log_{10}PFU$) value of rJEV10 (E107, E138, E176/177, and E279; 1.43) was slightly lower than that of rJEV9 (E107, E138, and E176/177; 1.99). The rJEV11/SA14 virus, wherein the E protein of SA14-14-2 was replaced with the E protein of wild-type SA14, exhibited the lowest LD_{50} ($log_{10}PFU$) value (0.66) among all the revertants, although it was still higher than that of the virulent wild-type SA14 virus (-0.92). These findings suggested, that among the five tested amino acid residues, E138 and E107 played the most important roles in the attenuation of SA14-14-2 neurovirulence.

Table 2. Neurovirulence of the mutated viruses in 3-week-old mice inoculated by the i.c. route.

Viruses	LD_{50} ($log_{10}PFU$) *
rJEV (SA14-14-2)	$\geq$6.48
rJEV1 (E107)	3.97
rJEV2 (E138)	2.89
rJEV3 (E176/E177)	$\geq$6.43
rJEV4 (E279) [†]	$\geq$6.24
rJEV5 (E107/E138)	1.70
rJEV6 (E107/E176/E177)	5.69
rJEV7 (E138/E176/E177)	3.64
rJEV8 (E138/E279)	2.82
rJEV9 (E107/E138/E176/E177)	1.99
rJEV10 (E107/E138/E176/E177/E279)	1.43
rJEV11/SA14	0.66
SA14	-0.92

* LD_{50} ($log_{10}PFU$) represents the plaque titers that cause death in 50% of tested mice; [†] Virus rJEV4 did not cause neurovirulence.

3.4. Reverse Mutations in the E Protein Increased the Mortality and Decreased the AST of I.C.-Inoculated Mice

The virulence phenotype of the revertant viruses was further evaluated by determining the mortality and AST of mice inoculated by the i.c. route with 5.18 $log_{10}PFU$ revertant virus (Table 3). Mice inoculated with rJEV2, rJEV5, rJEV7, rJEV8, rJEV9, rJEV10, rJEV11, or SA14 exhibited 100% mortality, whereas rJEV1 or rJEV3 inoculation resulted in 83.3% and 16.7% mortality, respectively, and SA14-14-2 or rJEV4 (E279) inoculation resulted in 0% mortality. These results showed that the E138 and E107 residues were more important than E279 and E176/177 at effecting SA14-14-2 virulence. The wild-type SA14 group exhibited the shortest AST (4 days), followed by the rJEV11 group (E107/E138/E176/177/E264/E279/E315/E439), with an AST of 5 days ($p \leq 0.05$, compared with SA14). The AST of mice inoculated with rJEV2 (E138) exhibited an AST of 6 days, and the ASTs of the rJEV1 (E107) and the rJEV6 (E107/E176/177) groups were 6.6 and 9 days, respectively ($p \leq 0.05$, comparing rJEV1 to rJEV6). The rJEV3 group (E176/177) exhibited the longest AST of 11 days. These results demonstrated that the E138 residue was a greater determinant of attenuated SA14-14-2 virulence, as compared with other residues in the E protein.

Table 3. AST and mortality of mice inoculated with the viruses by the i.c. route.

Viruses	No. of Dead Mice/ Total No. of Mice (%)	AST (day) Mean ± SD
rJEV (SA14-14-2)	0/6 (0)	-
rJEV1 (E107)	5/6 (83.3%)	6.6 ± 0.9 [$]
rJEV2 (E138)	6/6 (100%)	6 ± 0 [#,$]
rJEV3 (E176/E177)	1/6 (16.7%)	11 ± 0
rJEV4 (E279)	0/6 (0)	-
rJEV5 (E107/E138)	6/6 (100%)	6 ± 0 [#]
rJEV6 (E107/E176/E177)	3/6 (50%)	9 ± 0
rJEV7 (E138/E176/E177)	6/6 (100%)	6 ± 0 [#]
rJEV8 (E138/E279)	6/6 (100%)	6 ± 0 [#]
rJEV9 (E107/E138/E176/E177)	6/6 (100%)	6 ± 0 [#]
rJEV10 (E107/E138/E176/E177/E279)	6/6 (100%)	6 ± 0 [#]
rJEV11/SA14	6/6 (100%)	5 ± 0 [*]
SA14	6/6 (100%)	4 ± 0 [*]

[#] $p = 1$, compared with each other; [*] $p \leq 0.05$, compared with each other; [$] $p \leq 0.05$, compared with each other.

3.5. Effects of Specific Reverse Mutations on JEV SA14-14-2 Neuroinvasiveness

The neuroinvasiveness of all the revertants was tested using the same protocol as that used to test neurovirulence, except that the mice were inoculated by the s.c. route (Table 4). The LD_{50} ($\log_{10}PFU$) values of all the revertants containing single amino acid substitutions were similar to that of the attenuated SA14-14-2 strain, whereas the other revertants showed lower LD_{50} ($\log_{10}PFU$) values than that of SA14-14-2. The LD_{50} ($\log_{10}PFU$) values of the mice inoculated with rJEV5 (E107 and E138), rJEV7 (E138 and E176/177), or rJEV8 (E138 and E279) were ≥ 6.54, 5.76, and 6.01, respectively, suggesting that the E107 revertant mutation combined with E138 did not show the same synergistic effect as observed in the neurovirulence test. The LD_{50} ($\log_{10}PFU$) value of mice inoculated with rJEV10 (E107/E138/E176/E177/E279) was slightly higher than that of rJEV9 (E107/E138/E176/E177)-inoculated mice, and rJEV9 (E107/E138/E176/E177)-, rJEV10 (E107/E138/E176/E177/E279)-, and rJEV11/SA14-inoculated mice exhibited low LD_{50} ($\log_{10}PFU$) values of 5.40, 5.53, and 3.17, respectively. Furthermore, the LD_{50} ($\log_{10}PFU$) value of mice inoculated with rJEV11 (rJEV11/SA14) was higher than that of mice infected with virulent SA14, indicating that other regions in the JEV genome also contributed to the neuroinvasive phenotype.

Table 4. Neuroinvasiveness of the revertant viruses in 3-week-old mice inoculated by the s.c. route.

Inocula	LD_{50} ($\log_{10}PFU$)
rJEV (SA14-14-2)	≥ 6.14
rJEV1 (E107)	≥ 7.32
rJEV2 (E138)	≥ 6.20
rJEV3 (E176/E177)	≥ 6.93
rJEV4 (E279)	≥ 6.74
rJEV5 (E107/E138)	≥ 6.54
rJEV6 (E107/E176/E177)	≥ 6.71
rJEV7 (E138/E176/E177)	5.76
rJEV8 (E138/E279)	6.01
rJEV9 (E107/E138/E176/E177)	5.40
rJEV10 (E107/E138/E176/E177/E279)	5.53
rJEV11/SA14	3.17
SA14	1.86

4. Discussion

Reverse genetics is a powerful tool for studying the replication, virulence attenuation, and gene functions of positive-strand RNA viruses. The key step in this strategy involves constructing stable cDNA clones containing the full-length viral sequence. However, constructing the full-length cDNA clone of the JEV SA14-14-2 strain was hindered by the instability and toxicity of some gene products in *Escherichia coli* [13–15]. Previous studies utilized two strategies to overcome these hurdles. One was an in vitro ligation approach [16] and the second involved using low-copy plasmids, such as those containing artificial bacterial chromosomes, to stabilize the full-length cDNA of JEV [17]. Here, a different low-copy plasmid (pACNR) was employed to stably maintain the full-length cDNA of the infectious JEV. To generate a marker in the recombinant viruses, a silent mutation was inserted at nucleotide 473 (from A to C) that also created a new restriction site (*Kas*I) for DNA cloning. This genetic marker allowed confirmation that the recovered viruses were derived from the infectious cDNA. Furthermore, this cDNA cloning system previously enabled the mechanistic study of the virulence attenuation of *Flaviviruses* and the development of other *Flavivirus* vaccines [18].

Several studies reported nucleotide changes potentially underlying the attenuated phenotype of the JEV SA14-14-2 strain through comparisons with its parental strain SA14 [3,4,19]. Major nucleotide changes in *Flavivirus* E proteins responsible for viral neurovirulence were also revealed by comparing the JEV AT31 strain with its attenuated derivative [20] and between YFV (the Asibi train) and its attenuated 17D strain [21]. The results of mouse-infection studies showed that single substitutions at amino acid positions E107, E138, E176/177, or E279 differentially modulated viral virulence [22]. Our results showed that inoculation with the rJEV2 (E138) mutation increased SA14-14-2 neurovirulence to the highest level, followed by that of the rJEV1 (E107) mutation, whereas the single revertant mutation of E279 had no effect on neurovirulence. A synergistic virulence effect was observed when the E138 revertant mutation was combined with E107, but not E176/177 or E279, whereas infection with rJEV5 (E138/E107) exhibited the lowest LD_{50} ($log_{10}PFU$) value (1.70). These results were consistent with observations that after five passages in the suckling mouse brain, the revertant mutations at E138 and E107, from JEV SA14-14-2 to those of the parental SA14 strain, increased the neurovirulence of the resulting virus [22].

Residue E107 is located within a highly-conserved hairpin motif spanning amino acids 98 to 111 in domain II [23]. This region contains a fusogenic peptide according to studies of the tick-borne encephalitis virus, the Murray Valley encephalitis virus, and the dengue type 2 virus [24,25]. Mutations in close proximity to this region alter the fusion properties of the E protein in cell culture and are associated with changes in the neurovirulence of the dengue virus and the tick-borne encephalitis virus [26,27]. Residue E138 is located in the 'hinge' region at the interface of domains I and II of the E protein, and mutation at this position alters E protein conformation and function. Previous studies of *Flaviviruses* indicated that mutations within this region modulate viral virulence in mice [28–32], thereby supporting the results of this study.

The effect of the E176/177 cluster on viral neurovirulence is interesting. In contrast to other reverse mutations that increased SA14-14-2 virulence, single substitutions in the E176/177 cluster elevated viral virulence to a lesser extent than other substitutions; however, when combined with mutations at E107 or E138, E176/177 mutations significantly decreased viral virulence. Additionally, the virulence of rJEV6 (E107/E176/E177) was lower than that observed for rJEV1 (E107) (5.69 vs. 3.97), the virulence of rJEV7 (E138/E176/E177) was lower than that observed for rJEV2 (E138) (3.64 vs. 2.89), and the virulence of rJEV9 (E107/E138/E176/E177) was lower than that of rJEV5 (E107/E138) (1.99 vs. 1.77). Therefore, we concluded that the E176/177 mutations significantly neutralized the function of E107 and E138.

Residue E279 is located in the hinge region of the E protein, suggesting a possible regulatory role in E protein function, similar to that of E138. Previous studies showed that reverse mutation of residue E279 from methionine to lysine significantly increased neurovirulence [33]. Additionally, mutations in close proximity to E279 in the Murray Valley encephalitis virus impair hemagglutination and fusion properties of the E protein and reduce neuroinvasiveness in mice [28,34]. By contrast, reverse mutation of E279 in this study did not affect SA14-14-2 neurovirulence. The neutral effect of the E279 mutation might be explained by the decreased ability of the virus harboring the mutation to infect host cells, given that inoculation with rJEV4 (E279) resulted in that formation of the smallest plaques among all tested viruses, including SA14-14-2.

A previous study reported that the molecular determinants associated with the prM-E region of the attenuated JE SA14-14-2 virus are insufficient to confer an attenuated phenotype upon the JE Nakayama virus [6]. This suggested a role for determinants located in the 5′ untranslated region and/or the capsid protein of the JE SA 14-14-2 viral genome in influencing the virulent properties of the JE Nakayama virus in mouse models. Here, we observed that the revertant rJEV11 virus, having the same eight amino acids in the E protein as the parental virulent SA14 strain, exhibited significantly lower neurovirulence and neuroinvasiveness in mice as compared with JEV SA14 (Tables 2 and 4). These data demonstrated that mutations in viral proteins (including nonstructural protein) other than the E protein in JEV SA14-14-2 may also contribute to attenuated neurovirulence and neuroinvasiveness.

A previous report utilized a chimeric YFV/JEV SA14-14-2 virus to characterize the attenuation mechanism [35]. This chimeric virus contained the backbone of YFV and the prM and E protein sequences from the attenuated JEV SA14-14-2 strain. Consistent with our findings, Arroyo et al. [35] reported the importance of the E138 amino acid together with amino acid residues at other positions in neurovirulence attenuation; however, other results from that study differed from our findings. Arroyo et al. [35] reported that inoculation with variants harboring single reverse mutations of E107, E138, and E176/177 did not cause sickness or death in mice and that the single reverse mutation of E279 caused death in only 13% of mice. By contrast, we observed that inoculation with each of the three single reverse mutations (E107, E138, and E176/177) resulted in sickness or death in some of the mice and isolation of the revertant viruses in the brain. Additionally, the single reverse mutation of E279 did not cause sickness or death in mice. These discrepancies might be explained by the different inoculation doses used between the two studies, given that our inoculated mice became sick or died only when inoculated with >4.0 $\log_{10}$PFU of the revertants rJEV1 (E107), rJEV3 (E176/E177), and rJEV6 (E107/E176/E177), whereas the previous study used inoculation doses via i.c. of <4.0 $\log_{10}$PFU (10,000 PFU) [35]. Additionally, the results of that study did not suggest an important role for the single mutation of E107 in JEV attenuation, whereas our results provided a more detailed account of the roles of E176/177 and E279 in JEV attenuation, in combination with E138. Furthermore, Arroyo et al. [35] reported that the single substitution at E176/177 greatly enhanced virulence in combination with other mutations, whereas we observed decreased virulence associated with this mutation in combination with others. One explanation for these discrepancies might be use of a chimeric virus with YFV as the backbone in the previous study [35], whereas the JEV SA14 virus was used in this study to investigate the attenuation mechanism.

Yun et al. [36] showed that the passage of the JEV SA14-14-2 strain in the mouse brain was selected for mutations at the E244 position, which drastically altered the viral phenotype [36]. In this study, this amino acid position was not tested because the E244 position in the SA14 viral population harbors two different amino acids in the mouse brain (Figure 3), with glutamic acid at this position in wild-type SA14/USA and glycine at this position in the SA14/CDC and SA14/JAP strains (Table 1 and Figure 3).

The mechanisms associated with viral attenuation are complicated and involve membrane fusion [24,25], replication capacity [12], and heparin-binding activity [37]. In addition to the SA-14 E protein, our findings suggested that other regions of the viral genome likely also contribute to the attenuated phenotype. This was supported by our results showing that the virulence of the revertant rJEV11/SA14 strain remained lower than that of the parental SA14 strain, despite substitution with the intact E protein from wild-type SA14.

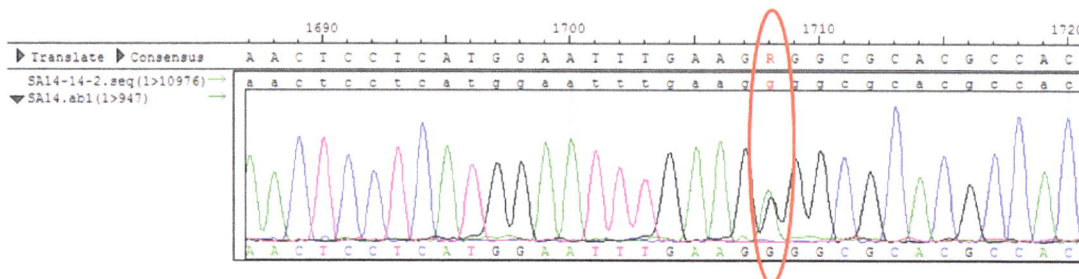

Figure 3. An equal proportion of two different amino acids exists at position E244 in the SA14 virus.

5. Conclusions

In summary, this study demonstrated that amino acids at positions E107, E138, E176/177, and E279 differentially contributed to virulence attenuation in the SA14-14-2 virus. Our findings indicated that the E138 position played the most important role in sustaining neurovirulence, but not the attenuated neuroinvasive phenotype associated with JEV SA14-14-2. Additionally, the role of the E107 position in attenuating virulence was revealed by its synergistic effect with the E138 position, although E107 alone also contributed to virulence attenuation. Compared with the E107 and E138 positions, E176/177 and E279 exhibited relatively minor roles in virulence attenuation. These results identified the key residues in the E protein involved in regulating attenuated JEV SA14-14-2 virulence, thereby elucidating the molecular mechanisms of JEV attenuation. The data presented in this study supported JEV vaccine guidelines stating that the stability of the E protein sequence should be used as the main safety indicator for the attenuated live JE vaccine (SA14-14-2 strain).

Acknowledgments: We would like to thank all members of the Animal Biotechnological Center for their contributions to this study. This study was supported by the national key projects of "863" High Technology, China (No. 2012AA02A401) and the Chinese Mega Project of Science Research for major new drug innovation and developmental research for quality control of the JE live attenuated vaccine and polio vaccine (Grant No. 2014ZX09304316-003). The funders had no role in study design, data collection and analysis, decision to publish, or manuscript preparation.

Author Contributions: Y.L. and H.Y. conceived and designed the experiments. J.Y., H.Y., Z.L., W.W., H.L., L.L., X.L., and Q.N. performed the experiments. X.Z. and Y.W. analyzed the data. J.Y. and H.Y. wrote the paper. J.Y. and H.Y. contributed equally to this work.

References

1. Wills, M.R.; Sil, B.K.; Cao, J.X.; Yu, Y.X.; Barrett, A.D. Antigenic characterization of the live attenuated Japanese encephalitis vaccine virus SA14-14-2: A comparison with isolates of the virus covering a wide geographic area. *Vaccine* **1992**, *10*, 861–872. [CrossRef]

2. Liu, Z.L.; Hennessy, S.; Strom, B.L.; Tsai, T.F.; Wan, C.M.; Tang, S.C.; Xiang, C.F.; Bilker, W.B.; Pan, X.P.; Yao, Y.J.; et al. Short-term safety of live attenuated Japanese encephalitis vaccine (SA14-14-2): Results of a randomized trial with 26,239 subjects. *J. Infect. Dis.* **1997**, *176*, 1366–1369. [CrossRef] [PubMed]

3. Ni, H.; Burns, N.J.; Chang, G.J.; Zhang, M.J.; Wills, M.R.; Trent, D.W.; Sanders, P.G.; Barrett, A.D. Comparison of nucleotide and deduced amino acid sequence of the 5′ non-coding region and structural protein genes of the wild-type Japanese encephalitis virus strain SA14 and its attenuated vaccine derivatives. *J. Gen. Virol.* **1994**, *75*, 1505–1510. [CrossRef] [PubMed]

4. Ni, H.; Chang, G.J.; Xie, H.; Trent, D.W.; Barrett, A.D. Molecular basis of attenuation of neurovirulence of wild-type Japanese encephalitis virus strain SA14. *J. Gen. Virol.* **1995**, *76*, 409–413. [CrossRef] [PubMed]

5. Yu, Y. Phenotypic and genotypic characteristics of Japanese encephalitis attenuated live vaccine virus SA14-14-2 and their stabilities. *Vaccine* **2010**, *28*, 3635–3641. [CrossRef] [PubMed]

6. Chambers, T.J.; Droll, D.A.; Jiang, X.; Wold, W.S.; Nickells, J.A. JE Nakayama/JE SA14-14-2 virus structural region intertypic viruses: Biological properties in the mouse model of neuroinvasive disease. *Virology* **2007**, *366*, 51–61. [CrossRef] [PubMed]

7. Jennings, A.D.; Gibson, C.A.; Miller, B.R.; Mathews, J.H.; Mitchell, C.J.; Roehrig, J.T.; Wood, D.J.; Taffs, F.; Sil, B.K.; Whitby, S.N.; et al. Analysis of a yellow fever virus isolated from a fatal case of vaccine-associated human encephalitis. *J. Infect. Dis.* **1994**, *169*, 512–518. [CrossRef] [PubMed]

8. Taffs, R.E.; Chumakov, K.M.; Rezapkin, G.V.; Lu, Z.; Douthitt, M.; Dragunsky, E.M.; Levenbook, I.S. Genetic stability and mutant selection in Sabin 2 strain of oral poliovirus vaccine grown under different cell culture conditions. *Virology* **1995**, *209*, 366–373. [CrossRef] [PubMed]

9. Chambers, T.J.; Liang, Y.; Droll, D.A.; Schlesinger, J.J.; Davidson, A.D.; Wright, P.J.; Jiang, X. Yellow fever virus/dengue-2 virus and yellow fever virus/dengue-4 virus chimeras: Biological characterization, immunogenicity, and protection against dengue encephalitis in the mouse model. *J. Virol.* **2003**, *77*, 3655–3668. [CrossRef] [PubMed]

10. Chambers, T.J.; Nestorowicz, A.; Mason, P.W.; Rice, C.M. Yellow fever/Japanese encephalitis chimeric viruses: Construction and biological properties. *J. Virol.* **1999**, *73*, 3095–3101. [PubMed]

11. Chambers, T.J.; Jiang, X.; Droll, D.A.; Liang, Y.; Wold, W.S.; Nickells, J. Chimeric Japanese encephalitis virus/dengue 2 virus infectious clone: Biological properties, immunogenicity, and protection against dengue encephalitis in mice. *J. Gen. Virol.* **2006**, *87*, 3131–3140. [CrossRef] [PubMed]

12. Muylaert, I.R.; Chambers, T.J.; Galler, R.; Rice, C.M. Mutagenesis of the N-linked glycosylation sites of the yellow fever virus NS1 protein: Effects on virus replication and mouse neurovirulence. *Virology* **1996**, *222*, 159–168. [CrossRef] [PubMed]

13. Mishin, V.P.; Cominelli, F.; Yamshchikov, V.F. A "minimal" approach in design of flavivirus infectious DNA. *Virus Res.* **2001**, *81*, 113–123. [CrossRef]

14. Ruggli, N.; Rice, C.M. Functional cDNA clones of the Flaviviridae: Strategies and applications. *Adv. Virus Res.* **1999**, *53*, 183–207. [PubMed]

15. Zhang, F.; Huang, Q.; Ma, W.; Jiang, S.; Fan, Y.; Zhang, H. Amplification and cloning of the full-length genome of Japanese encephalitis virus by a novel long RT-PCR protocol in a cosmid vector. *J. Virol. Methods* **2001**, *96*, 171–182. [CrossRef]

16. Sumiyoshi, H.; Hoke, C.H.; Trent, D.W. Infectious Japanese encephalitis virus RNA can be synthesized from in vitro-ligated cDNA templates. *J. Virol.* **1992**, *66*, 5425–5431. [PubMed]

17. Yun, S.I.; Kim, S.Y.; Rice, C.M.; Lee, Y.M. Development and application of a reverse genetics system for Japanese encephalitis virus. *J. Virol.* **2003**, *77*, 6450–6465. [CrossRef] [PubMed]

18. Lai, C.J.; Monath, T.P. Chimeric flaviviruses: Novel vaccines against dengue fever, tick-borne encephalitis, and Japanese encephalitis. *Adv. Virus Res.* **2003**, *61*, 469–509. [PubMed]

19. Nitayaphan, S.; Grant, J.A.; Chang, G.J.; Trent, D.W. Nucleotide sequence of the virulent SA-14 strain of Japanese encephalitis virus and its attenuated vaccine derivative, SA-14-14-2. *Virology* **1990**, *177*, 541–552. [CrossRef]

20. Zhao, Z.; Date, T.; Li, Y.; Kato, T.; Miyamoto, M.; Yasui, K.; Wakita, T. Characterization of the E-138 (Glu/Lys) mutation in Japanese encephalitis virus by using a stable, full-length, infectious cDNA clone. *J. Gen. Virol.* **2005**, *86*, 2209–2220. [CrossRef] [PubMed]

21. Galler, R.; Freire, M.S.; Jabor, A.V.; Mann, G.F. The yellow fever 17D vaccine virus: Molecular basis of viral attenuation and its use as an expression vector. *Braz. J. Med. Biol. Res.* **1997**, *30*, 157–168. [CrossRef] [PubMed]

22. Wu, Y.L.; Liu, J.; Yang, H.Q.; Zhao, Y.; Wang, W.; Mu, J.C.; Huang, Y.X.; Liu, R.; Sun, Y.; Yu, Y.X.; et al. Genetic property of attenuated Japanese encephalitis virus strain SA14-14-2 after subculture in suckling mouse brain. *Chin. J. Biol.* **2007**, *20*, 19–21.

23. Kolaskar, A.S.; Kulkarni-Kale, U. Prediction of three-dimensional structure and mapping of conformational epitopes of envelope glycoprotein of Japanese encephalitis virus. *Virology* **1999**, *261*, 31–42. [CrossRef] [PubMed]

24. Roehrig, J.T.; Hunt, A.R.; Johnson, A.J.; Hawkes, R.A. Synthetic peptides derived from the deduced amino acid sequence of the E-glycoprotein of Murray Valley encephalitis virus elicit antiviral antibody. *Virology* **1989**, *171*, 49–60. [CrossRef]

25. Roehrig, J.T.; Johnson, A.J.; Hunt, A.R.; Bolin, R.A.; Chu, M.C. Antibodies to dengue 2 virus E-glycoprotein synthetic peptides identify antigenic conformation. *Virology* **1990**, *177*, 668–675. [CrossRef]

26. Despres, P.; Frenkiel, M.P.; Deubel, V. Differences between cell membrane fusion activities of two dengue type-1 isolates reflect modifications of viral structure. *Virology* **1993**, *196*, 209–219. [CrossRef] [PubMed]

27. Rey, F.A.; Heinz, F.X.; Mandl, C.; Kunz, C.; Harrison, S.C. The envelope glycoprotein from tick-borne encephalitis virus at 2 A resolution. *Nature* **1995**, *375*, 291–298. [CrossRef] [PubMed]

28. Chen, L.K.; Lin, Y.L.; Liao, C.L.; Lin, C.G.; Huang, Y.L.; Yeh, C.T.; Lai, S.C.; Jan, J.T.; Chin, C. Generation and characterization of organ-tropism mutants of Japanese encephalitis virus in vivo and in vitro. *Virology* **1996**, *223*, 79–88. [CrossRef] [PubMed]

29. Gualano, R.C.; Pryor, M.J.; Cauchi, M.R.; Wright, P.J.; Davidson, A.D. Identification of a major determinant of mouse neurovirulence of dengue virus type 2 using stably cloned genomic-length cDNA. *J. Gen. Virol.* **1998**, *79*, 437–446. [CrossRef] [PubMed]

30. Hasegawa, H.; Yoshida, M.; Shiosaka, T.; Fujita, S.; Kobayashi, Y. Mutations in the envelope protein of Japanese encephalitis virus affect entry into cultured cells and virulence in mice. *Virology* **1992**, *191*, 158–165. [CrossRef]

31. McMinn, P.C.; Marshall, I.D.; Dalgarno, L. Neurovirulence and neuroinvasiveness of Murray Valley encephalitis virus mutants selected by passage in a monkey kidney cell line. *J. Gen. Virol.* **1995**, *76*, 865–872. [CrossRef] [PubMed]

32. Sumiyoshi, H.; Tignor, G.H.; Shope, R.E. Characterization of a highly attenuated Japanese encephalitis virus generated from molecularly cloned cDNA. *J. Infect. Dis.* **1995**, *171*, 1144–1151. [CrossRef] [PubMed]

33. Monath, T.P.; Arroyo, J.; Levenbook, I.; Zhang, Z.X.; Catalan, J.; Draper, K.; Guirakhoo, F. Single mutation in the flavivirus envelope protein hinge region increases neurovirulence for mice and monkeys but decreases viscerotropism for monkeys: Relevance to development and safety testing of live, attenuated vaccines. *J. Virol.* **2002**, *76*, 1932–1943. [CrossRef] [PubMed]

34. McMinn, P.C.; Weir, R.C.; Dalgarno, L. A mouse-attenuated envelope protein variant of Murray Valley encephalitis virus with altered fusion activity. *J. Gen. Virol.* **1996**, *77*, 2085–2088. [CrossRef] [PubMed]

35. Arroyo, J.; Guirakhoo, F.; Fenner, S.; Zhang, Z.X.; Monath, T.P.; Chambers, T.J. Molecular basis for attenuation of neurovirulence of a yellow fever virus/Japanese encephalitis virus chimera vaccine (ChimeriVax-JE). *J. Virol.* **2001**, *75*, 934–942. [CrossRef] [PubMed]

36. Yun, S.I.; Song, B.H.; Kim, J.K.; Yun, G.N.; Lee, E.Y.; Li, L.; Kuhn, R.J.; Rossmann, M.G.; Morrey, J.D.; Lee, Y.M. A molecularly cloned, live-attenuated Japanese encephalitis vaccine SA14–14–2 virus: A conserved single amino acid in the ij hairpin of the viral E glycoprotein determines neurovirulence in mice. *PLoS Pathog.* **2014**, *10*, e1004290. [CrossRef] [PubMed]

37. Silva, L.A.; Khomandiak, S.; Ashbrook, A.W.; Weller, R.; Heise, M.T.; Morrison, T.E.; Dermody, T.S. A single-amino-acid polymorphism in Chikungunya virus E2 glycoprotein influences glycosaminoglycan utilization. *J. Virol.* **2014**, *88*, 2385–2397. [CrossRef] [PubMed]

Novel Approach for Isolation and Identification of Porcine Epidemic Diarrhea Virus (PEDV) Strain NJ Using Porcine Intestinal Epithelial Cells

Wen Shi, Shuo Jia, Haiyuan Zhao, Jiyuan Yin, Xiaona Wang, Meiling Yu, Sunting Ma, Yang Wu, Ying Chen, Wenlu Fan, Yigang Xu * and Yijing Li *

College of Veterinary Medicine, Northeast Agricultural University, Harbin 150030, China;
wenshi_china@163.com (W.S.); jiashuo0508@163.com (S.J.); zhywxn1925@163.com (H.Z.); neauyjy@126.com (J.Y.);
xiaonawang0319@163.com (X.W.); yu19890130@126.com (M.Y.); masunting@163.com (S.M.);
wuyang_neau@126.com (Y.W.); chenying_neau@163.com (Y.C.); fanwenlu1230@163.com (W.F.)
* Correspondence: yigangxu_china@sohu.com (Y.X.); yijingli@163.com (Y.L.)

Academic Editors: Linda Dixon and Simon Graham

Abstract: Porcine epidemic diarrhea virus (PEDV), which is the causative agent of porcine epidemic diarrhea in China and other countries, is responsible for serious economic losses in the pork industry. Inactivated PEDV vaccine plays a key role in controlling the prevalence of PEDV. However, consistently low viral titers are obtained during the propagation of PEDV in vitro; this represents a challenge to molecular analyses of the virus and vaccine development. In this study, we successfully isolated a PEDV isolate (strain NJ) from clinical samples collected during a recent outbreak of diarrhea in piglets in China, using porcine intestinal epithelial cells (IEC). We found that the isolate was better adapted to growth in IECs than in Vero cells, and the titer of the IEC cultures was $10^{4.5}$ TCID$_{50}$/0.1 mL at passage 45. Mutations in the S protein increased with the viral passage and the mutations tended towards attenuation. Viral challenge showed that the survival of IEC-adapted cultures was higher at the 45th passage than at the 5th passage. The use of IECs to isolate and propagate PEDV provides an effective approach for laboratory-based diagnosis of PEDV, as well as studies of the epidemiological characteristics and molecular biology of this virus.

Keywords: porcine epidemic diarrhea virus (PEDV) NJ strain; porcine intestinal epithelial cells; isolation and identification

1. Introduction

Porcine epidemic diarrhea (PED), which is caused by the porcine epidemic diarrhea virus (PEDV), is an acute and highly contagious enteric viral disease in nursing pigs. PED is characterized by vomiting and lethal watery diarrhea; and is a global problem, especially in many swine-producing countries [1–7]. PED was first reported in feeder pigs and fattening swine in the United Kingdom in 1971 [8]; since then, it has emerged in numerous European and Asian countries, resulting in tremendous economic losses to the pork industry worldwide. In 2013, the first PED outbreak was reported in the U.S.; subsequently, the outbreak spread rapidly across the country, and similar outbreaks were also reported in Canada and Mexico [4–7]. In China, PED outbreaks have occurred infrequently with only sporadic incidents [9,10]. However, in late 2010, a remarkable increase in PED outbreaks was reported in the pork-producing provinces [11,12]. In 2014, an outbreak of severe acute diarrhea, with high morbidity and mortality, occurred in sucking piglets in Nanjing, China. Herds vaccinated with the CV777-inactivated vaccine were also infected.

During this period, the effectiveness of the CV777-based vaccine was questioned as PED outbreaks also occurred in vaccinated herds [12]. PED has since become one of the most significant epidemics affecting pig farming in China [13]. PEDV is an enveloped, single-stranded, positive-sense RNA virus belonging to the genus *Alphacoronavirus*, family *Coronaviridae*, and order *Nidovirales* [14,15]. The size of its genome is approximately 28 kb, with $5'$- and $3'$- untranslated regions (UTRs) and seven open reading frames (ORFs) that encode four structural proteins, i.e., spike (S), envelope (E), membrane (M), and nucleocapsid (N), and three nonstructural proteins [10,16]. The S protein of PEDV is the major enveloped protein of the virion, associated with growth adaptation in vitro and attenuation in vivo [17]. In addition, the S glycoprotein is used to determine the genetic relatedness among PEDV isolates and for developing diagnostic assays and effective vaccines [18–20].

The ability to propagate the virus is critical for the diagnosis and molecular analysis of PEDV, particularly the development of inactivated or attenuated vaccine. However, propagation of PEDV in vitro is challenging. Even though PEDV may be isolated from clinical samples, it gradually loses its infectivity during further passages in cell culture [4]. Therefore, it is necessary to evaluate the disinfection efficiency in vitro viral isolates using a cell culture system that promotes growth of PEDV. Currently, several PEDV strains, such as CV777, KPEDV-9, and 83P-5, have been successfully propagated in Vero cells using media with added trypsin [21–23]. In recent years, new variants of PEDV have emerged that are difficult to isolate and propagate in Vero cells with trypsin. Researchers have attempted to use pig bladder and kidney cells to isolate PEDV, with the addition of trypsin to the medium; this is the first report of isolation of PED virus in porcine cell culture [24]. PEDV infects the epithelium of the small intestine, which is a protease-rich environment, and causes atrophy of the villi resulting in diarrhea and dehydration; this indicates that porcine intestinal epithelial cells are the target cells of this virus. In 2014, Wang et al. established a porcine intestinal epithelial cell line (ZYM-SIEC02) by introducing the human telomerase reverse transcriptase (hTERT) gene into small intestinal epithelial cells derived from a neonatal, unsuckled piglet [25]. Several studies have used this established porcine intestinal epithelial cell (IEC) line [26,27]; however, the characteristics of PEDV cultured in this cell line have not been reported.

The present study aimed to confirm and identify PEDV in samples collected from piglets with suggestive clinical signs, using the IEC line established by Wang et al. [25]. A PEDV isolate, named PEDV strain NJ, was successfully isolated. Our results show that the PEDV strain NJ is adapted to growth in IECs with media containing trypsin, suggesting a new approach for the propagation of PEDV. Furthermore, the phylogeny and mutations of the *S* gene during serial passages were analyzed to determine its genetic homology and molecular variability. A virulence experiment for IEC-adapted NJ also confirmed that the virus had a tendency towards attenuation at 45 passages.

2. Materials and Methods

All applicable international and national guidelines for the care and use of animals were followed. Approval (2016NEFU-315) was obtained from the Institutional Committee of Northeast Agricultural University for the animal experiments.

2.1. Cells and Clinical Samples

The swine intestinal epithelial cell (IEC) line established by Wang et al. was kindly provided by Prof. Yanming Zhang, College of Veterinary Medicine, Northwest A&F University, Yangling, Shaanxi, China. IEC and Vero cells (ATCC CCL-81) were cultured in Dulbecco's modified Eagle's medium (DMEM; Gibco, Grand Island, NY, USA), and supplemented with 10% fetal bovine serum (Gibco). The clinical samples (small intestine tissues) used in this study were collected from a pig farm in Nanjing, China, at which an outbreak of acute diarrhea among piglets had been reported. The virus isolated from samples was identified as PEDV by *M* gene-based reverse transcription PCR (RT-PCR). The small intestine tissue was homogenized with serum-free DMEM, and then centrifuged (Thermo Scientific Sorvall Legend Micro 17, Waltham, MA, USA) at $5000 \times g$ at $4\,^{\circ}C$ for 10 min. The supernatant was filtered using 0.22-μm pore-size cellulose acetate (Merck Millipore, Darmstad, Germany), and used for virus isolation.

2.2. RNA Extraction and RT-PCR Assay

Total RNA was extracted from the clinical samples and virus cultures using the TRIzol® Plus RNA Purification Kit (Invitrogen Corp., Carlsbad, CA, USA) according to the manufacturer's instructions. Complementary DNA (cDNA) was produced via reverse transcription using the Superscript Reverse Transcriptase Reagent Kit (Takara, Tokyo, Japan) according to the manufacturer's instructions. The primer pairs of the partial M gene (316 bp) for identification of PEDV, 5'-TATGGCTTGCATCACTCTTA-3' (forward) and 5'-TTGACTGAACGACCAACACG-3' (reverse), were designed based on PEDV strain CV777, using the cDNA as a template. The PCR reaction system in a total volume of 50 μL was as follows: 5 μL of $10\times$ buffer, 3 μL of cDNA, 1 μL of LA Taq polymerase (TaKaRa, Tokyo, Japan), 2 μL of forward primer (10 μM), 2 μL of reverse primer (10 μM), 4 μL of dNTPs mix-ture (2.5 μM), and sterile water added up to 50 μL. The cycling parameters for PCR included 95 °C for 5 min, followed by 30 cycles at 94 °C for 30 s, 55.5 °C for 30 s followed by 72 °C for 30 s, and a final extension at 72 °C for 10 min. The PCR purified products were cloned into pMD-19T vector and sequenced by Comate Bioscience Company Limited (Jilin, China).

2.3. Virus Isolation

In this study, Vero cells and IECs were used to propagate PEDV. For the propagation of PEDV using Vero cells, the confluent cell monolayer was washed once with sterile phosphate-buffered saline (PBS; pH 7.2), and incubated with 1 mL of inoculum for 1 h in a T25 flask supplemented with 21 μg/mL of trypsin (Gibco) at 37 °C under 5% CO_2. Then, the inoculum was removed and the cells were washed twice with PBS, and 4 mL of maintenance medium (DMEM, Gibco) without fetal bovine serum supplemented with 5 μg/mL trypsin was added to the flask. The propagation of PEDV using IECs was performed according to the method described above, but with the use of 10 μg/mL of trypsin during adsorption. In parallel, cells mock-inoculated with DMEM were used as control. The PEDV infected cells and viral control cells were cultured at 37 °C under 5% CO_2. The cytopathic effect (CPE) was monitored daily, and cells were harvested until the CPE exceeded 80%. After one freeze-thaw cycle, the supernatants were collected, packed separately, and stored at −80 °C until required. Virus titer was measured in 96-well plates by 10-fold serial dilution of samples at five-passage intervals. The 50% tissue culture infective dose ($TCID_{50}$) was expressed as the reciprocal of the highest dilution showing CPE by the Reed–Muench method.

2.4. Electron Microscopy Assay

The supernatants of PEDV-infected IEC cultures were centrifuged at $3000\times g$ for 45 min, followed by ultracentrifugation through a 25% sucrose cushion at $30,000\times g$ for 2 h at 4 °C. Virus particles were resuspended in 100 μL of DMEM and observed by transmission electron microscopy (H-7650, Hitachi, Tokyo, Japan). For imaging of virions in infected IECs, PEDV-infected cells were fixed using 2.5% glutaraldehyde at 4 °C for 8 h, washed twice with PBS, and post-fixed with 2% osmium tetroxide at room temperature (20 °C–25 °C) for 50 min. After three washes with PBS, cells were dehydrated through a graded ethanol propylene oxide series and embedded. Then, ultra-thin sections were prepared and imaged via transmission electron microscopy.

2.5. Immunofluorescence Assay

After inoculation for 24 h, mock-infected IECs and IECs infected with PEDV, at multiplicity of infection (MOI) of 0.1 were fixed with 4% paraformaldehyde at room temperature (20 °C–25 °C) for 15 min, permeabilized with 0.2% Triton X-100 in PBS at room temperature (RT) for 10 min, and blocked with 0.3% bovine serum albumin in PBS at 37 °C for 30 min. Next, mouse anti-PEDV S protein monoclonal antibody (developed in our laboratory) and fluorescein isothiocyanate (FITC)-conjugated goat anti-mouse immunoglobulin G (IgG) (ZSGB-BIO, Beijing, China) were incubated as first and second antibodies, respectively, followed by counterstaining with 4,6-diamidino-2-phenylindole (DAPI, Beyotime, Shanghai, China). The coverslips were mounted on microscope glass slides in mounting buffer and cell staining was examined using a fluorescence microscope (Leica, Wetzlar, Germany).

2.6. Sequence Alignment and Phylogenetic Analysis of the S Gene

To monitor the amino acid variation of the S protein during the serial passaging, the parental NJ strain and the strain at different passages, i.e., NJ (15th), NJ (30th) and NJ (45th), were evaluated by RT-PCR. The S gene was amplified in four fragments using KOD-Plus-Neo (Toyobo, Osaka, Japan). The primers used were previously described by Zhao et al. [28]. The four fragments were amplified under same program of 2 min at 94 °C, 30 cycles of 10 s at 98 °C, 30 s at 52 °C and 1.5 min at 68 °C, and a final extension at 68 °C for 7 min. The purified PCR products were cloned into pMD-19T vector and sequenced by Comate Bioscience Company Limited (Jilin, China). The sequence analysis was performed using MegAlign in DNAStar Lasergene V 7.10 (DNAstar, Madison, WI, USA). To determine the relationships among the S gene of the representative PEDV isolates, phylogenetic analysis of the parent NJ strain was performed by the neighbor-joining method using molecular evolutionary genetics analysis (MEGA) software (version 4.0). Bootstrap values were estimated for 1000 replicates. The S gene sequences of PEDV strain NJ and the sequences of 33 known PEDV strains (listed in Table 1) retrieved from GenBank were subjected to comparative analysis.

Table 1. Reference strains of porcine epidemic diarrhea virus (PEDV) used in this study.

Strain	ID	Country	Strain	ID	Country
83P-5 (parent)	AB548618	Japan	HLJ-2012	JX512907	China
83P-5 (34th)	AB548619	Japan	IA1	KF468753	USA
83P-5 (100th)	AB548621	Japan	IA2	KF468754	USA
AD02	KC879281	South Korea	ISU13-19338E-IN-homogenate	KF650370	USA
AH2012	KC210145	China	ISU13-19338E-IN-passage3	KF650371	USA
BJ-2012-1	JX435299	China	ISU13-19338E-IN-passage9	KF650372	USA
Br1/87	Z25483	France	JS2008	KC109141	China
CH9-FJ	JQ979287	China	JS-HZ2012	KC210147	China
CH22-JS	JQ979290	China	KH	AB548622	Japan
CHGD-01	JN980698	China	LJB/03	DQ985739	China
Chinju99	AY167585	Korea	MK	AB548624	Japan
CH/S	JN547228	China	MN	KF468752	USA
CV777 vaccine	JN599150	China	NK	AB548623	Japan
CV777	AF353511	Belgium	YN1	KT021227	China
DR13	JQ023161	Korea	YN15	KT021228	China
DR13 (Attenuated)	JQ023162	Korea	YN144	KT021232	China
DX	JN104080	China			

2.7. Virulence Experiment for IEC-Adapted NJ

The in vivo swine studies were performed in a biosafety level-2 (BSL-2) laboratory. For the identification of attenuation, the pathogenicity of the lower and higher generations of NJ should be compared. PEDV NJ cultures propagated in IECs at passages 5 and 45 were used in this study. The PEDV-negative piglets as confirmed by RT-PCR method were neonatal landraces obtained from a pig farm without a PED outbreak or vaccination with PEDV vaccine. We randomly selected 20 healthy piglets as experimental animals. The piglets were divided into three groups. The two infected groups (eight pigs for each group) received an oral dose of $10^{4.5}$ median tissue culture infective dose (TCID$_{50}$)/mL of IEC-adapted NJ (5 mL) at passages 5 and 45, and the control group ($n = 4$) was orally administered virus-free cell culture media. The clinical signs and survival percentage of the piglets were monitored daily over a 10-day observation period, and stool samples were collected daily. The small intestine tissue samples were collected and stored at -80 °C until required. RT-PCR and immunofluorescence assays were performed to detect PEDV in the stool samples.

Necropsy was performed when the challenged piglets died post inoculation. The piglets were handled and maintained under strict ethical conditions according to international recommendations for animal welfare.

3. Results

3.1. Virus Isolation

PEDV was isolated from PEDV-positive samples collected from pig farms in China, using Vero cells and IECs, in which a severe outbreak of acute diarrhea had been reported in sucking piglets. The genomic RNA of the serially propagated virus was extracted and identified by RT-PCR. Vero cell cultures were negative for the *M* gene after two passages (Figure 1A), and no CPEs were observed in Vero cells during serial passaging (Figure 1C,D). IEC-adapted PEDV was successfully propagated (Figure 1B), and visible CPEs were observed at each passage; compared with uninfected IECs, the PEDV-infected IECs were characterized by cell fusion, syncytial and vacuole formation in the initial stage, then shrinkage, detachment, and amotic at 72 h post-inoculation (Figure 1E,F). The virus cultured in IECs and designated NJ was biologically cloned by three rounds of plaque purification in IECs prior to further virus characterization. These results demonstrate that the adaption of the PEDV strain NJ to growth in IECs was better than that in Vero cells, indicating that IECs are suitable for the isolation of PEDV from clinical samples.

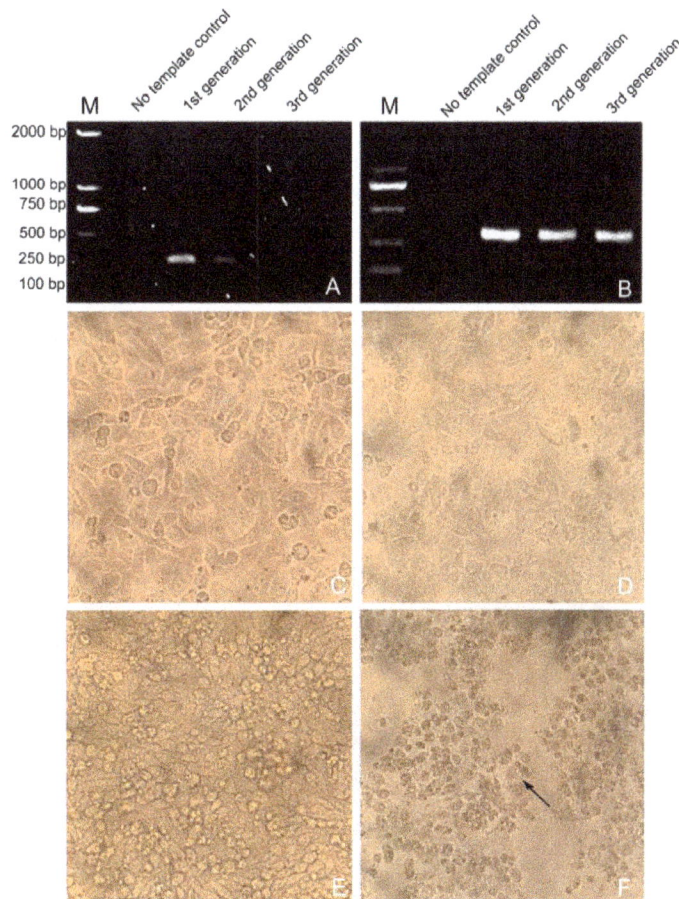

Figure 1. Isolation and identification of PEDV strain NJ in Vero cells and intestinal epithelial cell (IEC) cultures. (**A**) Identification of PEDV cultured in Vero cells at serial passages, by reverse transcription PCR (RT-PCR); (**B**) Identification of PEDV cultured in IEC at serial passages, by RT-PCR; (**C**) Control (uninfected) Vero cells; (**D**) PEDV-infected Vero cells; (**E**) Control (uninfected) IECs; (**F**) PEDV-infected IECs.

3.2. Determination of Viral Titer

The viral titer of the PEDV strain NJ propagated in IECs was determined at 5-passage intervals. The viral titer of the IEC-adapted PEDV strain NJ reached $10^{4.5}$ $TCID_{50}/0.1$ mL at passage 45 (Figure 2), suggesting that the use of an IEC culture is a promising approach for propagating PEDV.

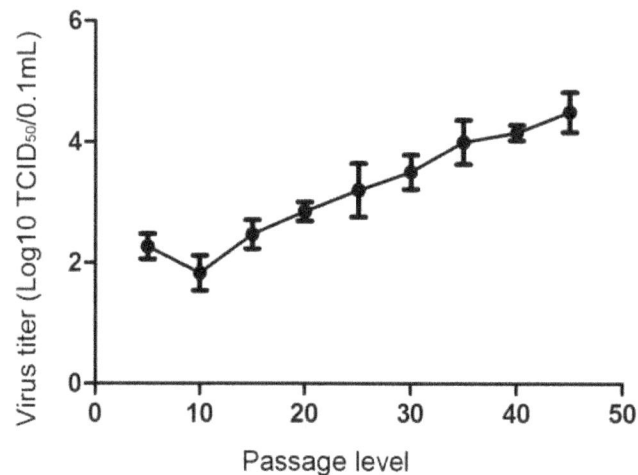

Figure 2. Viral titers of PEDV strain NJ propagated in IECs post-serial passages. All the results of a representative experiment performed with triplicate samples are shown.

3.3. Electron Microscopy

As shown in Figure 3A, the virion was circular in shape and 80–120 nm in diameter, with surface projections characteristic of coronaviruses. Thin sections of the PEDV strain NJ-infected IECs showed some of the virus particles appeared, many of the virus particles possessed a dense core, and masses of virus particles gathered in the cytoplasm at 24 h post-infection (Figure 3B).

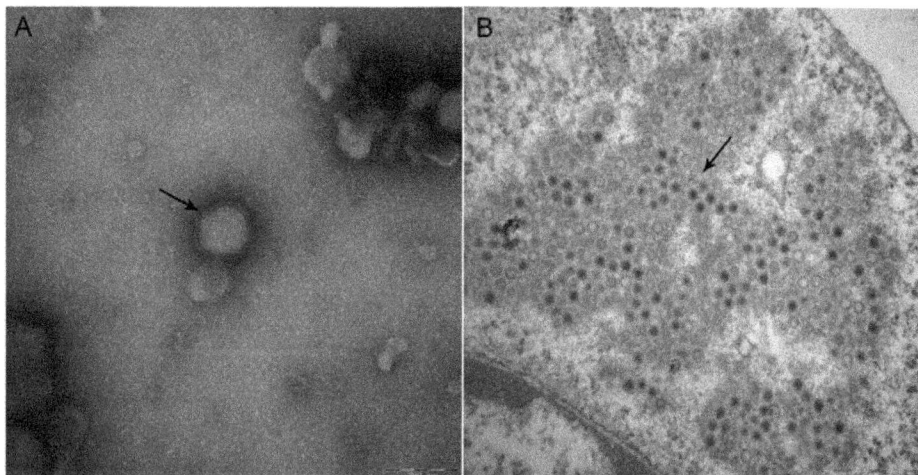

Figure 3. Images of PEDV strain NJ particles and PEDV strain NJ-infected IEC produced by electron microscopy. (**A**) Virions in culture media of IECs infected with PEDV strain NJ, as shown by the arrow; Bar = 200 nm. Magnification, ×100,000; (**B**) Thin section of IECs infected with PEDV strain NJ 24 h post-infection; many of the virus particles possessed a dense core and gathered in the cytoplasm as shown by the arrow; Bar = 500 nm. Magnification, ×50,000.

3.4. Immunofluorescence

Infection of IECs with the PEDV strain NJ was confirmed by immunofluorescence assay (Figure 4). PEDV strain NJ and cell nuclei were detected using mouse anti-PEDV S protein monoclonal antibody and 4,6-diamidino-2-phenylindole (DAPI), respectively. Specific green signals were observed in the PEDV strain NJ-infected IECs, but not in the mock-infected IECs. However, because the immunofluorescence assay was performed after inoculation for 24 h, the CPE was hard to observe during this period.

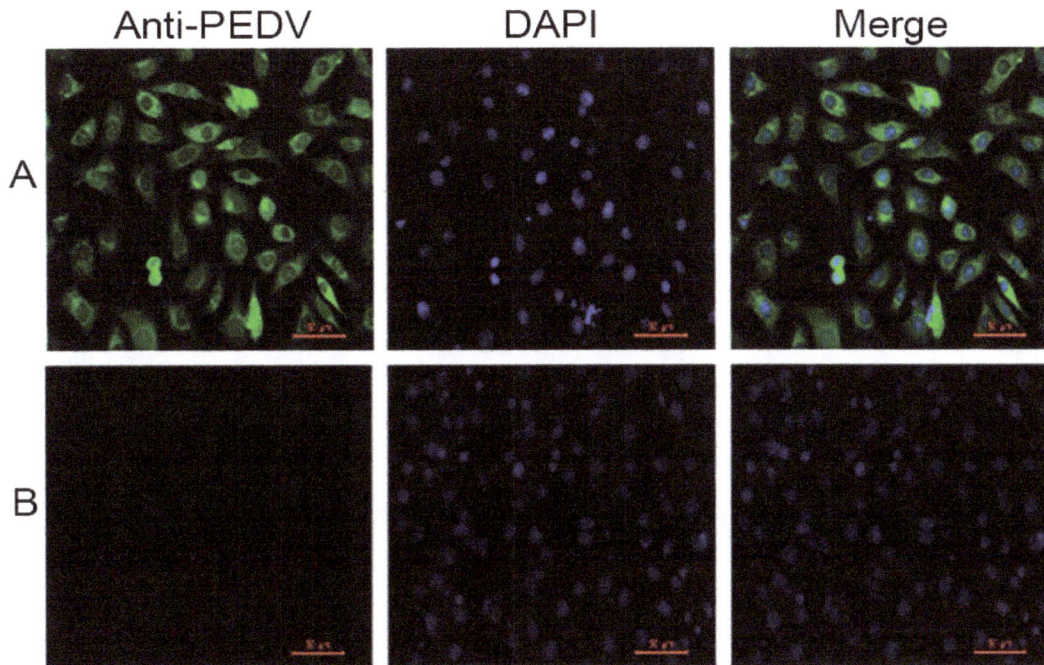

Figure 4. Detection of PEDV strain NJ in IECs by immunofluorescence assay at 24 h post-infection; mouse anti-PEDV S protein monoclonal antibody and fluorescein isothiocyanate (FITC)-conjugated goat anti-mouse immunoglobulin G (IgG) were respectively used as primary and secondary antibodies, followed by counterstaining with 4,6-diamidino-2-phenylindole (DAPI). (**A**) PEDV strain NJ-infected IECs; (**B**) Non-infected IECs.

3.5. Phylogenetic Analysis of the S Gene

The *S* gene of PEDV strain NJ was amplified by RT-PCR. Phylogenetic analysis based on the *S* gene was performed between parental PEDV strain NJ and other PEDV strains listed in Table 1. Phylogenetic analyses of *S* gene sequences revealed that all PEDV strains in this study could be separated into two groups: the NJ strain belonged to Group 1, which also contained classical PEDV CV777 and some strains isolated from China, Japan, and South Korea. As shown in Figure 5, the *S* gene of the parent NJ strain exhibited high sequence similarity with the epidemic strains CH9-FJ, CH22-JS, and DX isolated from southern China in recent years. These results suggest that the Chinese southern epidemic PEDV strains were likely derived from the same source. In addition, we did not find any insertions and deletions (INDEL) in the *S* gene compared with the Chinese PEDV *S* gene recombinant variants like IA1, IA2, and MN identified in U.S.

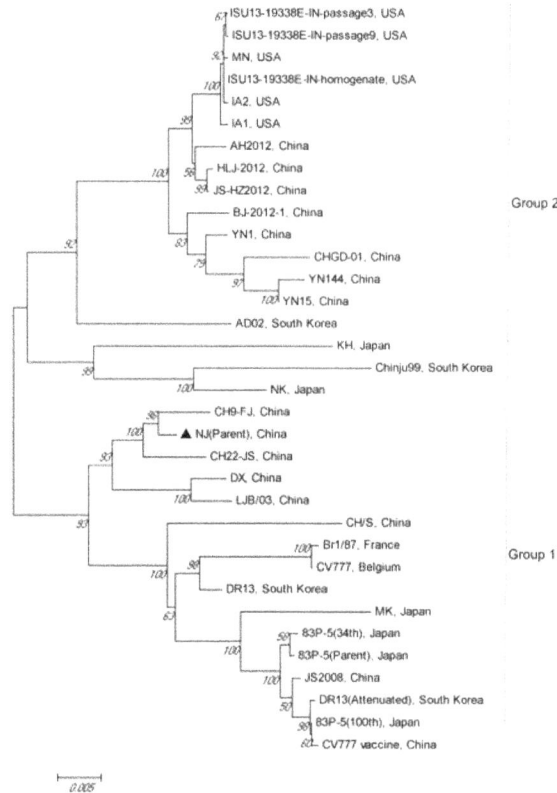

Figure 5. Phylogenetic analysis of PEDV strain NJ based on the *S* gene.

3.6. Amino Acid Variability of the S Protein of IEC-Adapted NJ after Serial Passaging

To evaluate the amino acid variability of the IEC-adapted PEDV strain NJ during serial passagings, the sequences of the *S* gene of the parent NJ and those of the virus at the 15th, 30th and 45th passage were amplified by RT-PCR. These amino acid sequences were then compared with the corresponding S protein sequences of the classical PEDV CV777 and its vaccine strain. The positions of amino acid changes were 3, 15, 70, 114, 282, 324, 378, 438, 973, 1023 and 1167 respectively (Table 2). In total, four mutations occurred at passage 15, three and four more mutations occurred at passages 30 and 45. The eight (8/11) mutations were the same as the ones occurring in the transition from classical PEDV CV777 to the CV777 vaccine strain during serial passaging, which suggested that the NJ might exhibit a tendency towards attenuation.

Table 2. The amino acid variation of S proteins during the serial passaging compared with CV777 and its vaccine strain.

Strain	Amino Acid Position										
	3	15	70	114	282	324	378	438	973	1023	1167
NJ(Parent)	S	S	A	N	L	S	N	I	Y	K	A
NJ(15th)	S	S	D	N	L	S	N	I	H	N	D
NJ(30th)	A	S	D	S	L	S	K	I	H	N	D
NJ(45th)	A	L	D	S	W	R	K	L	H	N	D
CV777	S	P	A	N	L	S	N	I	Y	K	A
CV777 vaccine	P	L	D	S	W	F	K	V	H	N	D

3.7. Pathogenicity Analysis of PEDV Strain NJ

To compare the pathogenicity of the lower-generation NJ with that of the higher-generation NJ, 5 mL of the IEC-adapted NJ culture at the 5th and 45th passage ($10^{4.5}$ TCID$_{50}$/mL) and volume-matched virus-free cell culture medium were administrated orally to the piglets. The clinical signs and survival percentage of the piglets was monitored daily over a 10-day observation period. All piglets infected with 5th passage cultures and three (3/8) piglets infected with 45th passage cultures showed diarrheic feces, significantly emaciated body condition, and experienced severe watery diarrhea with vomiting. Abundant yellow, watery, and foul-smelling stools were also observed around the perianal region of the piglets infected with 45th passage virus (Figure 6A,B). The remaining five (5/8) piglets infected with 45th passage showed signs of mild diarrhea after inoculation that seemed to be transient. However, in the control group, the piglets were healthy without watery diarrhea (data not shown). As shown in Figure 6E, one piglet death occurred within 3 days and in total seven piglets died after the 10-day challenge in the group that received the 5th passage. However, only three piglets died in the group that received the 45th passage; although the remaining piglets had the significant clinical signs of PED, they survived the 10-day challenge. The small intestines of one of the dead piglets infected with 45th passage IEC-adapted NJ developed severe clinical symptoms, the same as those in the piglets infected with the 5th passage virus. The small intestines were typically distended with accumulation of yellow fluid and mesenteric congestion, and the small intestinal wall was thin and transparent (Figure 6C,D). The diarrheal feces and intestinal tissues collected from these piglets were analyzed by RT-PCR targeting the *M* gene, and the PCR products were found to be consistent with the expected result. Moreover, immunofluorescence analysis of the small intestine tissue revealed that the viral antigen was predominantly present in the small intestines (data not shown).

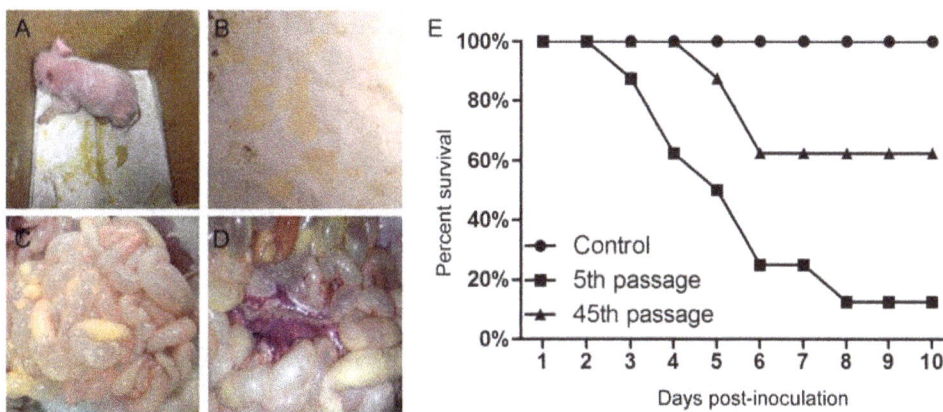

Figure 6. The clinical signs and necropsies of results of pigs infected with IEC-adapted NJ at 45th passage and the survival percentage of piglets after challenge. (**A**) Piglet with diarrhea and significantly dispirited status; (**B**) Watery diarrhea; (**C**) The intestinal tracts of infected piglets were thin and transparent; (**D**) Mesenteric congestion; (**E**) Survival percentage of piglets after challenge.

4. Discussion

The spread of PED to the U.S. and Canada has established that this viral infection represents a global epidemic [29] that has resulted in enormous economic losses to pig production. PEDV has caused similar economic losses to the pork industry in China, where frequent outbreaks have been recently reported [30]. Following the development of a CV777-based vaccine, based on the PEDV strain, and its wide application in the pig industry in China, only a limited number of incidents occurred before 2010; however, PED outbreaks have subsequently increased in frequency, particularly in pig-farming provinces. Notable, even pigs vaccinated with the CV777-based vaccine were found to be infected [31], indicating the need for the development of effective PEDV-based vaccines for the control of PED outbreaks.

To date, the propagation of PEDV remains challenging. Although Vero cells are commonly used to propagate PEDV, viral infectivity exhibits a gradual decline during serial passage in these cells. Many PEDV strains isolated from clinical samples were difficult to culture in Vero cells in recent years [4,32,33]. The adaptability of PEDV isolate infected with cell lines was the first step to successful isolation in vitro. Thus, it is urgent to develop new methods or cell lines to improve the tropism of PEDV cultured in cells. In this study, we successfully isolated and propagated the epidemic PEDV strain using porcine intestinal epithelial cells in vitro, and demonstrated that these cells are more suitable than Vero cells for the isolation of PEDV. This is the first report on the characteristics of PEDV cultured in this cell line which was established by Wang et al. [25].

The porcine intestinal epithelial cell is recognized as the target cell of PEDV [27]. Porcine aminopeptidase N (pAPN), a functional receptor of PEDV that is highly expressed in the small intestinal mucosa, plays a critical role in PEDV infection [31]. In addition, endogenous protease in the small intestine of porcine can cleave the S protein of PEDV in vivo and facilitate entry of PEDV virions into intestinal epithelial cells, resulting in massive propagation [34]. PEDV strain NJ isolated from clinical samples was as difficult to culture in Vero cells as other isolated prevalent PEDV strains; this is mainly attributed to mutations that arise in genes encoding spike proteins during serial passage. Moreover, the low levels of pAPN in Vero cells are thought to limit the attachment and entry of variant viruses. Before this study, we compared the isolation rate of five different positive PEDV strains isolated from clinical samples cultured in IECs and Vero cells. We found all the CPEs in Vero cells were invisible in three generations, and four PEDV isolates could not be detected by RT-PCR after two generations. In contrast, porcine IECs are more suitable than Vero cells for propagating PEDV isolated from clinical samples in vitro. The CPEs in IECs could be observed at the first or the third passage, and at each passage, cultures could be detected. Due to the characteristics of coronavirus, the degradation of the virus impacts the efficiency of the infection. The S glycoprotein forms peplomers on the virion envelope and contains receptor-binding regions and four major antigenic sites [35,36]; therefore, to minimize damage to the surface projections of the virus, the infected IECs were freeze-thawed only once before inoculation. The PEDV strain NJ was detected in cell culture during serial passages, by RT-PCR as well as immunofluorescence assay, indicating that this strain adapted to infection of IECs. Owing to the sensitivity of IECs, the trypsin concentration used in IEC culture was lower than that used in Vero cells during PEDV absorption. The viral titer of the IEC-adapted PEDV strain NJ propagated in IECs increased gradually and reached $10^{4.5}$ TCID$_{50}$/0.1 mL at 45 passages.

The S protein of PEDV is known to play pivotal roles in viral entry and in inducing the neutralizing antibodies in natural hosts, which makes it a primary target for development of effective vaccines against PEDV [23,37–39]. In this study, the S gene of the PEDV strain NJ was amplified by RT-PCR, and its genetic diversity and phylogenetic relationships were analyzed. S gene-based phylogenetic analysis showed that the PEDV strain NJ is closely related to the CH9-FJ, CH22-JS, and DX strains isolated from southern China in recent years. In addition, we did not find any insertions or deletions (INDEL) in the S gene compared to the Chinese PEDV S gene recombinant variants like IA1, IA2, and MN identified in the U.S.

The differences in virulence and genome sequences between the parental strain and derived attenuated strains have been extensively studied [40,41]. To investigate the genetic variability of the S protein of the PEDV strain NJ, the S gene was amplified during various serial passages and analyzed. Compared with the parent NJ strain, four mutations in the S protein occurred after the 15th passage, and then another seven mutations occurred after the 30th passage. In total, 11 amino acids changed during 45 passages. The eight (8/11) mutations were the same as those occurring in the transition from the classical PEDV CV777 strain to CV777 vaccine strain during serial passaging, which suggests that the high-passage IEC-adapted NJ might show attenuation. However, the molecular basis of viral adaptation to IECs remains a subject for future investigation.

Furthermore, to identify attenuation, the pathogenicity was compared for the lower- and higher-generation NJ. The piglets were inoculated with 5 mL of the IEC-adapted PEDV strain NJ ($10^{4.5}$ TCID$_{50}$/mL) at the 5th and 45th passages. We found that all of the piglets infected with 5th passage IEC-adapted cultures showed severe watery diarrhea with vomiting, and seven (7/8) piglets died by day 10 post inoculation. However, only three (3/8) piglets infected with viral cultures at the

45th passage showed significant PED signs, such as diarrheic feces and emaciated body condition; the remaining five (5/8) piglets showed mild diarrhea after inoculation that seemed to be transient, and only three piglets died by day 10 post inoculation. The in vivo challenge experiment implied that IEC-adapted NJ at a high passage number had a tendency towards attenuation. The mutations in the S protein during serial passaging might play a major role in the attenuation of virulence, and might suggest potential genetic changes for a candidate attenuated vaccine. However, the genetic mutations in the whole genome of the NJ strain during passages needs to be investigated further.

In conclusion, we successfully isolated and identified a novel PEDV, strain NJ, from clinical samples using IECs. The adaption of the PEDV strain NJ to growth in IECs was better than that in Vero cells. To our knowledge, this is the first report to describe the isolation and characterization of the IEC-adapted PEDV strain NJ. Furthermore, the present work reveals a novel approach for the propagation of PEDV in vitro. These findings are expected to be of importance for the development and evaluation of the efficacy of vaccines against PED.

Acknowledgments: We thank Yan-Ming Zhang, College of Veterinary Medicine, Northwest A&F University, for providing the IEC line. This work was supported by the General Program of National Natural Science Foundation of China (31472226) and the National Key R & D Program (2016YFD0500100).

Author Contributions: Yijing Li and Yigang Xu conceived and designed the study. Wen Shi, Shuo Jia, Haiyuan Zhao, Jiyuan Yin, Xiaona Wang, Meiling Yu, and Sunting Ma performed the experiments. Yang Wu, Ying Chen and Wenlu Fan analyzed and interpreted the data. Wen Shi wrote the paper. All authors read and approved the manuscript.

References

1.　Debouck, P.; Pensaert, M. Experimental infection of pigs with a new porcine enteric coronavirus, CV 777. *Am. J. Vet. Res.* **1980**, *41*, 219–223. [PubMed]

2.　Pensaert, M.B.; de Bouck, P. A new coronavirus-like particle associated with diarrhea in swine. *Arch. Virol.* **1978**, *58*, 243–247. [CrossRef] [PubMed]

3.　Wood, E.N. An apparently new syndrome of porcine epidemic diarrhoea. *Vet. Rec.* **1977**, *100*, 243–244. [CrossRef] [PubMed]

4.　Chen, Q.; Li, G.; Stasko, J.; Thomas, J.T.; Stensland, W.R.; Pillatzki, A.E.; Gauger, P.C.; Schwartz, K.J.; Madson, D.; Yoon, K.-J. Isolation and characterization of porcine epidemic diarrhea viruses associated with the 2013 disease outbreak among swine in the United States. *J. Clin. Microbiol.* **2014**, *52*, 234–243. [CrossRef] [PubMed]

5.　Kim, S.H.; Kim, I.J.; Pyo, H.M.; Tark, D.S.; Song, J.Y.; Hyun, B.H. Multiplex real-time RT-PCR for the simultaneous detection and quantification of transmissible gastroenteritis virus and porcine epidemic diarrhea virus. *J. Virol. Methods* **2007**, *146*, 172–177. [CrossRef] [PubMed]

6.　Ojkic, D.; Hazlett, M.; Fairles, J.; Marom, A.; Slavic, D.; Maxie, G.; Alexandersen, S.; Pasick, J.; Alsop, J.; Burlatschenko, S. The first case of porcine epidemic diarrhea in Canada. *Can. Vet. J.* **2015**, *56*, 149–152. [PubMed]

7.　Trujillo-Ortega, M.E.; Beltran-Figueroa, R.; Garcia-Hernandez, M.E.; Juarez-Ramirez, M.; Sotomayor-Gonzalez, A.; Hernandez-Villegas, E.N.; Becerra-Hernandez, J.F.; Sarmiento-Silva, R.E. Isolation and characterization of porcine epidemic diarrhea virus associated with the 2014 disease outbreak in Mexico: Case report. *BMC Vet. Res.* **2016**, *12*, 132. [CrossRef] [PubMed]

8.　Oldham, J. Letter to the editor. *Pig Farming* **1972**, *10*, 72–73. [CrossRef]

9.　Bi, J.; Zeng, S.L.; Xiao, S.B.; Chen, H.C.; Fang, L.R. Complete Genome Sequence of Porcine Epidemic Diarrhea Virus Strain AJ1102 Isolated from a Suckling Piglet with Acute Diarrhea in China. *J. Virol.* **2012**, *86*, 10910–10911. [CrossRef] [PubMed]

10.　Chen, J.; Wang, C.; Shi, H.; Qiu, H.; Liu, S.; Chen, X.; Zhang, Z.; Feng, L. Molecular epidemiology of porcine epidemic diarrhea virus in China. *Arch. Virol.* **2010**, *155*, 1471–1476. [CrossRef] [PubMed]

11.　Li, W.; Li, H.; Liu, Y.; Pan, Y.; Deng, F.; Song, Y.; Tang, X.; He, Q. New variants of porcine epidemic diarrhea virus, China, 2011. *Emerg. Infect. Dis.* **2012**, *18*, 1350–1353. [CrossRef] [PubMed]

12. Luo, Y.; Zhang, J.; Deng, X.; Ye, Y.; Liao, M.; Fan, H. Complete genome sequence of a highly prevalent isolate of porcine epidemic diarrhea virus in South China. *J. Virol.* **2012**, *86*, 9551. [CrossRef] [PubMed]

13. Sun, R.Q.; Cai, R.J.; Chen, Y.Q.; Liang, P.S.; Chen, D.K.; Song, C.X. Outbreak of porcine epidemic diarrhea in suckling piglets, China. *Emerg. Infect. Dis.* **2012**, *18*, 161–163. [CrossRef] [PubMed]

14. Duarte, M.; Gelfi, J.; Lambert, P.; Rasschaert, D.; Laude, H. Genome organization of porcine epidemic diarrhoea virus. *Adv. Exp. Med. Biol.* **1993**, *342*, 55–60. [PubMed]

15. Spaan, W.; Cavanagh, D.; Horzinek, M.C. Coronaviruses: Structure and genome expression. *J. Gen. Virol.* **1988**, *69*, 2939–2952. [CrossRef] [PubMed]

16. Song, D.; Park, B. Porcine epidemic diarrhoea virus: A comprehensive review of molecular epidemiology, diagnosis, and vaccines. *Virus Genes* **2012**, *44*, 167–175. [CrossRef] [PubMed]

17. Sato, T.; Takeyama, N.; Katsumata, A.; Tuchiya, K.; Kodama, T.; Kusanagi, K. Mutations in the spike gene of porcine epidemic diarrhea virus associated with growth adaptation in vitro and attenuation of virulence in vivo. *Virus Genes* **2011**, *43*, 72–78. [CrossRef] [PubMed]

18. Lee, S.; Lee, C. Outbreak-related porcine epidemic diarrhea virus strains similar to US strains, South Korea, 2013. *Emerg. Infect. Dis.* **2014**, *20*, 1223–1226. [CrossRef] [PubMed]

19. Lee, D.K.; Park, C.K.; Kim, S.H.; Lee, C. Heterogeneity in spike protein genes of porcine epidemic diarrhea viruses isolated in Korea. *Virus Res.* **2010**, *149*, 175–182. [CrossRef] [PubMed]

20. Gerber, P.F.; Gong, Q.; Huang, Y.W.; Wang, C.; Holtkamp, D.; Opriessnig, T. Detection of antibodies against porcine epidemic diarrhea virus in serum and colostrum by indirect ELISA. *Vet. J.* **2014**, *202*, 33–36. [CrossRef] [PubMed]

21. Hofmann, M.; Wyler, R. Propagation of the virus of porcine epidemic diarrhea in cell culture. *J. Clin. Microbiol.* **1988**, *26*, 2235–2239. [PubMed]

22. Kusanagi, K.; Kuwahara, H.; Katoh, T.; Nunoya, T.; Ishikawa, Y.; Samejima, T.; Tajima, M. Isolation and serial propagation of porcine epidemic diarrhea virus in cell cultures and partial characterization of the isolate. *J. Vet. Med. Sci.* **1992**, *54*, 313–318. [CrossRef] [PubMed]

23. Kweon, C.H.; Kwon, B.J.; Lee, J.G.; Kwon, G.O.; Kang, Y.B. Derivation of attenuated porcine epidemic diarrhea virus (PEDV) as vaccine candidate. *Vaccine* **1999**, *17*, 2546–2553. [CrossRef]

24. Shibata, I.; Tsuda, T.; Mori, M.; Ono, M.; Sueyoshi, M.; Uruno, K. Isolation of porcine epidemic diarrhea virus in porcine cell cultures and experimental infection of pigs of different ages. *Vet. Microbiol.* **2000**, *72*, 173–182. [CrossRef]

25. Wang, J.; Hu, G.; Lin, Z.; He, L.; Xu, L.; Zhang, Y. Characteristic and functional analysis of a newly established porcine small intestinal epithelial cell line. *PLoS ONE* **2014**, *9*, e110916. [CrossRef] [PubMed]

26. Li, W.; Wang, G.; Liang, W.; Kang, K.; Guo, K.; Zhang, Y. Integrin beta3 is required in infection and proliferation of classical swine fever virus. *PLoS ONE* **2014**, *9*, e110911.

27. Xu, X.; Zhang, H.; Zhang, Q.; Dong, J.; Liang, Y.; Huang, Y.; Liu, H.J.; Tong, D. Porcine epidemic diarrhea virus E protein causes endoplasmic reticulum stress and up-regulates interleukin-8 expression. *Virol. J.* **2013**, *10*, 26. [CrossRef] [PubMed]

28. Zhao, P.D.; Tan, C.; Dong, Y.; Li, Y.; Shi, X.; Bai, J.; Jiang, P. Genetic variation analyses of porcine epidemic diarrhea virus isolated in mid-eastern China from 2011 to 2013. *Can. J. Vet. Res.* **2015**, *79*, 8–15. [PubMed]

29. Kim, S.H.; Lee, J.M.; Jung, J.; Kim, I.J.; Hyun, B.H.; Kim, H.I.; Park, C.K.; Oem, J.K.; Kim, Y.H.; Lee, M.H.; et al. Genetic characterization of porcine epidemic diarrhea virus in Korea from 1998 to 2013. *Arch. Virol.* **2015**, *160*, 1055–1064. [CrossRef] [PubMed]

30. Gao, Y.; Kou, Q.; Ge, X.; Zhou, L.; Guo, X.; Yang, H. Phylogenetic analysis of porcine epidemic diarrhea virus field strains prevailing recently in China. *Arch. Virol.* **2013**, *158*, 711–715. [CrossRef] [PubMed]

31. Li, B.X.; Ge, J.W.; Li, Y.J. Porcine aminopeptidase N is a functional receptor for the PEDV coronavirus. *Virology* **2007**, *365*, 166–172. [CrossRef] [PubMed]

32. Oka, T.; Saif, L.J.; Marthaler, D.; Esseili, M.A.; Meulia, T.; Lin, C.M.; Vlasova, A.N.; Jung, K.; Zhang, Y.; Wang, Q. Cell culture isolation and sequence analysis of genetically diverse US porcine epidemic diarrhea virus strains including a novel strain with a large deletion in the spike gene. *Vet. Microbiol.* **2014**, *173*, 258–269. [CrossRef] [PubMed]

33. Lee, S.; Kim, Y.; Lee, C. Isolation and characterization of a Korean porcine epidemic diarrhea virus strain KNU-141112. *Virus Res.* **2015**, *208*, 215–224. [CrossRef] [PubMed]

34. Shirato, K.; Matsuyama, S.; Ujike, M.; Taguchi, F. Role of proteases in the release of porcine epidemic diarrhea virus from infected cells. *J. Virol.* **2011**, *85*, 7872–7880. [CrossRef] [PubMed]

35. Correa, I.; Jiménez, G.; Suñé, C.; Bullido, M.J.; Enjuanes, L. Antigenic structure of the E2 glycoprotein from transmissible gastroenteritis coronavirus. *Virus Res.* **1988**, *10*, 77–93. [CrossRef]

36. Gebauer, F.; Posthumus, W.; Correa, I.; Suné, C.; Sánchez, C.; Smerdou, C.; Lenstra, J.; Meloen, R.; Enjuanes, L. Residues involved in the formation of the antigenic sites of the S protein of transmissible gastroenteritis coronavirus. *Virology* **1991**, *183*, 225–238. [CrossRef]

37. Sun, D.; Feng, L.; Shi, H.; Chen, J.; Cui, X.; Chen, H.; Liu, S.; Tong, Y.; Wang, Y.; Tong, G. Identification of two novel B cell epitopes on porcine epidemic diarrhea virus spike protein. *Vet. Microbiol.* **2008**, *131*, 73–81. [CrossRef] [PubMed]

38. Park, S.J.; Moon, H.J.; Yang, J.S.; Lee, C.S.; Song, D.S.; Kang, B.K.; Park, B.K. Sequence analysis of the partial spike glycoprotein gene of porcine epidemic diarrhea viruses isolated in Korea. *Virus Genes* **2007**, *37*, 321–332. [CrossRef] [PubMed]

39. Chang, S.H.; Bae, J.L.; Kang, T.J.; Kim, J.; Chung, G.H.; Lim, C.W.; Laude, H.; Yang, M.S.; Jang, Y.S. Identification of the epitope region capable of inducing neutralizing antibodies against the porcine epidemic diarrhea virus. *Mol. Cells* **2002**, *14*, 295–299. [PubMed]

40. Song, D.S.; Yang, J.S.; Oh, J.S.; Han, J.H.; Park, B.K. Differentiation of a Vero cell adapted porcine epidemic diarrhea virus from Korean field strains by restriction fragment length polymorphism analysis of ORF 3. *Vaccine* **2003**, *21*, 1833–1842. [CrossRef]

41. Song, D.S.; Oh, J.S.; Kang, B.K.; Yang, J.S.; Moon, H.J.; Yoo, H.S.; Jang, Y.S.; Park, B.K. Oral efficacy of Vero cell attenuated porcine epidemic diarrhea virus DR13 strain. *Res. Vet. Sci.* **2007**, *82*, 134–140. [CrossRef] [PubMed]

12

A Glimpse of Nucleo-Cytoplasmic Large DNA Virus Biodiversity through the Eukaryotic Genomics Window

Lucie Gallot-Lavallée [1] and Guillaume Blanc [1,2,*]

[1] Structural and Genomic Information Laboratory (IGS), Aix-Marseille Université, CNRS UMR7256 (IMM FR3479), 13288 Marseille cedex 09, France; lucie.gallot-lavallee@igs.cnrs-mrs.fr

[2] Mediterranean Institute of Oceanography (MIO), Aix Marseille Université, Université de Toulon, CNRS/INSU, IRD, UM 110, 13288 Marseille cedex 09, France

* Correspondence: guillaume.blanc@igs.cnrs-mrs.fr

Academic Editor: Bernard La Scola

Abstract: The nucleocytoplasmic large DNA viruses (NCLDV) are a group of extremely complex double-stranded DNA viruses, which are major parasites of a variety of eukaryotes. Recent studies showed that certain eukaryotes contain fragments of NCLDV DNA integrated in their genome, when surprisingly many of these organisms were not previously shown to be infected by NCLDVs. We performed an update survey of NCLDV genes hidden in eukaryotic sequences to measure the incidence of this phenomenon in common public sequence databases. A total of 66 eukaryotic genomic or transcriptomic datasets—many of which are from algae and aquatic protists—contained at least one of the five most consistently conserved NCLDV core genes. Phylogenetic study of the eukaryotic NCLDV-like sequences identified putative new members of already recognized viral families, as well as members of as yet unknown viral clades. Genomic evidence suggested that most of these sequences resulted from viral DNA integrations rather than contaminating viruses. Furthermore, the nature of the inserted viral genes helped predicting original functional capacities of the donor viruses. These insights confirm that genomic insertions of NCLDV DNA are common in eukaryotes and can be exploited to delineate the contours of NCLDV biodiversity.

Keywords: nucleo-cytoplasmic large DNA virus; lateral gene transfer; virus insertion

1. Introduction

Viruses have long been viewed only under the angle of human, animal, and plant diseases, which considerably restrained our vision of the viral world and its role in global ecology. In this age of virus discovery, we are beginning to appreciate the enormous diversity of viruses, far beyond what we originally thought. Nucleo-cytoplasmic large DNA viruses (NCLDVs) [1,2] form a monophyletic clade of eukaryotic viruses with a large double-stranded DNA (dsDNA) genome ranging from 100 kbp in the smallest iridoviruses up to 2.50 Mbp in the gigantic pandoraviruses [3]. Their hosts show a remarkably wide taxonomic spectrum from microscopic unicellular eukaryotes to larger animals, including humans [2]. The biodiversity of NCLDVs is thought to be immense however we still do not know how many major clades do exist [4]. Seven taxonomic families have been defined so far including *Ascoviridae, Asfarviridae, Iridoviridae, Marseilleviridae, Mimiviridae, Phycodnaviridae,* and *Poxviridae,* but new viral isolates, such as pandoravirus, pithovirus, and mollivirus, are likely to become founding members of new families. Historically, isolation of large DNA viruses infecting eukaryotic algae or protists has proceeded by co-culturing a host together with a virus sampled from the environment. In this experimental approach, a eukaryotic host is chosen a priori for its capacity of being infected by a virus and adapted to lab culture prior to virus isolation. Recently, the metagenomic approach has accelerated the rate at which new viruses are brought to light [5,6]. However, this approach suffers from two main shortcomings: first, viral sequences assembled from metagenomic data are generally short, encompassing often only a few genes at best. Second, the hosts of the identified viruses remain unknown. Yet, host information is an absolutely essential component in the study of viruses, since viral replication is dependent on host organisms [7]. Thus, drawing the contours of virus/host diversities calls for a development of new approaches that can circumvent limitations of the co-culturing and metagenomics methods.

Recent studies have identified NCLDV-related sequences in genomic and transcriptomic datasets generated from eukaryotic organisms [8–13]. Some of these viral sequences were shown to originate from virus genome fragments integrated into the nuclear genome of their presumed eukaryotic hosts, including protists [10,13], land plant [9], and algae [8,11,12,14]. These fragments encompass up to several hundreds of kbp and can contain hundreds of viral genes, including common NCLDV phylogenetic markers. It is currently unclear how and by which mechanisms these viral DNAs became integrated into eukaryotic genomes. DNA integration may result from an active process (i.e., as a result of a virus-encoded integrase activity) or from an accidental incorporation of viral DNA freely floating inside the cell (i.e., as a result of an aborted infection). Phaeoviruses are the only members of NCLDVs to show evidence of a lysogenic cycle. Presumably, they integrate into the genome of their host by means of an integrase encoded by the virus [14,15]. In addition to reports of NCLDV DNA inserts in eukaryotic hosts, NCLDV-like sequences were also found in some algal transcriptomes [11]. These transcripts may originate from viral genes integrated in the host genome or from infected host cells present in the culture from which RNA were extracted. Altogether, these studies suggested that viral DNA insertion in the host genome is a common feature of NCLDVs. However, the frequency at which this phenomenon occurs across eukaryotic lineages, and the short- and long-term evolutions of inserted viral sequences are still poorly understood. Whether they have a potential role in defense mechanisms against infecting viruses based on sequence recognition and/or RNA silencing is also an open question.

Interesting information has come out from the discovery of viral inserts: many of the organisms in which NCLDV sequences were identified were not previously known to be infected by NCLDVs. Moreover, phylogenetic markers harbored by viral genomic inserts or transcripts suggested that certain virus donors were distantly related to known NCLDVs [8,9,16]. Thus, assuming that the NCLDV sequences identified in eukaryotic datasets result from infecting viruses or lateral gene transfers, these sequences may be used as a tool to better describe the realm of NCLDVs. Importantly, identification of NCLDV genes may allow predicting novel virus/host associations and shedding new light on the biodiversity of NCLDVs. With the vertiginous throughput and dropping cost of DNA sequencing, new eukaryotic genomes are nowadays sequenced at an unprecedented pace. Thus, since the pioneering studies performed over the last couple of years [8–13,16], many new eukaryotic genomes have been released in public databases. This prompted us to perform an update survey of NCLDV genes hidden in eukaryotic sequences to measure the incidence of this phenomenon in common public databases. Here, we show that sequences generated from 66 eukaryotes contained NCLDV core genes, most of which have never been reported so far. Phylogenetic reconstruction showed that many of these sequences originated from members of existing NCLDV families, but also possibly from as yet unknown NCLDV clades, thus extending the range of the NCLDV biodiversity.

2. Materials and Methods

Sequences from the five NCLDV core proteins were retrieved from the NCLDV clusters of orthologous gene (NCVOG) database [17,18] and aligned against protein databases using BLASTP (E-value $< 1 \times 10^{-5}$). BLAST searches against RefSeq and 1KP databases were performed on the dedicated website at NCBI and [19]. Unannotated genome assemblies were downloaded from the NCBI Assembly database. We only downloaded the eukaryotic fraction of the assembly database to the exclusion of very large genomes (i.e., >1 Gbp) to limit computational time; however, annotated proteins of a majority of very large genomes were already available for search in the RefSeq database. Open reading frames >100 codons were extracted from the genome assemblies prior to BLAST searches. Predicted proteins from the Marine Microbial Eukaryote Transcriptome Sequencing Project (MMETSP) transcriptomes were downloaded from the iMicrobe server [20]. The documentation on the experimental conditions used during transcriptome acquisitions was obtained at the following internet adresses: [21] (MMETSP) and [22] (1KP).

Significant similarity and phylogenetic intertwining exist between packaging ATPase of NCLDVs and polintons, a family of large self-synthesizing transposons encoding up to 10 open reading frames [23]. To sort packaging ATPases between NCLDVs and polintons, a phylogenetic tree was constructed with all identified ATPases, NCLDV ATPases and polinton ATPases (reference polinton sequences were retrieved from the relevant supplementary data file of the Yutin et al. paper [23]). Identified ATPases that grouped with reference polinton homologs were removed from further study.

General phylogenetic analyses were performed as follows: additional homologous sequences were first searched in the RefSeq database using the BLAST EXPLORER tool [24]. Multiple-sequence alignment of homologous proteins was then performed using the MAFFT program [25]. We removed alignment positions containing >90% gaps before maximum likelihood phylogenetic reconstruction, which was performed using the FastTree program [26] with the LG + Gamma model of amino acid substitution. Statistical support for branches was estimated with the SH-like local support method. Sequences, alignments and phylogenetic trees are available in Dataset S1. The lengths of multiple-alignments used for phylogenetic reconstruction were 411, 1,198, 2,117, 692, and 565 amino acid positions (including position containing <90% gaps) for the ATPase, D5 primase-helicase, B-family DNA polymerase (DNAP), major capsid protein (MCP), and Very Late Transcription Factor 3 (VLTF3), respectively.

3. Results and Discussion

3.1. NCLDV Protein Markers in Eukaryotes

Although they typically encode hundreds of proteins, NCLDVs were reported to only share five universally-conserved core genes, including genes for MCP, D5 primase-helicase, DNAP, A32-like packaging ATPase, andVLTF3 [17]. These core viral proteins were used as query in BLAST searches against four eukaryotic sequence databases. The databases queried in this study included Genbank Refseq, which contained all annotated proteins from 680 sequenced eukaryotic species. In addition, we screened the Genbank Assembly database which contained 602 raw eukaryotic genome sequences that were not annotated and, therefore, not referenced in RefSeq. Open reading frames >100 codons were extracted from the non-annotated genome assemblies prior to their mining by BLASTP. Altogether, the RefSeq and Assembly databases comprised 1282 fully sequenced eukaryotic genomes. Because preliminary analysis revealed that NCLDV insertions were most frequent in aquatic unicellular eukaryotes, we also downloaded transcriptomic data from sequencing initiatives specifically targeting these organisms. The MMETSP database contained 679 assembled transcriptomes from 413 distinct marine unicellular eukaryotes, including some of the more abundant and ecologically significant species in the oceans such as diatoms [27]. The "1000 plants" (1KP) initiative database contained transcriptomic data from over 1000 plant species, including 214 unicellular eukaryotic algae from the Archaeplastida and Chromista groups [28].

Protein homologous to the five NCLDV core proteins were identified in 48 eukaryotic genomic sequencing projects, including 12 annotated genomes from RefSeq and 36 non-annotated genomes (Table 1). Nine of these genome assemblies contained the five core genes, while 14 genome assemblies contained only one of them. Some of the viral sequences arose from larger viral inserts that have already been described in the genomes of *Ectocarpus siliculosus* [12], *Bigelowiella natans* [8], *Physcomitrella patens* [9], *Acanthamoeba* spp. [13,16], *Hydra vulgaris* [10,13] and *Phytophthora parasitica* [10]. Most of the genomic datasets associated with NCLDV core protein homologs correspond to organisms living in soil or aquatic environments. The working genome dataset was highly dominated by Metazoa ($n = 495$), Fungi ($n = 392$) and land plants ($n = 110$), collectively representing 80% of the analyzed genomes. However, only 15% (7/48) of the eukaryotes positive for NCLDV proteins belonged to one of these groups, including three metazoans (*Daphnia pulex*, *H. vulgaris*, *Echinacea pallida*), three fungi (*Gonapodya prolifera*, *Rhizophagus irregularis*, *Allomyces macrogynus*), and a land plant (moss *P. patens*). Thus, a majority of eukaryotes associated with NCLDV proteins have a unicellular or simple multicellular structure and are members of less studied clades. The most impacted eukaryotic groups in terms of frequency are (i) brown algae (*Phaeophyceae*) for which all three genomes contained NCLDV homologs, Amoebozoa (11 out of 32 genomes = 34%), green algae (9/28 = 32%; i.e., *Chlorophyta* + *Streptophyta*), and *Oomycetes* (10/40 = 25%). As a matter of fact, small eukaryotes living either constantly (aquatic) or transiently (soil or swimming gametes (i.e., moss *P. patens*)) in waters appear to more frequently have integrated NCLDV sequences in their genome. This host bias may be a consequence of the relative large size of NCLDV particles, which makes their propagation more difficult out of fluidic environments. Also, the large virion size may limit propagation in complex multicellular organisms that have thick cell walls (i.e., terrestrial plants), giving them less chance to access the germ-line cells where lateral gene transfers must occur to be transmitted to the next host generation.

In addition, 18 transcriptomes generated from eukaryotic microalgae or aquatic protistans encode homologs to at least one of the NCLDV core proteins. In contrast to genomic datasets, none of the transcriptomes encode the five core proteins; 14 contained only one core protein sequence. DNAP was the most frequently identified NCLDV core gene among transcriptomes (10 species). This observation is consistent with a previous study reporting that a supernumerary DNAP subunit of possible NCLDV origin was transcribed in the rhizarian alga *B. natans*, whereas most of the other inserted NCLDV-like genes were transcriptionally silent [8]. The NCLDV-like DNAP of *B. natans* has been shown to be targeted to the nucleomorph where it might be involved in the nucleomorph genome replication [29]. Thus, some of the NCLDV-like DNAPs identified in these transcriptomes might also originate from lateral gene transfer from viruses and have acquired a functional role in their respective eukaryotes. Another possibility is that these transcripts were produced by viruses replicating in infected cells in the cultures used for sequencing. This is most likely the case for the *Emiliania huxleyii* viral transcripts because the corresponding transcriptomes have been reportedly acquired during a viral infection experiment (see information on the experimental conditions in Materials and Methods). In addition, the *Pleurochrysis carterae* strain sequenced in the MMETSP was suspected to contain a persistent virus, and our analysis gives credit to this hypothesis.

All in all, our study reveals many more potential NCLDV hosts than previously thought. Out of the 66 sequence datasets positive for NCLDV core genes, only four were generated from species already known for being infected by NCLDVs (i.e., *Acanthamoeba castellanii*, *Acanthamoeba polyphaga*, *E. siliculosus*, *E. huxleyii*). Two other species are closely related to organisms hosting NCLDVs. This is the case for the marine flagellate *Halocafeteria seosinensis* that is closely related to *Cafeteria roenbergensis* [30], a host for giant viruses and virophage [31,32]. The freshwater green alga *Chlorella vulgaris* is also closely related to *Craspedia variabilis* infected by *Paramecium bursaria* Chlorella viruses [33]. Overall NCLDV core proteins were identified in virtually all major groups of algae, including Chlorophyta, Streptophyta, Stramenopiles, *Cryptophyta*, Euglenozoa, Haptophyceae, and Rhizaria. Remarkably, NCLDV core proteins were identified from multiple species of a same genus such as *Acanthamoeba* spp., *Sphaeroforma* spp., *Phytophthora* spp., *Pythium* spp., *Klebsormidium* spp. and *Chlamydomonas* spp.

Table 1. Nucleocytoplasmic large DNA viruses (NCLDV) core protein homologs in eukaryotic sequence datasets.

Eukaryotic Clade	Species	Habitat	Database	DNAP *	MCP *	ATPase *	D5 *	VLTF3 *
Genomic datasets								
Amoebozoa (Discosea)	Acanthamoeba astronyxis	terrestrial and aquatic	Assembly	√		√		√
	Acanthamoeba castellanii		RefSeq		√	√	√	√
	Acanthamoeba divionensis		Assembly	√		√		√
	Acanthamoeba healyi		Assembly		√			
	Acanthamoeba lenticulata		Assembly		√	√	√	√
	Acanthamoeba lugdunensis		Assembly		√	√		
	Acanthamoeba mauritaniens		Assembly	√		√		√
	Acanthamoeba pearcei		Assembly			√	√	√
	Acanthamoeba polyphaga		Assembly		√	√	√	√
	Acanthamoeba quina		Assembly		√	√	√	√
	Acanthamoeba rhysodes		Assembly		√			
Cryptophyta (Pyrenomonadales)	Guillardia theta	sea	RefSeq					MI
Euglenozoa	Euglena gracilis	freshwater	Assembly					
Fungi (Chytridiomycota)	Gonapodya prolifera	freshwater	RefSeq	Phy	√	√	Phy	√
Fungi (Glomeromycota)	Rhizophagus irregularis	terrestrial	Assembly	Asf	√			
Fungi (Blastocladiomycota)	Allomyces macrogynus	freshwater	RefSeq				Phy	
Metazoa (Arthropoda)	Daphnia pulex	freshwater	RefSeq		√			
Metazoa (Cnidaria)	Exaiptasia pallida	sea	RefSeq		Mi		Mi	Asf
	Hydra vulgaris	freshwater	RefSeq			√		√
Opisthokonta (Ichthyosporea)	Sphaeroforma arctica	sea	RefSeq	Irma		Irma		
	Sphaeroforma sirkka	sea	Assembly	Irma	Irma	Irma	√	Irma
Rhizaria (Cercozoa)	Bigelowiella natans	sea	RefSeq	Phy + √	√	√	Phy	√
Stramenopiles (Bicosoecida)	Halocafeteria seosinensis	saltern pond	Assembly					√
Stramenopiles (Eustigmatophyceae)	Nannochloropsis limnetica	freshwater	Assembly		Pha			Pha
Stramenopiles (Hyphochytriomycetes)	Hyphochytrium catenoides	terrestrial	Assembly	Asf	Asf	Asf	Asf	Asf
Stramenopiles (Oomycetes)	Phytophthora sp. totara	soilborne plant pathogen	Assembly		Asf			
	Phytophthora agathidicida		Assembly		Asf			
	Phytophthora alni		Assembly					Asf
	Phytophthora cambivora		Assembly					Asf
	Phytophthora cryptogea		Assembly		Asf			
	Phytophthora nicotianae		Assembly		Asf			Asf
	Phytophthora parasitica		RefSeq		Asf			Asf
	Pythium irregulare		Assembly					Asf
	Pythium oligadrum		Assembly		Asf			Asf
	Pythium ultimum		Assembly		Asf			Asf

Table 1. *Cont.*

Eukaryotic Clade	Species	Habitat	Database	DNAP*	MCP*	ATPase*	D5*	VLTF3*
Stramenopiles (Phaeophyceae)	*Cladosiphon okamuranus*	sea	Assembly	Pha	Pha	Pha	Pha	Pha
	Ectocarpus siliculosus	sea	RefSeq	Pha	Pha	Pha	Pha	Pha
	Saccharina japonica	sea	RefSeq		Pha			
Viridiplantae (Chlorophyta)	*Asterochloris glomerata*	lichen photobiont	Assembly	Phy	Phy	Phy	Phy	
	Chlamydomonas applanata	terrestrial	Assembly	Mi	Mi	Mi	Mi + Phy	Mi
	Chlamydomonas asymmetrica	freshwater	Assembly	Mi	Mi	Mi	Mi + Phy	Mi + √
	Chlamydomonas sphaeroides	freshwater	Assembly	Mi	Mi			
	Chlorella vulgaris	freshwater	Assembly		√			
	Coccomyxa sp. LA000219	unknown	Assembly	Mi	Mi	Mi	Mi	Mi
	Cymbomonas tetramitiformis	sea	Assembly	Phy	Phy	Phy	Phy	Phy
	Haematococcus pluvialis	freshwater	Assembly	Mi	Mi	Mi	Mi + Phy	Mi
Viridiplantae (Streptophyta)	*Klebsormidium flaccidum*	terrestrial	RefSeq	Phy	Phy	Phy	√	Phy
Viridiplantae (Streptophyta)	*Physcomitrella patens*	terrestrial	RefSeq	Pitho			Pitho	
Transcriptomic datasets								
Cryptophyta (Cryptomonadales)	*Hemiselmis andersenii*	sea	MMETSP	√				
Cryptophyta (Pyrenomonadales)	*Hanusia phi*	sea	MMETSP	√				
Haptophyceae (Coccolithales)	*Pleurochrysis carterae*	sea	MMETSP		√	√		√
Haptophyceae (Isochrysidales)	*Chrysochromulina polylepis*	sea	MMETSP			√		
	Isochrysis galbana	sea	MMETSP			√		
Haptophyceae (Phaeocystales)	*Phaeocystis antarctica*	sea	MMETSP	Coc	√			
Haptophyceae (Prymnesiales)	*Emiliania huxleyi*	sea	MMETSP		Coc		Coc	
Rhizaria (Cercozoa)	*Lotharella globosa*	sea	MMETSP	Phy				
Stramenopiles (Labyrinthulomycetes)	*Aurantiochytrium limacinum*	sea	MMETSP	Pha				
	Schizochytrium aggregatum	sea	MMETSP		√			√
	Thraustochytrium sp.	sea	MMETSP		√			√
Undescribed Strain	CCMP2135	sea	MMETSP				√	√
Undescribed Strain	CCMP2436	sea	MMETSP					
Viridiplantae (Chlorophyta)	*Carteria crucifera*	freshwater	1KP	Mi			Mi	
	Cylindrocapsa geminella	freshwater	1KP	Mi				
Viridiplantae (Streptophyta)	*Entransia fimbriat*	freshwater	1KP	Phy				
	Interfilum paradoxum	terrestrial	1KP	Phy				
	Klebsormidium subtile	terrestrial	1KP	Phy				

* putative phylogenetic grouping of the NCLDV core protein homologs based on the phylogenetic trees presented in Figure 1 and Figure S1-S4: √ = unknown clade, Phy = *Phycodnaviridae*, Mi = *Mimiviridae*, Pha = phaeoviruses, Pitho = putative Pithoviridae, Coc = coccolothoviruses, Asf = *Asfarviridae* and IrMa = *Iridoviridae/Marseilleviridae* cluster. Column names: DNAP, DNA polymerase; MCP, major capsid protein; ATPase, DNA packaging ATPase; D5, D5 helicase; VLTF3, very late transcription factor 3.

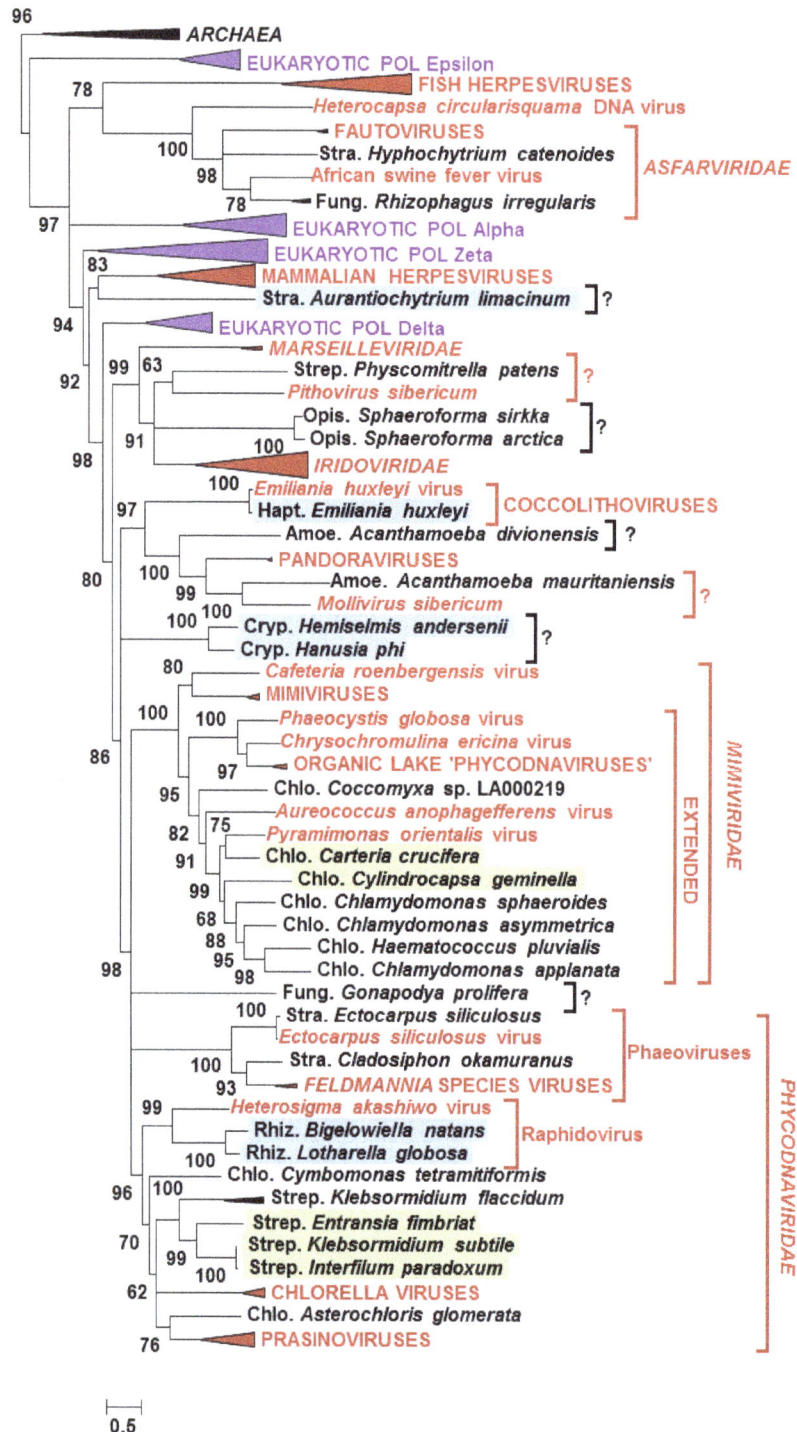

Figure 1. Maximum likelihood phylogenetic tree of DNA polymerase proteins. Statistical supports for branch (SH-like local support test) are given above or below nodes in percent. Branches with support less than 50% were collapsed. Species names with colored background indicate transcribed genes: green, 1KP transcriptomes; blue, MMETSP transcriptomes. Red and black question marks show potential extension of recognized viral groups or new viral clades, respectively. The scale bar indicates the number of substitution per site. Sequences, alignments, and phylogenetic trees are available in Dataset S1.

3.2. Phylogeny of Eukaryotic NCLDV-Like Proteins

To investigate the phylogenetic relationships between NCLDV-like proteins identified in eukaryotic datasets and their homologs in extant viruses, maximum likelihood phylogenetic trees were constructed for each of the five NCLDV core proteins (Figure 1 and Figure S1–S4). Overall, the resulting phylogenetic trees revealed several general characteristics of the NCLDV-like sequences. First, most of the NCLDV core proteins identified in eukaryotic datasets branched close to or within existing viral clades, further supporting the hypothesis of their viral origin. Second, 20 eukaryotes listed in Table 1, containing two or more NCLDV core genes, occupied consistent positions across phylogenetic trees (i.e., grouped within the same viral clade in each phylogenetic tree). This observation suggests that the viral sequences in each eukaryote arose from a single unique virus rather than multiple unrelated viral sources. Lastly, closely related eukaryotes tended to share closely-related viral sequences. This involved organisms beyond the genus rank such as for example chlorophytan or streptophytan species which had sequences forming subtrees within the *Phycodnaviridae* or *Mimiviridae* clades, or Stramenopile species branching within the *Asfarviridae* clade. This phylogenetic "correlation" between virus sequences and potential hosts can occur if closely related virus-like sequences originate from a single viral genome integration event in an ancestral eukaryotic host—the transferred genes could then spread across the host progeny up to the extant species. It has been suggested that most *Acanthamoeba* inserted viral genes became nonfunctional and decayed by accumulation of mutations [16]; it is, therefore, possible that sequence homology can no longer be recognized between viral inserts after a sufficiently long period of divergence. Alternatively, some of these sequences may originate from closely related contaminating viruses. In fact, some viral clades are apparently specific to certain eukaryotic groups, such as for example chloroviruses that infect *Chlorella* species [33], prasinoviruses that infect prasinophyte algae [34], or phaeoviruses that infect brown algae [35], possibly because their speciation and diversification co-occurred with those of their hosts. Under this co-speciation scenario, closely related eukaryotes may be infected by closely related viral species, which can also result in the observations made in our phylogenetic trees.

The B-family DNAP gene is a core NCLDV gene traditionally used as a reference phylogenetic marker to establish the taxonomy of large DNA viruses [36,37]. The DNAPs of NCLDVs are phylogenetically related to those of eukaryotes [38]. Our phylogenetic tree is largely in agreement with previous studies and shows that the eukaryotic DNAP delta emerged as a sister group to most NCLDV DNAPs (Figure 1). Furthermore, *Asfarviridae*, together with fish herpesviruses and *Heterocapsa circularisquama* DNA virus, form a separate group from the other NCLDV DNAPs [38,39]. As expected, some of the virus-like DNAPs identified in algae branched within the *Phycodnaviridae* family, which is a group of NCLDVs exclusively infecting phytoplanktonic species [35]. Among these sequences were proteins from streptophytan and rhizarian species, two major algal clades for which no NCLDV has ever been reported so far. Moreover, virus-like DNAPs from seven chlorophytan species grouped within the "extended *Mimiviridae*" clade, another group of large DNA viruses infecting a variety of microalgae but more closely related to giant mimiviruses [40]. Interestingly these chlorophytes include three species of *Chlamydomonas*. This genus of green algae also contains *Chlamydomonas reinhardtii,* a model organism for molecular and chloroplast biology. Although 25 putative viral genes were identified in this species [11], none of the five NCLDV core protein genes were found in the *C. reinhardtii* genome. Finally, the non-photosynthtic stramenopile *Hyphochytrium catenoides* and the fungus *Rhizophagus irregularis* contained DNAP sequences branching within the *Asfarviridae*. This result suggests that viruses from the *Asfarviridae* family have a much wider host range than currently thought. Fungi have never been reported as a potential host for NCLDVs and the presence of a viral DNAP in *R. irregularis* and in another fungus *Gonapodya prolifera* (as well as a D5 primase-helicase in the fungus *Allomyces macrogynus*; Figure S3 and Table 1) suggest that these organisms may have been infected by members of NCLDVs. Note that *G. prolifera* and *A. macrogynus* are members of two ancestral fungal lineages, which contain species feeding on algae. Furthermore, the fungal D5 primase-helicases branched close to the *Heterosigma akashiwo* virus, which is a member of the *Phycodnaviridae*. Thus we cannot rule out that the NCLDV-like sequences identified in these two lower fungi may in fact originate from viruses infecting their algal preys.

Interestingly a number of virus-like DNAPs branched outside recognized taxonomic clades [41] suggesting that they belong to yet unknown taxa (e.g., indicated by black question marks in Figure 1). Other sequences grouping close to single, unclassified viral isolates (e.g., indicated by red question marks in Figure 1), might originate from members of the extended putative *Pithoviridae* family (e.g., represented by a sequence from the moss *P. patens*) and *Molliviridae* family (e.g., represented by a sequence from the amoeba *A. mauritaniensis*). Thus, our data, and more generally the approach of searching viral sequences in eukaryotic sequence data, make it possible to consolidate and even improve our knowledge of the NCLDV biodiversity. It is likely that some of the donor viruses encoded original functions and have developed new ways of interacting with their host that are radically different from the mechanisms already characterized.

3.3. Genomic Context around NCLDV-Like Genes

The discovery of viral sequences in eukaryotic genomes naturally poses the question of their origin, which can be a viral contamination, a provirus or a lateral gene transfer. A viral infection was suspected to be at the origin of the viral transcripts identified in the *E. huxleyii* and *P. carterae* transcriptomes. Filée suggested to examine the genomic environment around the virus-like genes to decipher whether they come from inserted viral DNA or contamination with free viral DNA during sequencing [13]. According to the author, insertion is the most likely hypothesis when virus-like genes are surrounded by intron-rich genes highly similar to eukaryotic homologs. We investigated the nature of genes surrounding NCLDV-like sequences in genomes that have a publicly available annotation, to the exception of organisms for which viral inserts have already been studied in details (i.e., *E. siliculosus* [12], *B. natans* [8], *P. patens* [9], *Acanthamoeba* spp. [13,16], *H. vulgaris* [10,13] and *P. parasitica* [10]). Figure S5 show the gene organization in contigs containing NCLDV core genes in seven eukaryotes, namely *A. macrogynus*, *D. pulex* [42], *E. pallida* [43], *Sphaeroforma arctica*, *Klebsormidium flaccidum* [44], *Saccharina japonica* [45], and *G. prolifera* [46].

Contigs with homologs to NCLDV core genes had sizes ranging from 1.2 kbp to 1.4 Mbp, so most of them contained more than one gene. The origin of the neighboring genes, inferred from the taxonomic information of their best match, generally indicates that other viral genes are present in the immediate vicinity of NCLDV core genes (Figure S5). Thus, most of the NCLDV core genes do not seem to result from horizontal transfer of a single isolated gene. However, this is not the case for the two metazoans, *D. pulex* and *E. pallida*, that each carries copies of a single NCLDV core gene (respectively, MCP and VLTF3) isolated in the midst of typical metazoan genes (Figure S5). Furthermore, the viral genes of the fungi *G. prolifera and A. macrogynus*, the brown alga *S. japonica* and the protist *S. arctica* are grouped in small genomic islands amid genes of eukaryotic origin, suggesting that they result from an insertion of a larger viral genome fragment. The viral sequences of *S. japonica* are closely related to phaeoviruses, which have a lysogenic reproduction and exist as provirus elements incorporated into the genomes of the brown algae *Ectocarpus siliculosus* and *Feldmannia* species [12,14,15]. Thus, the viral genes identified in the *S. japonica* genome might be remnants of an ancient provirus.

In contrast, NCLDV core genes of *K. flaccidum* are contained in contigs that have a dominance of viral genes, intermingled with a minority of genes most closely related to bacterial or eukaryotic homologs. Such a gene mosaicism is typical of NCLDV genomes [47]. Except for contig DF237168, there is no apparent juncture between a eukaryotic genomic region and a viral genomic region (i.e., a region containing a majority of eukaryotic genes followed by a region containing a majority of viral genes). Although this observation is compatible with a contaminating virus, this hypothesis is unlikely because a single viral genotype would be expected in the case of an infected culture. In contrast, we found five contigs containing a DNAP gene while NCLDVs only contain a single DNAP gene per genome. We also found six packaging ATPase genes, six MCP genes, and five VLTF3 genes. The levels of protein similarity between the DNAPs ranged from 40% to 92%, which excluded the possibility that the homologous regions originate from variants of a same initial viral genotype. Furthermore the gene order was highly rearranged between homologous regions, further refuting the hypothesis of a single viral genotype. These data suggest the *K. flaccidum* genome contains distinct viral insertions. These inserts may result from duplication of an original viral insert followed by sequence divergence and rearrangements of the duplicated copies. Alternatively, they may result from independent acquisitions from multiple viruses.

Altogether, our analysis of seven annotated eukaryotic genomes supports the hypothesis of lateral gene transfer from viruses rather than contamination with free viral DNA during sequencing. The same conclusion was drawn in other studies of various organisms including *Acanthamoeba* spp., *P. patens* and *B. natans* [8,9,13,16]. Thus, there is a general consensus indicating that when viral sequences are identified in a given eukaryotic genome assembly, they are likely to result from bona fide viral genomic insertions rather than an alternative source. Given the substantial number of eukaryotic genomes concerned by inserted viral sequences (Table 1), viruses, and especially NCLDVs, may soon take center stage in our understanding of eukaryotic genome evolution. Viral insertions may turn out to be a major force driving lateral gene transfers between viruses and eukaryotes or between eukaryotes. The wide phylogenetic spectrum of eukaryotes containing viral sequences also suggest that these inserts might serve as DNA template in an evolutionarily conserved defense mechanism against viruses based on sequence recognition such as the RNA interference (RNAi) pathway [48].

3.4. Hints on Viral Functions

Another interesting aspect of viral inserts is that genes contained within viral regions can provide hints on unexpected functional capabilities of the original viruses. For instance, we found two highly similar expansin genes in two NCLDV-like contigs of the *K. flaccidum* genome assembly (DF237168.1 and DF237869.1; Figure S5). Although most similar to plant homologs, these expansin genes are both surrounded by a VLTF3 gene and a hypothetical protein gene that only has homologs in phycodnaviruses. This suggests that the expansin genes had been captured by the original donor virus from a plant or algal cell, before lending to the *K. flaccidum* genome through integration of viral DNA. Expansins mediate cell wall extension in plants by disrupting non-covalent binding of wall polysaccharides [49,50]. Lateral transfers of expansin genes from plants toward their fungal and bacterial parasites have been described, and their functional similarity suggests that these proteins mediate plant-microbial interaction [51]. Thus, in analogy to cellular plant parasites, this gene could also have a role during viral infection by enabling the virus to cross the host cell wall barrier. This would consist in a case of functional convergence between eukaryotic, bacterial and viral plant pathogens.

Additionally, two *K. flaccidum* viral regions contained a gene encoding a U-type cyclin domain (DF237607.1 and DF237785.1; Figure S5). In the available *K. flaccidum* genome annotation the cyclin domain is predicted to be fused with a MCP domain. However, this protein structure is likely an annotation error resulting from the merging of two independent exons each containing one of the two domains. In fact, no transcript sequence supports the junction between the two introns in a *K. flaccidum* RNAseq study [44]. On the other side of the cyclin gene, we found a viral DNA packaging ATPase gene. Although the cyclin domain is more closely related to plant homologs, the viral origin of the surrounding genes suggests a cyclin gene captured from a plant cell was present in the original donor viral genome. Viral encoded cyclins have been identified in several viral families including herpesviruses, retroviruses, and baculoviruses, where they drive cell cycle transitions of the host [52]. Many DNA viruses induce quiescent cells to enter the cell cycle; this is thought to increase pools of deoxynucleotides and, thus, facilitate viral replication. In contrast, some viruses can arrest cells in a particular phase of the cell cycle that is favorable for replication of the specific virus [53]. If the existence of the predicted *K. flaccidum* virus is confirmed in future studies, it would represent the first instance of a cyclin gene in a NCLDV.

A contig of the fungus *G. prolifera* (KQ965906.1; Figure S5) contained a chitinase gene, together with three other viral genes encoding a MCP, a VATPase_H domain containing protein and a protein of unknown function. Chitinases are enzymes that degrade chitin, which is one of the most abundant biopolymers in nature. Chitin occurs in various contexts across a broad range of species and is the main constituent of fungal cell wall [54]. Remarkably, the *G. prolifera* chitinase is most closely related to homologs in chloroviruses (phycodnavirus) which are presumably involved in degradation of the cell wall of green algae of the *Chlorella* genus [55]. Interestingly the same *G. prolifera* viral region contains another chitinase-like protein surrounded by 2 viral genes, but this one is more similar to bacterial homologs. Thus, the putative *G. prolifera* virus might use a similar enzyme apparatus as chloroviruses to pass through the chitin-rich fungal cell wall.

4. Conclusions

We are only beginning to appreciate the extraordinary diversity of NCLDVs, which are among the most intriguing viruses on the planet. Here we show that substantial progress in the description of the NCLDV biodiversity can be made by mining potential host sequences in order to identify genetic markers of NCLDVs. Using this approach, both a virus and its putative host can be brought to light, a significant advantage over metagenomics, which cannot directly identify the two partners. This approach takes advantage of what appears to be an important, but as yet poorly understood, feature of NCLDVs: they leave footprints of their passage in the cell in the form of viral inserts in the host genome. Here we chose to search virus-like sequences using the five most consistently conserved genes in NCLDVs. Others suggested to use the RNA polymerase subunit 2 to identify giant viruses sequences in (meta)genomic data [10]. However, some of these genes are not universally conserved among NCLDVs. For instance, a RNA polymerase gene is absent in most *Phycodnaviridae* genomes, whereas a MCP gene could not be detected in the genomes of pandoraviruses and *P. sibericum* [3,56,57]. *P. sibericum* is also apparently lacking a gene for packaging ATPase. Thus additional combinations of NCLDV reference markers may lead to an increasing number of eukaryotic datasets positive for NCLDV sequences [11]. It is also worth noting that the abundance of viruses in environmental samples is sometimes estimated by quantitative PCR using primers specific for virus genes [58] or by the number of metagenomic reads overlapping viral genes [6]. However, given the apparent ease with which NCLDV genes find themselves integrated into host genomes, these approaches may lead to over estimating the viral abundances if the surveyed samples also contain hosts harboring viral HGTs.

In this study we could predict the existence of new members of NCLDVs, some of which are apparently distantly related from already characterized viruses and may define new viral clades. Examination of the gene content of viral regions also helped us predicting some potential functional capabilities of the original viruses. We also predicted a wide range of potential hosts, most of whom have never had an association described with NCLDVs. The validity of all these predictions must now be evaluated through experimental approaches. Most of the organisms in which viral sequences were found are cultivable in laboratory conditions. This offers a favorable experimental framework to prospect environmental samples in order to isolate viral strains by co-culture with a eukaryote. A potential host may be chosen according to the phylogenetic reconstruction of its viral sequences in order to target the isolation of novel NCLDVs of special scientific interest. If such an approach proves to be successful, it may help in improving our understanding of the NCLDV world.

Acknowledgments: We thank Daniele Armaleo and Olivier Vallon for providing the *A. glomerata* and *H. pluvialis* sequences ahead of publication.

Author Contributions: G.B. conceived and designed the experiments; L.G.-L. and G.B. analyzed the data; L.G.-L. and G.B. wrote the paper.

References

1. Iyer, L.M.; Balaji, S.; Koonin, E.V.; Aravind, L. Evolutionary genomics of nucleo-cytoplasmic large DNA viruses. *Virus Res.* **2006**, *117*, 156–184. [CrossRef] [PubMed]

2. Koonin, E.V.; Yutin, N. Nucleo-cytoplasmic Large DNA Viruses (NCLDV) of Eukaryotes. *eLS* **2012**.

3. Philippe, N.; Legendre, M.; Doutre, G.; Couté, Y.; Poirot, O.; Lescot, M.; Arslan, D.; Seltzer, V.; Bertaux, L.; Bruley, C.; et al. Pandoraviruses: Amoeba Viruses with Genomes Up to 2.5 Mb Reaching That of Parasitic Eukaryotes. *Science* **2013**, *341*, 281–286. [CrossRef] [PubMed]

4. Fischer, M.G. Giant viruses come of age. *Curr. Opin. Microbiol.* **2016**, *31*, 50–57. [CrossRef] [PubMed]

5. Wommack, K.E.; Nasko, D.J.; Chopyk, J.; Sakowski, E.G. Counts and sequences, observations that continue to change our understanding of viruses in nature. *J. Microbiol. Seoul Korea* **2015**, *53*, 181–192. [CrossRef] [PubMed]

6. Hingamp, P.; Grimsley, N.; Acinas, S.G.; Clerissi, C.; Subirana, L.; Poulain, J.; Ferrera, I.; Sarmento, H.; Villar, E.; Lima-Mendez, G.; et al. Exploring nucleo-cytoplasmic large DNA viruses in Tara Oceans microbial metagenomes. *ISME J.* **2013**, *7*, 1678–1695. [CrossRef] [PubMed]

7. Mihara, T.; Nishimura, Y.; Shimizu, Y.; Nishiyama, H.; Yoshikawa, G.; Uehara, H.; Hingamp, P.; Goto, S.; Ogata, H. Linking Virus Genomes with Host Taxonomy. *Viruses* **2016**, *8*, 66. [CrossRef] [PubMed]

8. Blanc, G.; Gallot-Lavallée, L.; Maumus, F. Provirophages in the Bigelowiella genome bear testimony to past encounters with giant viruses. *Proc. Natl. Acad. Sci. USA* **2015**, *112*, E5318–E5326. [CrossRef] [PubMed]

9. Maumus, F.; Epert, A.; Nogué, F.; Blanc, G. Plant genomes enclose footprints of past infections by giant virus relatives. *Nat. Commun.* **2014**, *5*. [CrossRef] [PubMed]

10. Sharma, V.; Colson, P.; Giorgi, R.; Pontarotti, P.; Raoult, D. DNA-Dependent RNA Polymerase Detects Hidden Giant Viruses in Published Databanks. *Genome Biol. Evol.* **2014**, *6*, 1603–1610. [CrossRef] [PubMed]

11. Wang, L.; Wu, S.; Liu, T.; Sun, J.; Chi, S.; Liu, C.; Li, X.; Yin, J.; Wang, X.; Yu, J. Endogenous viral elements in algal genomes. *Acta Oceanol. Sin.* **2014**, *33*, 102–107. [CrossRef]

12. Delaroque, N.; Boland, W. The genome of the brown alga *Ectocarpus siliculosus* contains a series of viral DNA pieces, suggesting an ancient association with large dsDNA viruses. *BMC Evol. Biol.* **2008**, *8*, 110. [CrossRef] [PubMed]

13. Filée, J. Multiple occurrences of giant virus core genes acquired by eukaryotic genomes: The visible part of the iceberg? *Virology* **2014**, *466–467*, 53–59. [CrossRef] [PubMed]

14. Meints, R.H.; Ivey, R.G.; Lee, A.M.; Choi, T.-J. Identification of Two Virus Integration Sites in the Brown Alga Feldmannia Chromosome. *J. Virol.* **2008**, *82*, 1407–1413. [CrossRef] [PubMed]

15. Delaroque, N.; Maier, I.; Knippers, R.; Müller, D.G. Persistent virus integration into the genome of its algal host, *Ectocarpus siliculosus* (Phaeophyceae). *J. Gen. Virol.* **1999**, *80*, 1367–1370. [CrossRef] [PubMed]

16. Maumus, F.; Blanc, G. Study of gene trafficking between *Acanthamoeba* and giant viruses suggests an undiscovered family of amoeba-infecting viruses. *Genome Biol. Evol.* **2016**. [CrossRef] [PubMed]

17. Yutin, N.; Wolf, Y.I.; Raoult, D.; Koonin, E.V. Eukaryotic large nucleo-cytoplasmic DNA viruses: Clusters of orthologous genes and reconstruction of viral genome evolution. *Virol. J.* **2009**, *6*, 223. [CrossRef] [PubMed]

18. Available online: ftp://ftp.ncbi.nih.gov/pub/wolf/COGs/NCVOG/ (accessed on 30 November 2016).

19. Available online: https://www.bioinfodata.org/Blast4OneKP (accessed on 30 November 2016).

20. Available online: ftp://ftp.imicrobe.us/projects/104/CAM_P_0001000.pep.fa.gz (accessed on 30 November 2016).

21. Available online: ftp://ftp.imicrobe.us/projects/104/Callum_FINAL_biosample_ids.xls (accessed on 30 November 2016).

22. Available online: http://www.onekp.com/samples/list.php (accessed on 30 November 2016).

23. Yutin, N.; Raoult, D.; Koonin, E.V. Virophages, polintons, and transpovirons: a complex evolutionary network of diverse selfish genetic elements with different reproduction strategies. *Virol. J.* **2013**, *10*, 158. [CrossRef] [PubMed]

24. Dereeper, A.; Audic, S.; Claverie, J.-M.; Blanc, G. BLAST-EXPLORER helps you building datasets for phylogenetic analysis. *BMC Evol. Biol.* **2010**, *10*, 8. [CrossRef] [PubMed]

25. Katoh, K.; Kuma, K.; Toh, H.; Miyata, T. MAFFT version 5: improvement in accuracy of multiple sequence alignment. *Nucleic Acids Res.* **2005**, *33*, 511–518. [CrossRef] [PubMed]

26. Price, M.N.; Dehal, P.S.; Arkin, A.P. FastTree 2 – Approximately Maximum-Likelihood Trees for Large Alignments. *PLoS ONE* **2010**, *5*, e9490. [CrossRef] [PubMed]

27. Keeling, P.J.; Burki, F.; Wilcox, H.M.; Allam, B.; Allen, E.E.; Amaral-Zettler, L.A.; Armbrust, E.V.; Archibald, J.M.; Bharti, A.K.; Bell, C.J.; et al. The Marine Microbial Eukaryote Transcriptome Sequencing Project (MMETSP): Illuminating the Functional Diversity of Eukaryotic Life in the Oceans through Transcriptome Sequencing. *PLoS Biol* **2014**, *12*, e1001889. [CrossRef] [PubMed]

28. Matasci, N.; Hung, L.-H.; Yan, Z.; Carpenter, E.J.; Wickett, N.J.; Mirarab, S.; Nguyen, N.; Warnow, T.; Ayyampalayam, S.; Barker, M.; et al. Data access for the 1,000 Plants (1KP) project. *GigaScience* **2014**, *3*, 17. [CrossRef] [PubMed]

29. Suzuki, S.; Ishida, K.-I.; Hirakawa, Y. Diurnal transcriptional regulation of endosymbiotically derived genes in the chlorarachniophyte *Bigelowiella natans*. *Genome Biol. Evol.* **2016**. [CrossRef] [PubMed]

30. Park, J.S.; Cho, B.C.; Simpson, A.G.B. *Halocafeteria seosinensis* gen. et sp. nov. (Bicosoecida), a halophilic bacterivorous nanoflagellate isolated from a solar saltern. *Extrem. Life Extreme Cond.* **2006**, *10*, 493–504. [CrossRef] [PubMed]

31. Fischer, M.G.; Allen, M.J.; Wilson, W.H.; Suttle, C.A. Giant virus with a remarkable complement of genes infects marine zooplankton. *Proc. Natl. Acad. Sci. USA* **2010**, *107*, 19508–19513. [CrossRef] [PubMed]

32. Fischer, M.G.; Suttle, C.A. A virophage at the origin of large DNA transposons. *Science* **2011**, *332*, 231–234. [CrossRef] [PubMed]

33. Van Etten, J.L.; Dunigan, D.D. Chloroviruses: not your everyday plant virus. *Trends Plant Sci.* **2012**, *17*, 1–8. [CrossRef] [PubMed]

34. Clerissi, C.; Grimsley, N.; Ogata, H.; Hingamp, P.; Poulain, J.; Desdevises, Y. Unveiling of the Diversity of Prasinoviruses (*Phycodnaviridae*) in Marine Samples by Using High-Throughput Sequencing Analyses of PCR-Amplified DNA Polymerase and Major Capsid Protein Genes. *Appl. Environ. Microbiol.* **2014**, *80*, 3150–3160. [CrossRef] [PubMed]

35. Wilson, W.H.; Van Etten, J.L.; Allen, M.J. The *Phycodnaviridae*: The Story of How Tiny Giants Rule the World. *Curr. Top. Microbiol. Immunol.* **2009**, *328*, 1–42. [PubMed]

36. Hanson, L.A.; Rudis, M.R.; Vasquez-Lee, M.; Montgomery, R.D. A broadly applicable method to characterize large DNA viruses and adenoviruses based on the DNA polymerase gene. *Virol. J.* **2006**, *3*, 28. [CrossRef] [PubMed]

37. Chen, F.; Suttle, C.A. Evolutionary relationships among large double-stranded DNA viruses that infect microalgae and other organisms as inferred from DNA polymerase genes. *Virology* **1996**, *219*, 170–178. [CrossRef] [PubMed]

38. Takemura, M.; Yokobori, S.; Ogata, H. Evolution of Eukaryotic DNA Polymerases via Interaction Between Cells and Large DNA Viruses. *J. Mol. Evol.* **2015**, *81*, 24–33. [CrossRef] [PubMed]

39. Yutin, N.; Koonin, E.V. Hidden evolutionary complexity of Nucleo-Cytoplasmic Large DNA viruses of eukaryotes. *Virol. J.* **2012**, *9*, 161. [CrossRef] [PubMed]

40. Yutin, N.; Colson, P.; Raoult, D.; Koonin, E.V. *Mimiviridae*: clusters of orthologous genes, reconstruction of gene repertoire evolution and proposed expansion of the giant virus family. *Virol. J.* **2013**, *10*, 106. [CrossRef] [PubMed]

41. King, A.M.Q.; Adams, M.J.; Carstens, E.B.; Lefkowitz, E.J. *Virus Taxonomy: Classification and Nomenclature of Viruses: Ninth Report of the International Committee on Taxonomy of Viruses*; Elsevier Academic Press: San Diego, CA, USA, 2011.

42. Colbourne, J.K.; Pfrender, M.E.; Gilbert, D.; Thomas, W.K.; Tucker, A.; Oakley, T.H.; Tokishita, S.; Aerts, A.; Arnold, G.J.; Basu, M.K.; et al. The ecoresponsive genome of *Daphnia pulex*. *Science* **2011**, *331*, 555–561. [CrossRef] [PubMed]

43. Baumgarten, S.; Simakov, O.; Esherick, L.Y.; Liew, Y.J.; Lehnert, E.M.; Michell, C.T.; Li, Y.; Hambleton, E.A.; Guse, A.; Oates, M.E.; et al. The genome of Aiptasia, a sea anemone model for coral symbiosis. *Proc. Natl. Acad. Sci. USA* **2015**, *112*, 11893–11898. [CrossRef] [PubMed]

44. Hori, K.; Maruyama, F.; Fujisawa, T.; Togashi, T.; Yamamoto, N.; Seo, M.; Sato, S.; Yamada, T.; Mori, H.; Tajima, N.; et al. *Klebsormidium flaccidum* genome reveals primary factors for plant terrestrial adaptation. *Nat. Commun.* **2014**, *5*, 3978. [CrossRef] [PubMed]

45. Ye, N.; Zhang, X.; Miao, M.; Fan, X.; Zheng, Y.; Xu, D.; Wang, J.; Zhou, L.; Wang, D.; Gao, Y.; et al. Saccharina genomes provide novel insight into kelp biology. *Nat. Commun.* **2015**, *6*, 6986. [CrossRef] [PubMed]

46. Chang, Y.; Wang, S.; Sekimoto, S.; Aerts, A.L.; Choi, C.; Clum, A.; LaButti, K.M.; Lindquist, E.A.; Yee Ngan, C.; Ohm, R.A.; et al. Phylogenomic Analyses Indicate that Early Fungi Evolved Digesting Cell Walls of Algal Ancestors of Land Plants. *Genome Biol. Evol.* **2015**, *7*, 1590–1601. [CrossRef] [PubMed]

47. Filée, J.; Chandler, M. Gene Exchange and the Origin of Giant Viruses. *Intervirology* **2010**, *53*, 354–361. [CrossRef] [PubMed]

48. Stram, Y.; Kuzntzova, L. Inhibition of viruses by RNA interference. *Virus Genes* **2006**, *32*, 299–306. [CrossRef] [PubMed]

49. McQueen-Mason, S.; Durachko, D.M.; Cosgrove, D.J. Two endogenous proteins that induce cell wall extension in plants. *Plant Cell* **1992**, *4*, 1425–1433. [CrossRef] [PubMed]

50. Yennawar, N.H.; Li, L.-C.; Dudzinski, D.M.; Tabuchi, A.; Cosgrove, D.J. Crystal structure and activities of EXPB1 (Zea m 1), a beta-expansin and group-1 pollen allergen from maize. *Proc. Natl. Acad. Sci. USA* **2006**, *103*, 14664–14671. [CrossRef] [PubMed]

51. Nikolaidis, N.; Doran, N.; Cosgrove, D.J. Plant expansins in bacteria and fungi: evolution by horizontal gene transfer and independent domain fusion. *Mol. Biol. Evol.* **2014**, *31*, 376–386. [CrossRef] [PubMed]

52. Hardwick, J.M. Cyclin' on the viral path to destruction. *Nat. Cell Biol.* **2000**, *2*, E203–E204. [CrossRef] [PubMed]

53. Bagga, S.; Bouchard, M.J. Cell cycle regulation during viral infection. *Methods Mol. Biol. Clifton NJ* **2014**, *1170*, 165–227.

54. Bowman, S.M.; Free, S.J. The structure and synthesis of the fungal cell wall. *BioEssays* **2006**, *28*, 799–808. [CrossRef] [PubMed]

55. Yamada, T.; Onimatsu, H.; Van Etten, J.L. *Chlorella* viruses. *Adv. Virus Res.* **2006**, *66*, 293–336. [PubMed]

56. Klose, T.; Rossmann, M.G. Structure of large dsDNA viruses. *Biol. Chem.* **2014**, *395*, 711–719. [CrossRef] [PubMed]

57. Legendre, M.; Bartoli, J.; Shmakova, L.; Jeudy, S.; Labadie, K.; Adrait, A.; Lescot, M.; Poirot, O.; Bertaux, L.; Bruley, C.; et al. Thirty-thousand-year-old distant relative of giant icosahedral DNA viruses with a pandoravirus morphology. *Proc. Natl. Acad. Sci. USA* **2014**, *111*, 4274–4279. [CrossRef] [PubMed]

58. Short, S.M. The ecology of viruses that infect eukaryotic algae. *Environ. Microbiol.* **2012**, *14*, 2253–2271. [CrossRef] [PubMed]

13

A Point Mutation in a Herpesvirus Co-Determines Neuropathogenicity and Viral Shedding

Mathias Franz [1],*, Laura B. Goodman [2], Gerlinde R. Van de Walle [3], Nikolaus Osterrieder [4] and Alex D. Greenwood [1,5],*

[1] Department of Wildlife Diseases, Leibniz Institute for Zoo and Wildlife Research, Berlin 10315, Germany

[2] Department of Population Medicine and Diagnostic Sciences, Cornell University, Ithaca, NY 14850, USA; laura.goodman@cornell.edu

[3] Baker Institute for Animal Health, Cornell University, Ithaca, NY 14850, USA; grv23@cornell.edu

[4] Institut für Virologie, Freie Universität Berlin, Robert Von Ostertag-Str. 7 – 13, Berlin 14163, Germany; no.34@fu-berlin.de

[5] Department of Veterinary Medicine, Freie Universität Berlin, Oertzenweg 19b, Berlin 14163, Germany

* Correspondence: m.franz@izw-berlin.de (M.F.); greenwood@izw-berlin.de (A.D.G.)

Academic Editor: Joanna Parish

Abstract: A point mutation in the DNA polymerase gene in equine herpesvirus type 1 (EHV-1) is one determinant for the development of neurological disease in horses. Three recently conducted infection experiments using domestic horses and ponies failed to detect statistically significant differences in viral shedding between the neuropathogenic and non-neuropathogenic variants. These results were interpreted as suggesting the absence of a consistent selective advantage of the neuropathogenic variant and therefore appeared to be inconsistent with a systematic increase in the prevalence of neuropathogenic strains. To overcome potential problems of low statistical power related to small group sizes in these infection experiments, we integrated raw data from all three experiments into a single statistical analysis. The results of this combined analysis showed that infection with the neuropathogenic EHV-1 variant led to a statistically significant increase in viral shedding. This finding is consistent with the idea that neuropathogenic strains could have a selective advantage and are therefore systematically increasing in prevalence in domestic horse populations. However, further studies are required to determine whether a selective advantage indeed exists for neuropathogenic strains.

Keywords: equine herpesvirus type 1; neuropathogenicity; viral shedding; trade-off hypothesis

1. Introduction

Equine herpesvirus type 1 (EHV-1) is a ubiquitous alphaherpesvirus that can cause respiratory disease, abortion, neonatal foal death and equine herpes myeloencephalopathy (EHM), a neurological disease that can be lethal [1,2]. Although it is not well understood how EHM develops following EHV-1 infection, it has been shown that neuropathogenicity is significantly associated with a single nucleotide polymorphism in the viral DNA polymerase, resulting in a specific amino acid change; asparagine to aspartic acid (N/D_{752}) [1–3]. To demonstrate the causality of this association, mutations were introduced in the backgrounds of representative D_{752} and N_{752} strains from Europe and North America, cloned as bacterial artificial chromosomes and then characterized thoroughly both in vitro and in vivo [4,5]. Although involvement of a variety of host, viral, and environmental risk factors are known or suspected, infections with D_{752} strains strongly increase the risk of developing EHM [1]. Based on two retrospective large-scale studies, it has been estimated that the odds of having EHM with a D_{752} strain were 490 times greater than those with a N_{752} strain [2].

Whether the neuropathogenic D_{752} strains are increasing in prevalence and what could enable such an increase are important and as yet unresolved questions. The D_{752} or the N_{752} at synonymous positions in DNA polymerases of other herpesviruses is highly conserved [3], although EHM has been considered rare in the past. Following worldwide EHM outbreaks in 2006–2007, the US Department of Agriculture classified EHM as a potentially emerging disease [6]. Related concerns that neuropathogenic D_{752} strains are spreading in domestic horse populations have been supported by a limited number of large-scale retrospective studies. These studies indicated that while N_{752} strains are more common [2,7], D_{752} strains have increased in prevalence in recent decades [8], which may suggest the existence of a selective advantage of neuropathogenic D_{752} strains. However, what could mediate such a selective advantage remains unclear. In contrast to the potential fitness benefits, D_{752} strains might suffer from fitness costs related to increased host death rates that result from neuropathogenicity [9].

D_{752} strains might have nevertheless a selective advantage if they have sufficiently increased transmission rates compared to N_{752} strains. EHV-1 transmission occurs via direct contact, fomites, or by inhalation of respiratory secretions [2]. Primary infections caused by EHV-1 occur at the respiratory epithelium and result in nasal shedding for 10 to 14 days. These infections typically result in life-long latent infections with subsequent viral reactivation and viral shedding during periods of stress [1]. Differences in transmission rates among strains can therefore have at least three causes: (1) the differences in the amount of nasal virus shedding during primary infections; (2) differences in the amount of nasal virus shedding at reactivation; and (3) differences in the rate of reactivation. Among these sources, previous studies have been limited to investigations of viral shedding during primary infections. Specifically, two experimental infection studies, which document that the D_{752} variant is more likely to cause neuropathogenicity, failed to find evidence that infections with the D_{752} variant cause statistically significant increased amounts of nasal virus shedding compared to the N_{752} variant [4,5]. These results were interpreted as suggesting the absence of a consistent selective advantage of the D_{752} variant and therefore were thought to indicate the absence of a systematic increase in the prevalence of neuropathogenic D_{752} strains.

Here, we re-examined data from the two previous studies to address the potential problem that the absence of statistically significant differences in shedding amounts reflected an actual absence of differences; and the absence of statistical significance could be related to low statistical power. To overcome potential problems of low statistical power related to small sample sizes, we integrated the original raw data from all three experimental infection studies into a single statistical analysis and re-assessed whether increased neuropathogenicity of the D_{752} variant coincides with increased viral shedding.

2. Materials and Methods

In our analysis, we integrated data on nasal virus shedding from infection experiments conducted by Goodman et al. [4] and Van de Walle et al. [5]. Both studies used a similar experimental design in which two groups of animals were infected with the N_{752} or D_{752} EHV-1 variant (Table 1). However, the performed experiments differed in specific details that include infection dose, breed of studied animals and number of infected animals. Goodman et al. conducted two separate experiments: (1) two groups of four Welsh Mountain ponies infected with 7×10^6 plaque-forming units (PFUs) of the virus, and (2) two groups of seven mixed-breed horses infected with 1×10^7 PFUs [4]. Van de Walle et al. conducted one experiment on a group of three and six horses (infected with the N_{752} and D_{752} variant, respectively), which were each inoculated with 1.5×10^7 PFUs [5]. Following infection, nasal swabs were obtained from each animal at a daily or near-daily basis (Figure 1). For all three experiments, quantitative polymerase chain reaction was used to measure the amount of nasal virus shedding and estimate the viral genome copy number in nasal swabs. All three experiments were performed in accordance with relevant guidelines and regulations. The experimental protocols on horses were approved by the Cornell University Institutional Animal Care and Use Committee [4,5]. The experimental protocol on ponies was approved by the UK Home Office [4].

Table 1. Overview of analyzed data.

Experiment	Number of Data Points	Number of Sampling Days	Number of Animals Infected		Estimated Viral Genome Copy Numbers [1]	
			N_{752}	D_{752}	N_{752}	D_{752}
Horses 2007 [4]	178	13	7	7	$5.2 \times 10^6 \pm 3.1 \times 10^7$	$1.2 \times 10^7 \pm 3.6 \times 10^7$
Horses 2009 [5]	108	12	6	3	$3.5 \times 10^5 \pm 1.2 \times 10^6$	$5.2 \times 10^7 \pm 4.4 \times 10^8$
Ponies 2007 [4]	96	12	4	4	$9.5 \times 10^5 \pm 3.7 \times 10^6$	$1.4 \times 10^6 \pm 4.0 \times 10^6$

[1] Mean $\pm$ standard deviation (SD).

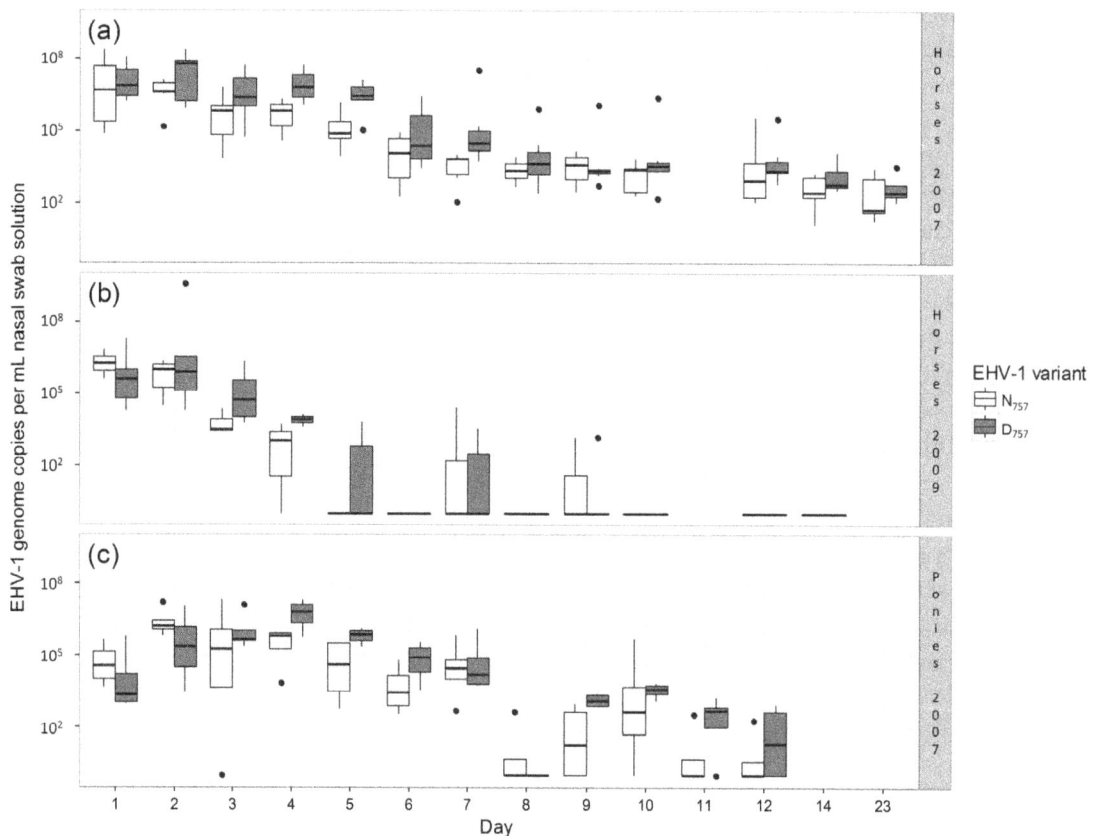

Figure 1. Boxplots illustrating the differences in amounts of nasal virus shedding between equine herpesvirus type 1 (EHV-1) variants (N/D_{752}) across all days and animals for each experiment. (**a**) Experiment on 14 horses (seven infected with the N_{752} and seven infected with the D_{752} variant) conducted by Goodman et al. [4]; (**b**) Experiment on nine horses (three infected with the N_{752} and six infected with the D_{752} variant) conducted by Van de Walle et al. [5]; (**c**) Experiment on eight ponies (four infected with the N_{752} and four infected with the D_{752} variant) conducted by Goodman et al. [4]. If sampling took place on a given day, then all infected animals were sampled, with the exception of day 23 in the experiment on horses conducted by Goodman et al. where samples were restricted to six individuals infected with the N_{752} variant and four individuals infected with the D_{752} variant. In each boxplot, the bottom and top of the box indicate the first and third quartiles, and the band inside the box indicates the median. Upper (and lower) whiskers extend to the highest (and lowest) value that is within 1.5 times the inter-quartile range (i.e., the distance between the first and third quartiles). Data points beyond the end of the whiskers are plotted as dots [10]. Due to highly skewed distributions, the amount of nasal virus shedding was plotted on a logarithmic scale.

In our statistical analysis, we aimed to test whether amounts of nasal virus shedding differed between the two EHV-1 variants. We focused only on shedding amount and did not analyze shedding duration because the majority of animals (17 out of 31) were still shedding on the last respective sampling day. To maximize statistical power in our analysis of nasal virus shedding, we pooled all available raw data from the three experiments. Accordingly, our data set consisted of 382 data points, where each data point was related to one individual measure of nasal virus shedding (Table 1, Figure 1). This approach required the analysis to appropriately control for non-independence of data points and related variation in shedding amounts that emerged from (1) consistent differences among animals, which might, for example, relate to physiological or genetic differences; (2) systematic differences among the three experiments, which could, for example, relate to differences in infection treatment or horse breeds; and (3) temporal variation in shedding over the course of each experiment. To address these issues, we used a linear mixed model, in which each nasal swab was treated as a single data point with the amount of nasal virus shedding as the response variable. The EHV-1 variant (N_{752} vs. D_{752}) was treated as a fixed effect, which tested whether shedding differed between the two EHV-1 variants. To account for variation in nasal virus shedding among experiments and over time within each experiment, we included a categorical predictor 'experiment_day' as a fixed effect in our model. This predictor combines information on the experiment and day after infection, which resulted in 37 unique levels (13 + 12 + 12 days). To account for potential differences among animals and non-independence of data points, the identity of each animal was included as a random effect. To prevent violations of model assumptions, we log-transformed the response variable. After this transformation, visual inspections of the model output did not indicate any violations of assumptions regarding the normality and homogeneity of error variances. Model fitting and analysis were performed using the package 'glmmADMB' [11] in the statistical software R [12]. Statistical significance was assessed based on a two-sided test with an α-level of 0.05. The datasets supporting this article are available from the Dryad Digital Repository [13].

3. Results

Results of our combined analysis showed that the amount of EHV-1 nasal shedding was significantly increased in animals infected with the D_{752} variant compared with infection with the N_{752} variant ($p = 0.001$). The model indicated that infections with the D_{752} variant led, on average, to a four-fold higher amount of nasal EHV-1 shedding compared to infections with the N_{752} variant (Figure 1). To control for potentially confounding effects of combining data on mixed-breed horses and Welsh Mountain ponies, we also performed an analysis that excluded the pony data (Figure 1c). The analysis of this reduced data set did not qualitatively change our results: the differences between both variants remained significant ($p = 0.004$) with an estimated 3.9-fold increase in the amount of nasal virus shedding by animals infected with the neuropathogenic D_{752} variant.

4. Discussion

In contrast to previously conducted and separately analyzed studies, we found support for the hypothesis that, compared to the non-neuropathogenic N_{752} variant, primary infection with neuropathogenic D_{752} results in increased nasal virus shedding. This finding indicates that the absence of statistical differences in shedding amounts in previous studies was related to low statistical power. Given the constraints of performing infection experiments on a large number of animals, our findings emphasize (1) the importance of using advanced statistical modeling approaches and (2) the usefulness of pooling raw data from individual experiments.

Our finding of increased nasal virus shedding of the D_{752} variant following experimental EHV-1 infections of equids is consistent with the idea that D_{752} strains could have a selective advantage and are therefore systematically increasing in prevalence in domestic horse populations [8]. However, our finding is not sufficient to determine whether a fitness advantage indeed exists for D_{752} variants, which would require more precise information on infection-induced host death rates and rates of viral reactivation during latent infections. In addition, due to effects on host death and transmission rates, the fitness of different EHV-1 strains might be strongly determined by within-host interactions. Co-infection with both variants is possible [1,7], which suggests that within-host interactions between different strains might occur, which could impact strain fitness and related evolutionary dynamics [14,15]. We are not aware of any study that has investigated within-host interactions between the two variants. Nevertheless, there is some indication that the neuropathogenic variant has a competitive advantage regarding within-host interactions. Compared to the N_{752} variant, infections with the D_{752} variant led to higher levels of viremia and virus-neutralizing antibody titers [4,5]. This finding is consistent with D_{752} eliciting a higher immune response and being more resistant to the response than the N_{752} variant [16,17].

A growing body of research on this polymorphism has attempted to elucidate the biological mechanisms connecting DNA polymerase activity to EHM. There is a general consensus that magnitude and duration of viremia are central aspects of the disease ([18]; reviewed by [19]), with secondary fevers commonly seen in neurologic cases. Replication kinetics in cultured fibroblasts and epithelial cells were similar between variants [4]. Studies in primary respiratory epithelial and brain endothelial model systems have elucidated the ability of D_{752} strains to modulate immune evasion [20,21]. In nasal mucosal explants, D_{752} was more effective at crossing epithelial boundaries [22], but less effective in vaginal mucosal explants [23]. The ability to infect migratory leukocytes may be a key aspect of EHM progression, with high levels of nasal virus shedding one of the consequences.

Our finding that neuropathogenicity is linked to nasal virus shedding is consistent with assumptions underlying the trade-off hypothesis—the currently dominant but still controversial hypothesis of virulence evolution [24–26]. This hypothesis assumes that virulence evolution is crucially influenced by a mechanistic link between virulence (defined as increased host death rate) and pathogen transmission: decreasing virulence constrains pathogen shedding and the related rate at which pathogens are transmitted to susceptible hosts. This interdependence generates an evolutionary trade-off for pathogens because the ideal state of low virulence and high transmission rate becomes impossible to reach. Instead, fitness benefits of reducing virulence come at a fitness cost of a reduced transmission rate. Although theoretically appealing, there is currently only limited empirical support for the underlying assumption of virulence–transmission trade-offs [24–34].

In the absence of reliable quantifications of virulence and transmission rates of different EHV-1 strains, it is not possible to directly assess the validity of the trade-off hypothesis for EHV-1. Nevertheless, our finding of increased and nasal virus shedding for the neuropathogenic EHV-1 variant is fully consistent with this hypothesis. Conceptual descriptions and formal implementations of the trade-off hypothesis usually assume that pathogens can evolve different levels of virulence (i.e., increased host death rate) and transmission along a continuous scale [24–26]. In contrast, the currently existing variation in EHV-1 neuropathogenicity seems to be mainly caused by a single nucleotide polymorphism, which suggests that a continuous variation in virulence and transmission might be impossible in this system. As a consequence, it could be possible that there is no single optimal virulence in this system and that instead both variants permanently coexist [5].

Acknowledgments: M.F., N.O. and A.D.G. were supported by a grant from the Leibniz Gemeinschaft, SAW-2015-IZW-1 440.

Author Contributions: M.F. and A.D.G. conceived the study; M.F. conducted the statistical analysis; L.B.G., G.R.V.d.W. and N.O. contributed the experimental data; M.F., L.B.G., G.R.V.d.W., N.O. and A.D.G. wrote the paper.

References

1.   Lunn, D.; Davis-Poynter, N.; Flaminio, M.; Horohov, D.; Osterrieder, K.; Pusterla, N.; Townsend, H. Equine Herpesvirus-1 Consensus Statement. *J. Vet. Internal Med.* **2009**, *23*, 450–461. [CrossRef] [PubMed]

2. Perkins, G.A.; Goodman, L.B.; Tsujimura, K.; Van de Walle, G.R.; Kim, S.G.; Dubovi, E.J.; Osterrieder, N. Investigation of the prevalence of neurologic equine herpes virus type 1 (EHV-1) in a 23-year retrospective analysis (1984–2007). *Vet. Microbiol.* **2009**, *139*, 375–378. [CrossRef] [PubMed]

3. Nugent, J.; Birch-Machin, I.; Smith, K.; Mumford, J.; Swann, Z.; Newton, J.; Bowden, R.; Allen, G.; Davis-Poynter, N. Analysis of equid herpesvirus 1 strain variation reveals a point mutation of the DNA polymerase strongly associated with neuropathogenic versus nonneuropathogenic disease outbreaks. *J. Virol.* **2006**, *80*, 4047–4060. [CrossRef] [PubMed]

4. Goodman, L.B.; Loregian, A.; Perkins, G.A.; Nugent, J.; Buckles, E.L.; Mercorelli, B.; Kydd, J.H.; Palù, G.; Smith, K.C.; Osterrieder, N. A point mutation in a herpesvirus polymerase determines neuropathogenicity. *PLoS Pathog.* **2007**, *3*, e160. [CrossRef] [PubMed]

5. Van de Walle, G.R.; Goupil, R.; Wishon, C.; Damiani, A.; Perkins, G.A.; Osterrieder, N. A single-nucleotide polymorphism in a herpesvirus DNA polymerase is sufficient to cause lethal neurological disease. *J. Infect. Dis.* **2009**, *200*, 20–25. [CrossRef] [PubMed]

6. U.S. Department of Agriculture Animal and Plant Health Inspection Service. *Equine Herpesvirus Myeloencephalopathy: A Potentially Emerging Disease*; US Department of Agriculture Animal and Plant Health Inspection Service, 2007. Available online: https://www.aphis.usda.gov/animal_health/emergingissues/downloads/ehv1final.pdf (accessed on 9 December 2008).

7. Allen, G.; Bolin, D.; Bryant, U.; Carter, C.; Giles, R.; Harrison, L.; Hong, C.; Jackson, C.; Poonacha, K.; Wharton, R. Prevalence of latent, neuropathogenic equine herpesvirus-1 in the Thoroughbred broodmare population of central Kentucky. *Equine Vet. J.* **2008**, *40*, 105–110. [CrossRef] [PubMed]

8. Smith, K.L.; Allen, G.P.; Branscum, A.J.; Cook, R.F.; Vickers, M.L.; Timoney, P.J.; Balasuriya, U.B. The increased prevalence of neuropathogenic strains of EHV-1 in equine abortions. *Vet. Microbiol.* **2010**, *141*, 5–11. [CrossRef] [PubMed]

9. Henninger, R.W.; Reed, S.M.; Saville, W.J.; Allen, G.P.; Hass, G.F.; Kohn, C.W.; Sofaly, C. Outbreak of Neurologic Disease Caused by Equine Herpesvirus-1 at a University Equestrian Center. *J. Vet. Internal Med.* **2007**, *21*, 157–165. [CrossRef]

10. McGill, R.; Tukey, J.W.; Larsen, W.A. Variations of box plots. *Am. Stat.* **1978**, *32*, 12–16. [CrossRef]

11. Fournier, D.A.; Skaug, H.J.; Ancheta, J.; Ianelli, J.; Magnusson, A.; Maunder, M.N.; Nielsen, A.; Sibert, J. AD Model Builder: Using automatic differentiation for statistical inference of highly parameterized complex nonlinear models. *Optim. Methods Softw.* **2012**, *27*, 233–249. [CrossRef]

12. R Core Team. *R: A Language and Environment for Statistical Computing*; R Core Team: Vienna, Austria, 2015.

13. Dryad Digital Repository. Available online: http://dx.doi.org/10.5061/dryad.61k7n (accessed on 5 January 2017).

14. Alizon, S.; de Roode, J.C.; Michalakis, Y. Multiple infections and the evolution of virulence. *Ecol. Lett.* **2013**, *16*, 556–567. [CrossRef] [PubMed]

15. Mideo, N.; Alizon, S.; Day, T. Linking within-and between-host dynamics in the evolutionary epidemiology of infectious diseases. *Trends Ecol. Evol.* **2008**, *23*, 511–517. [CrossRef] [PubMed]

16. Alizon, S.; van Baalen, M. Multiple infections, immune dynamics, and the evolution of virulence. *Am. Nat.* **2008**, *172*, E150–E168. [CrossRef] [PubMed]

17. Brown, S.P.; Grenfell, B.T. An unlikely partnership: Parasites, concomitant immunity and host defence. *Proc. R. Soc. Lond. B Biol. Sci.* **2001**, *268*, 2543–2549. [CrossRef] [PubMed]

18. Allen, G.; Breathnach, C. Quantification by real-time PCR of the magnitude and duration of leucocyte-associated viraemia in horses infected with neuropathogenic vs. non-neuropathogenic strains of EHV-1. *Equine Vet. J.* **2006**, *38*, 252–257. [CrossRef] [PubMed]

19. Pusterla, N.; Hussey, G.S. Equine herpesvirus 1 myeloencephalopathy. *Vet. Clin. N. Am. Equine Pract.* **2014**, *30*, 489–506. [CrossRef] [PubMed]

20. Hussey, G.S.; Ashton, L.V.; Quintana, A.M.; Lunn, D.P.; Goehring, L.S.; Annis, K.; Landolt, G. Innate immune responses of airway epithelial cells to infection with Equine herpesvirus-1. *Vet. Microbiol.* **2014**, *170*, 28–38. [CrossRef] [PubMed]

21. Goehring, L.; Hussey, G.; Ashton, L.; Schenkel, A.; Lunn, D. Infection of central nervous system endothelial cells by cell-associated EHV-1. *Vet. Microbiol.* **2011**, *148*, 389–395. [CrossRef] [PubMed]

22. Vandekerckhove, A.P.; Glorieux, S.; Gryspeerdt, A.; Steukers, L.; Duchateau, L.; Osterrieder, N.; Van de Walle, G.; Nauwynck, H. Replication kinetics of neurovirulent versus non-neurovirulent equine herpesvirus type 1 strains in equine nasal mucosal explants. *J. Gen. Virol.* **2010**, *91*, 2019–2028. [CrossRef] [PubMed]

23. Negussie, H.; Li, Y.; Tessema, T.S.; Nauwynck, H.J. Replication characteristics of equine herpesvirus 1 and equine herpesvirus 3: Comparative analysis using ex vivo tissue cultures. *Vet. Res.* **2016**, *47*, 1. [CrossRef] [PubMed]

24. Alizon, S.; Hurford, A.; Mideo, N.; Van Baalen, M. Virulence evolution and the trade-off hypothesis: History, current state of affairs and the future. *J. Evol. Biol.* **2009**, *22*, 245–259. [CrossRef] [PubMed]

25. Bull, J.J.; Lauring, A.S. Theory and empiricism in virulence evolution. *PLoS Pathog.* **2014**, *10*, e1004387. [CrossRef] [PubMed]

26. Cressler, C.E.; McLeod, D.V.; Rozins, C.; van den Hoogen, J.; Day, T. The adaptive evolution of virulence: A review of theoretical predictions and empirical tests. *Parasitology* **2016**, *143*, 915–930. [CrossRef] [PubMed]

27. Paul, R.; Lafond, T.; Müller-Graf, C.; Nithiuthai, S.; Brey, P.; Koella, J. Experimental evaluation of the relationship between lethal or non-lethal virulence and transmission success in malaria parasite infections. *BMC Evol. Biol.* **2004**, *4*, 30. [CrossRef] [PubMed]

28. Salvaudon, L.; Héraudet, V.; Shykoff, J.A.; Koella, J. Parasite-host fitness trade-offs change with parasite identity: Genotype-specific interactions in a plant-pathogen system. *Evolution* **2005**, *59*, 2518–2524. [CrossRef] [PubMed]

29. Mackinnon, M.J.; Gandon, S.; Read, A.F. Virulence evolution in response to vaccination: The case of malaria. *Vaccine* **2008**, *26*, C42–C52. [CrossRef] [PubMed]

30. Mackinnon, M.J.; Read, A.F. Genetic relationships between parasite virulence and transmission in the rodent malaria Plasmodium chabaudi. *Evolution* **1999**, *53*, 689–703. [CrossRef]

31. Ferguson, H.; Mackinnon, M.; Chan, B.; Read, A. Mosquito mortality and the evolution of malaria virulence. *Evolution* **2003**, *57*, 2792–2804. [CrossRef] [PubMed]

32. Atkins, K.E.; Read, A.F.; Savill, N.J.; Renz, K.G.; Walkden-Brown, S.W.; Woolhouse, M.E. Modelling Marek's Disease Virus (MDV) infection: Parameter estimates for mortality rate and infectiousness. *BMC Vet. Res.* **2011**, *7*, 1. [CrossRef] [PubMed]

33. Alizon, S.; Michalakis, Y. Adaptive virulence evolution: The good old fitness-based approach. *Trends Ecol. Evol.* **2015**, *30*, 248–254. [CrossRef] [PubMed]

34. Schmid Hempel, P.; Schmid-Hempel, P. *Evolutionary Parasitologythe Integrated Study of Infections, Immunology, Ecology, and Genetics*; Oxford University Press: Oxford, UK, 2011.

Permissions

List of Contributors

Tapas K. Nayak, Subhasis Chattopadhyay, Laishram Pradeep K. Singh and Subhransu S. Sahoo
School of Biological Sciences, National Institute of Science Education & Research, Bhubaneswar, HBNI, Jatni, Khurda, Odisha 752050, India

Prabhudutta Mamidi, Soma Chattopadhyay and Abhishek Kumar
Infectious Disease Biology, Institute of Life Sciences, (Autonomous Institute of Department of Biotechnology, Government of India), Nalco Square, Bhubaneswar, Odisha 751023, India

Hongtao Kang, Dafei Liu, Jin Tian, Xiaoliang Hu, Xiaozhan Zhang, Hongxia Wu, Chunguo Liu, Dongchun Guo, Zhijie Li, Qian Jiang, Jiasen Liu and Liandong Qu
Division of Zoonosis of Natural Foci, State Key Laboratory of Veterinary Biotechnology, Harbin Veterinary Research Institute, Chinese Academy of Agricultural Sciences, 678 Haping road, Xiangfang District, Harbin 150000, China

Hang Yin
College of Veterinary Medicine, Northeast Agricultural University, Harbin 150000, China

WeiminWu, Yan Liu, Yong Lin, Danzhen Pan and Yang Xu
Department of Pathogen Biology, School of Basic Medicine, Tongji Medical College, Huazhong University of Science and Technology, Wuhan 430030, China

Dongliang Yang
Department of Infectious Diseases, Union Hospital, Tongji Medical College, Huazhong University of Science and Technology, Wuhan 430022, China

Mengji Lu
Department of Pathogen Biology, School of Basic Medicine, Tongji Medical College, Huazhong University of Science and Technology, Wuhan 430030, China
Institute of Virology, University Hospital of Essen, 45147 Essen, Germany

Sarah E. Heath, Pedro F. Vale and Sinead Collins
Institute of Evolutionary Biology, School of Biological Sciences, University of Edinburgh, Ashworth Laboratories, The King's Buildings, Charlotte Auerbach Road, Edinburgh EH9 3FL, UK

Kirsten Knox
Institute of Molecular Plant Sciences, School of Biological Sciences, University of Edinburgh, Rutherford Building, Max Born Crescent, Edinburgh EH9 3BF, UK

Ben A. Wagstaff, Iulia C. Vladu, J. Elaine Barclay and Robert A. Field
Department of Biological Chemistry, John Innes Centre, Norwich Research Park, Norwich NR4 7UH, UK

Declan C. Schroeder
Marine Biological Association of the UK, Plymouth PL1 2PB, UK

Gill Malin
Centre for Ocean and Atmospheric Studies, School of Environmental Sciences, University of East Anglia, Norwich Research Park, Norwich NR4 7TJ, UK

Andrea Highfield, Ian Joint and Declan C. Schroeder
The Marine Biological Association, The Laboratory, Citadel Hill, Plymouth PL1 2PB, UK

Jack A. Gilbert
The Microbiome Centre, Department of Surgery, University of Chicago, Chicago, IL 60637, USA
Division of Bioscience, Argonne National Laboratory, 9700 South Cass Avenue, Argonne, IL 60439, USA

Katharine J. Crawfurd
Department of Biological Oceanography, NIOZ–Royal Netherlands Institute for Sea Research, 1790 AB Den Burg, Texel, The Netherlands

Steven W. Wilhelm, Jordan T. Bird, Kyle S. Bonifer, Benjamin C. Calfee, Tian Chen, Samantha R. Coy, P. Jackson Gainer, Eric R. Gann, Huston T. Heatherly, Jasper Lee, Xiaolong Liang, Jiang Liu, April C. Armes, Mohammad Moniruzzaman, J. Hunter Rice, Joshua M. A. Stough, Robert N. Tams, Evan P. Williams and Gary R. LeCleir
The Department of Microbiology, The University of Tennessee, Knoxville, TN 37996, USA

Anastasia N. Vlasova and Linda J. Saif
Food Animal Health Research Program, CFAES, Ohio Agricultural Research and Development Center
Department of Veterinary Preventive Medicine, The Ohio State University, Wooster, OH 44691, USA

Joshua O. Amimo
Department of Animal Production, Faculty of Veterinary Medicine, University of Nairobi, Nairobi 30197
Bioscience of Eastern and Central Africa, International Livestock Research Institute (BecA-ILRI) Hub, Nairobi 30709, Kenya

Leszek Błaszczyk, Marcin Biesiada and Katarzyna J. Purzycka
Institute of Bioorganic Chemistry, Polish Academy of Sciences, Poznan 61-704, Poland

Agniva Saha and David J. Garfinkel
Department of Biochemistry & Molecular Biology, University of Georgia, Athens, GA 30602, USA

Xianwu Zeng, Huiqiang Yang, Zhushi Li, Wei Wang, Hua Lin, Lina Liu and Qianzhi Ni
Department of Viral Vaccine, Chengdu Institute of Biological Products Co., Ltd., China National Biotech Group, Chengdu 610023, China

Jian Yang
Department of Viral Vaccine, Chengdu Institute of Biological Products Co., Ltd., China National Biotech Group, Chengdu 610023, China
Department of Microbiology and Immunology, North Sichuan Medical College, Nanchong 637007, China

Xinyu Liu
Department of Arbovirus Vaccine, National Institutes for Food and Drug Control, Beijing 100050, China

Yonglin Wu
China National Biotech Group, Beijing 100029, China

Yuhua Li
Department of Viral Vaccine, Chengdu Institute of Biological Products Co., Ltd., China National Biotech Group, Chengdu 610023, China

Department of Arbovirus Vaccine, National Institutes for Food and Drug Control, Beijing 100050, China
State Key Laboratory of Biotherapy and Cancer Center, West China Hospital, Sichuan University and Collaborative Innovation Center for Biotherapy, Chengdu 610000, China

Wen Shi, Shuo Jia, Haiyuan Zhao, Jiyuan Yin, Xiaona Wang, Meiling Yu, Sunting Ma, Yang Wu, Ying Chen, Wenlu Fan, Yigang Xu and Yijing Li
College of Veterinary Medicine, Northeast Agricultural University, Harbin 150030, China

Lucie Gallot-Lavallée
Structural and Genomic Information Laboratory (IGS), Aix-Marseille Université, CNRS UMR7256 (IMM FR3479), 13288 Marseille cedex 09, France

Guillaume Blanc
Structural and Genomic Information Laboratory (IGS), Aix-Marseille Université, CNRS UMR7256 (IMM FR3479), 13288 Marseille cedex 09, France
Mediterranean Institute of Oceanography (MIO), Aix Marseille Université, Université de Toulon, CNRS/INSU, IRD, UM 110, 13288 Marseille cedex 09, France

Mathias Franz
Department of Wildlife Diseases, Leibniz Institute for Zoo and Wildlife Research, Berlin 10315, Germany

Laura B. Goodman
Department of Population Medicine and Diagnostic Sciences, Cornell University, Ithaca, NY 14850, USA

Gerlinde R. Van de Walle
Baker Institute for Animal Health, Cornell University, Ithaca, NY 14850, USA; grv23@cornell.edu

Nikolaus Osterrieder
Institut für Virologie, Freie Universität Berlin, Robert Von Ostertag-Str. 7 – 13, Berlin 14163, Germany

Alex D. Greenwood
Department of Wildlife Diseases, Leibniz Institute for Zoo and Wildlife Research, Berlin 10315, Germany
Department of Veterinary Medicine, Freie Universität Berlin, Oertzenweg 19b, Berlin 14163, Germany

Index

A

Abortion, 194

Acanthamoeba, 64, 73, 90, 92-93, 95, 97, 100-102, 104, 182-184, 187-189, 191

Active and Passive Immunity, 108

Algal Bloom, 64, 98, 106

Algal Virus, 64, 70, 72, 102

Alphaherpesvirus, 194

Alphavirus, 1, 19

Apoptosis, 1-2, 7-9, 13-16, 20-21, 120

Attenuation Mechanism, 155, 163-164

C

Chikungunya Virus, 1, 6, 17-21, 166

Chimeric Genome, 34

Climate Change, 47, 60-61, 75

Coccolithovirus, 72, 75, 87

Cost of Resistance, 46-47, 56-57, 59-60, 62

Cv777-based Vaccine, 168, 175

E

Emiliania Huxleyi (EHV), 64

Epidemiology, 108, 110, 113, 115, 121, 123, 125, 128, 131-132, 177-178, 199

Equine Herpesvirus Type 1 (EHV-1), 194, 196

Eukaryotes, 72, 74, 90-91, 93, 95, 97-98, 101, 105, 107, 135-136, 154, 180-183, 187-189, 191-192

Evolution, 17, 23, 32, 46-49, 51-58, 60-63, 78, 90, 94-95, 100-101, 103, 124, 127, 131, 189, 191-193, 198-200

F

Feline Panleucopenia Virus (FPV), 22, 25

G

Gastroenteritis, 108, 119, 122-124, 127, 130, 132, 177, 179

Genetic Variability, 76, 108, 119, 176

Giant Viruses, 74, 90-91, 93-107, 183, 190-191, 193

H

Haptophyte, 64-65, 74

Hepatitis B Virus, 19, 34-36, 44-45

Herpesvirus, 23, 194, 196, 198-200

Hsp90, 1-3, 5, 11-16, 19-21

I

Identification, 97, 102, 110, 112, 118-119, 122-132, 166-167, 169-171, 179, 181, 191

Infectious Pathogen, 22

J

Japanese Encephalitis Virus (JEV), 155-156

L

Lateral Gene Transfer, 180, 183, 188-189

M

Macrophage, 1-3, 6-7, 10-11, 14-16, 18

Marine Viral Ecology, 46

Mcp Gene, 96, 190

Megaviridae, 64, 69-72, 74, 104

Mhc, 1, 10-11, 13, 15-16, 20-21

Mimiviridae, 90, 96-99, 107, 180, 185, 187, 192

N

Ncldv, 64, 74, 95-97, 102-104, 180-191

Neonatal Foal Death, 194

Neuroinvasiveness, 155-156, 158, 161, 163-164, 166

Neuropathogenicity, 194-195, 198-199

Neurovirulence, 155-156, 158, 160-166

Nonstructural Protein 2 (NS2), 22

O

Ocean Acidification, 61, 75, 83, 86-88

Ostreococcus Tauri, 46-47, 61-62

P

Phycodnavirus, 46, 69, 190

Picoplankton, 46

Point Mutation, 36, 40, 42, 194, 199

Porcine Epidemic Diarrhea Virus (PEDV), 167, 170, 178

Porcine Intestinal Epithelial Cells, 167-168, 176

Porcine Rotavirus, 108, 122-123, 125-131

Prasinovirus, 46

Prevalence, 23, 108-117, 119, 121, 123-126, 128, 132, 167, 194-195, 197, 199

Protein Markers, 182

Protein Mutations, 155

Prymnesium Parvum, 64-66, 68, 70, 73-74

R

Respiratory Disease, 194

Rna Structure, 135-136, 149, 153-154

Rotavirus Vaccines, 108, 133-134

Rotaviruses (RVS), 108

T

Tnf, 1-2, 9-10, 13-16, 21, 120
Trade-off Hypothesis, 194, 198, 200
Translation Regulation, 135
Ty1 Retrotransposon, 135, 152-154

V

Viral Shedding, 194-195
Viral Titer, 6, 172, 176
Virus Insertion, 180

Virus Isolation, 95, 169, 171, 181
Virus Resistance, 46, 61-62
Virus-host Interactions, 46, 85, 94, 104

W

Woodchuck Hepatitis Virus, 34-36, 45

Z

Zoonotic Potential, 108, 117, 119, 121, 126
Zooxanthellae, 90, 99